Cellulose Nanocrystal/Nanoparticles Hybrid Nanocomposites

Woodhead Publishing Series in Composites Science and Engineering

Cellulose Nanocrystal/ Nanoparticles Hybrid Nanocomposites

From Preparation to Applications

Edited by

Denis Rodrigue
Abou el Kacem Qaiss
Rachid Bouhfid

Woodhead Publishing is an imprint of Elsevier
The Officers' Mess Business Centre, Royston Road, Duxford, CB22 4QH, United Kingdom
50 Hampshire Street, 5th Floor, Cambridge, MA 02139, United States
The Boulevard, Langford Lane, Kidlington, OX5 1GB, United Kingdom

Notices

Knowledge and best practice in this field are constantly changing. As new research and experience broaden our understanding, changes in research methods, professional practices, or medical treatment may become necessary.

Practitioners and researchers must always rely on their own experience and knowledge in evaluating and using any information, methods, compounds, or experiments described herein. In using such information or methods they should be mindful of their own safety and the safety of others, including parties for whom they have a professional responsibility.

To the fullest extent of the law, neither the Publisher nor the authors, contributors, or editors, assume any liability for any injury and/or damage to persons or property as a matter of products liability, negligence or otherwise, or from any use or operation of any methods, products, instructions, or ideas contained in the material herein.

Library of Congress Cataloging-in-Publication Data
A catalog record for this book is available from the Library of Congress

British Library Cataloguing-in-Publication Data
A catalogue record for this book is available from the British Library

ISBN: 978-0-12-822906-4 (print)
ISBN: 978-0-12-823090-9 (online)

For information on all Woodhead publications
visit our website at https://www.elsevier.com/books-and-journals

Publisher: Matthew Deans (ELS-OXF)
Acquisitions Editor: Jones, Glyn (ELS-OXF)
Editorial Project Manager: Capille, Gabriela D. (ELS-CON)
Production Project Manager: Vignesh Tamil
Cover Designer: Studholme, Alan (ELS-OXF)

Typeset by SPi Global, India

Contents

Contributors

Numbers in parentheses indicate the pages on which the author's contributions begin.

Ayonbala Baral (205), Faculty of Metallurgy & Energy Engineering, Kunming University of Science & Technology, Kunming, China

Hala Bensalah (181), Equipe de Mécanique et des Matériaux (E.M.M), Centre de Recherche en Enérgie (C.R.E.), Faculty of Sciences, Mohammed V University in Rabat, Rabat, Morocco

Hanane Benzeid (115), Laboratoire de Chimie Analytique et de Bromatologie, Faculté de Médecine et de Pharmacie, Université Mohamed V de Rabat, Rabat, Morocco

Rachid Bouhfid (1, 27, 115, 181, 247), Moroccan Foundation for Advanced Science, Innovation and Research (MAScIR), Composites and Nanocomposites Center, Rabat Design Center, Rabat, Morocco

Hanane Chakhtouna (115), Moroccan Foundation for Advanced Science, Innovation and Research (MAScIR), Composites and Nanocomposites Center, Rabat Design Center; Laboratoire de Chimie Analytique et de Bromatologie, Faculté de Médecine et de Pharmacie, Université Mohamed V de Rabat, Rabat, Morocco

Likius Daniel (141), Faculty of Science, Department of Chemistry and Biochemistry, University of Namibia, Windhoek, Namibia

Mounir El Achaby (247), Materials Science and Nano-Engineering (MSN) Department, Mohammed VI Polytechnic University (UM6P), Ben Guerir, Morocco

Khadija El Bourakadi (1, 27), Moroccan Foundation for Advanced Science, Innovation and Research (MAScIR), Composites and Nanocomposites Center, Rabat Design Center, Rabat, Morocco

Hamid Essabir (181), Moroccan Foundation for Advanced Science, Innovation and Research (MAScIR), Composites and Nanocomposites Center, Rabat Design Center, Rabat; Mechanic, Materials, and Composites (MMC), Laboratory of Energy Engineering, Materials and Systems, National School of Applied Sciences of Agadir, Ibn Zohr University, Agadir, Morocco

Kamal Guerraoui (181), Equipe de Mécanique et des Matériaux (E.M.M), Centre de Recherche en Enérgie (C.R.E.), Faculty of Sciences, Mohammed V University in Rabat, Rabat, Morocco

G Hemath Kumar (223), Composite Research Center, Chennai, India

Kalsoom Jan (165), Department of Plastic Engineering, University of Massachusetts Lowell, Lowell, MA, United States

Zineb Kassab (247), Materials Science and Nano-Engineering (MSN) Department, Mohammed VI Polytechnic University (UM6P), Ben Guerir, Morocco

Hossein Kazemi (65), Department of Chemical Engineering, Université Laval, Quebec City, QC, Canada

Mohamed El Mehdi Mekhzoum (1), Moroccan Foundation for Advanced Science, Innovation and Research (MAScIR), Composites and Nanocomposites Center, Rabat Design Center, Rabat, Morocco

Frej Mighri (65), Department of Chemical Engineering, Université Laval, Quebec City, QC, Canada

H Mohit (223), Natural Composites Research Group Lab, Department of Materials and Production Engineering, The Siridhorn International Thai-German Graduate School of Engineering, King Mongkut's University of Technology North Bangkok, Bangkok, Thailand

V.S.R. Rajasekhar Pullabhotla (141), Department of Chemistry, University of Zululand, Richards Bay, South Africa

Abou el Kacem Qaiss (1, 27, 115, 181, 247), Moroccan Foundation for Advanced Science, Innovation and Research (MAScIR), Composites and Nanocomposites Center, Rabat Design Center, Rabat, Morocco

Ateeq Rahman (141), Faculty of Science, Department of Chemistry and Biochemistry, University of Namibia, Windhoek, Namibia

Marya Raji (181), Moroccan Foundation for Advanced Science, Innovation and Research (MAScIR), Composites and Nanocomposites Center, Rabat Design Center, Rabat, Morocco

P Ramesh (223), Department of Production Engineering, National Institute of Technology, Tiruchirappalli, India

Denis Rodrigue (65), Department of Chemical Engineering, Université Laval, Quebec City, QC, Canada

Aneeya K. Samantara (205), National Institute of Science Education and Research (NISER), Khordha, Odisha; Homi Bhabha National Institute (HBNI), Mumbai, India

MR Sanjay (223), Natural Composites Research Group Lab, Department of Materials and Production Engineering, The Siridhorn International Thai-German Graduate School of Engineering, King Mongkut's University of Technology North Bangkok, Bangkok, Thailand

Carlo Santulli (99), Geology Division, Università degli Studi di Camerino, School of Science and Technology, Camerino, Italy

Lakkoji Satish (205), Department of Chemistry, Ravenshaw University, Cuttack, Odisha, India

Fatima-Zahra Semlali Aouragh Hassani (247), Moroccan Foundation for Advanced Science, Innovation and Research (MAScIR), Composites and Nanocomposites Center, Rabat Design Center, Rabat, Morocco

Farnaz Shahamati Fard (65), Department of Chemical Engineering, Université Laval, Quebec City, QC, Canada

S Siengchin (223), Natural Composites Research Group Lab, Department of Materials and Production Engineering, The Siridhorn International Thai-German Graduate School of Engineering, King Mongkut's University of Technology North Bangkok, Bangkok, Thailand

Veikko Uahengo (141), Faculty of Science, Department of Chemistry and Biochemistry, University of Namibia, Windhoek, Namibia

Nadia Zari (115), Moroccan Foundation for Advanced Science, Innovation and Research (MAScIR), Composites and Nanocomposites Center, Rabat Design Center, Rabat, Morocco

Chapter 1

Cellulose nanocrystal/nanoparticles hybrid nanocomposites: From preparation to applications

Mohamed El Mehdi Mekhzoum, Khadija El Bourakadi, Abou el Kacem Qaiss, and Rachid Bouhfid
Moroccan Foundation for Advanced Science, Innovation and Research (MAScIR), Composites and Nanocomposites Center, Rabat Design Center, Rabat, Morocco

1.1 Introduction

During the past decades, the most used plastic materials are based on conventional petroleum-based polymer products in several areas leading to a wide range of environmental problems, as well as growing landfill space. Nevertheless, through the establishment of laws insisting on respect for the environment, the majority of industries substituted the petroleum-derived polymers with bio-based materials due to the efforts of scientists and engineers. In fact, the use of these bio-based polymers that are biocompatible, eco-friendly, and non-petroleum based are reducing carbon dioxide emissions into the atmosphere, as well as limiting the greenhouse effect. Therefore, the growing interest of biodegradable plastics and nanocomposites is justified by their renewable sources and could replace synthetic polymers, while decreasing our global dependence on fossil fuel sources [1, 2]. In order to reduce the effect of plastic consumption, most of the researchers and industries are focused on the use of cellulose as a reinforcing polymer at the micro- and nanoscale, which is by far one of the world's most abundant natural material and the world's oldest known biomaterial that can be obtained from virtually inexhaustible biomass feedstocks [3, 4]. More recent comprehensive review papers of this unique cellulosic structure have been published [5, 6].

Recently, the interest in the next generation of cellulose, or nanocellulose, has broadly increased owing to the specific chemical and physical properties of these materials such as high specific surface area, high elastic modulus, low

Cellulose Nanocrystal/Nanoparticles Hybrid Nanocomposites. http://doi.org/10.1016/B978-0-12-822906-4.00008-6

density, high transparency, high thermal stability, non-toxicity, lightweight, low cost, easy surface functionalization, biocompatibility, as well as biodegradability [7]. Interestingly, nanocellulose combines the important chemical properties of cellulose with attractive phenomena of nanoscale materials, originating from the very large surface area of these materials [8–11]. These fascinating properties make this material attractive for several applications such as mechanical reinforcing agent, thickening agent, rheological modifier, emulsion stabilizer, low-calorie food additive, scaffolds, carrier vehicles for controlled release systems, biosensor and bioimaging agent, and pharmaceutical binder [12]. In general, a nanocellulose is also a cellulose nanoparticle in the individual cellulose element with a diameter or a length on the order of a nanometer (<100 nm) [13]. Nanocellulose is typically divided into three major categories: cellulose nanofibrils (CNF), cellulose nanocrystals (CNC), and bacterial nanocellulose (BNC) [14]. In addition, the nomenclature of nanocellulose has not been fully standardized and some ambiguities are reported in the literature [15]. Fig. 1.1 shows the nomenclature, abbreviation, and dimensions applicable to each subgroup. Nevertheless, cellulose nanocrystal was recommended by the Technical Association of the Pulp and Paper Industry (TAPPI) as a standard nomenclature. A typical nanocellulose classification was reported in the review by [17]. These types of nanocellulose showed different properties which dictate their applicability and functionality; i.e., specific types of nanocellulose are better suited for specific applications than others. On the other hand, nanocellulose can be

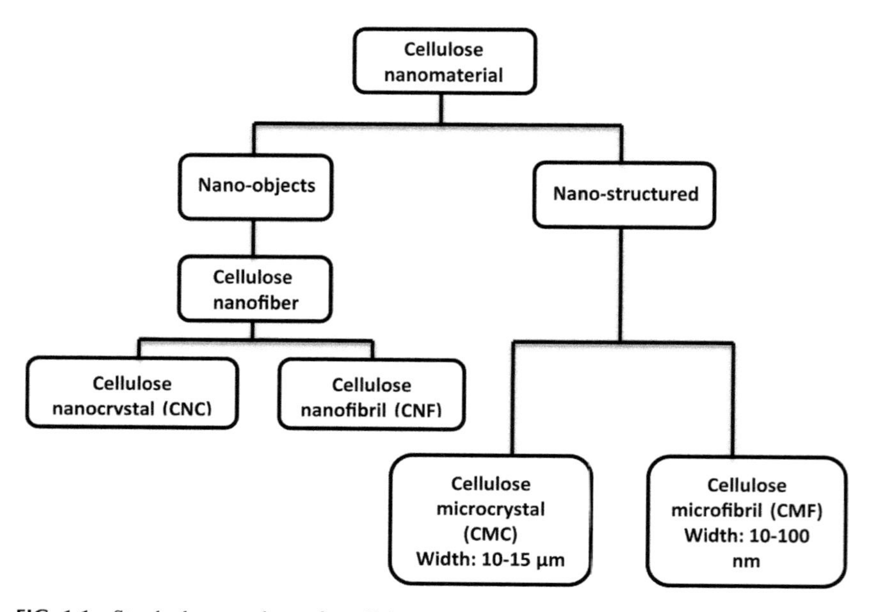

FIG. 1.1 Standard nomenclature for cellulose nanomaterials (TAPPI W13021) [16].

extracted from various natural resources by an appropriate method or combination thereof. The final properties, namely size, surface morphology, shape, and the yield of the nanocellulose materials, are dependent on the cellulose sources, pretreatment, and preparation method [18, 19].

Researchers have initially focused on isolating, characterizing, and using nanocellulose from cellulose to improve the performance of different products and develop materials with unique characteristics [20]. These nanocellulosic materials have the potential to reinforce a variety of polymer matrices either thermoplastic or thermoset [21], with low filler loadings from 1 to 10 wt% to improve the mechanical properties and other functional properties of the polymer nanocomposite materials [22]. These properties depend on the degree of dispersion of the nanocellulose in the polymer matrix and the nature of interfacial interactions [23–25]. A number of review papers focusing on the preparation and application of nanocellulose reinforced polymer composites have been published recently [26, 27]. Moreover, nanocellulose based polymer nanocomposites have enormous potential in several applications ranging from biomedical, packaging, and electronics to environmental and water treatment fields [28–30]. To broaden the field of application of nanocellulose based polymer nanocomposites, scientists focused on the hybridization of nanocellulose with other nanoparticles as a new promising way to develop high-performance polymer nanocomposites with excellent functional properties. By using a hybrid polymer nanocomposite containing two or more types of nanofillers, the advantages of one type of nanofiller could complement with what is lacking with the other. Therefore, a balance in cost and performance can be achieved through proper material design [31]. In this chapter, we provide an overview on this emerging nanocellulosic material, focusing on cellulose nanocrystal in terms of its structure, properties, extraction from lignocellulosic biomass, and some recent advances of cellulose nanocrystal/nanoparticles hybrid nanocomposites with their applications.

1.2 Cellulose nanocrystal: Structure, source, and properties

In general, cellulose is composed of two main domains that are crystalline and non-crystalline (amorphous). The latter acts as structural defects and can be easily removed by treating the cellulose under acid conditions. The obtained highly crystalline nano-structural cellulose from wood or plant cellulose biosynthesis processes which are frequently called cellulose nanocrystals (CNC), can be typically needle-like, ribbon-like, sphere-like, or rod-like shape nanoparticles with diameter less than 100 nm and length ranging from 10 to 100 nm [32]. Other synonymous terminologies are often found in the literature including cellulose nanowhiskers (CNW), or nanocrystalline cellulose [33]. These nomenclatures arise owing to their physical characteristics of stiffness, thickness, and length, with dimensions in the nanometer scale [34–36]. Fig. 1.2 depicts the location and extraction of cellulose nanocrystals.

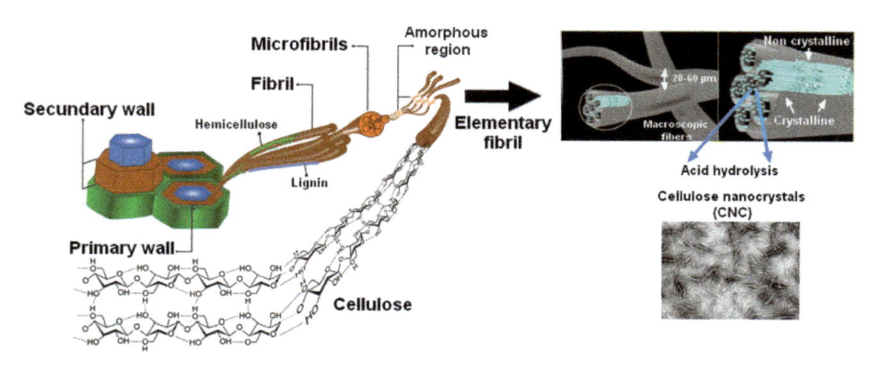

FIG. 1.2 Schematic illustration of cellulose nanocrystal from a vegetable source [37, 38].

Recently, CNC have been associated to a novel class of natural nanomaterials attracting much interest from both scientists and industrial engineers. It is now becoming an increasingly topical subject as confirmed by the increasing number of papers year after year [39, 40]. CNC are high-value green materials since they can convert the performance of existing products and help to generate new, unique, and improved products. They also have unique and appealing characteristics/properties that are highlighting them in relation to other types of nanocellulose namely their amphiphilic nature, chemical purity and reactivity, unique electrical and optical properties, nano-dimension, excellent mechanical properties, good renewability, biodegradability, and biocompatibility, high specific surface area, and aspect ratio, non- abrasive and non-toxic nature, low density and cost, self-assembling property, high crystallinity, rich hydroxyl groups for modification, excellent environmental benefits, high thermal and barrier properties [41, 42], CNC have been extensively used in various potential application including polymer reinforcement, antimicrobial and medical materials, biosensors, drilling fluids, drug delivery, green catalysts, barrier films, electronic devices, flexible substrates, emulsion stabilizers, coatings, textiles, aerospace, food packaging, cosmetics, paper making, pharmaceuticals, automotive industry, and building materials. Table 1.1 summarizes the main CNC properties and applications.

Interestingly, the surface hydroxyl groups and high specific surface offer an excellent platform for CNC chemical modification. Various hydrophobic groups with different functionality, such as fluorine, alkenyl, alkynyl, thiol groups, and pyridine moieties, have been attached to CNC [43–45]. This modified CNC, as precursors, can be used for further modifications including grafting smaller molecules (biomarkers or metal nanoparticles), and larger macromolecules namely proteins, thereby conferring the possibility of tailoring new properties and promoting their use in numerous practical applications. In addition, hydrophobization not only improves the poor dispersion of modified CNC in organic

TABLE 1.1 Properties and potential applications of CNC.

Properties	Potential applications
Few nm in diameter From 10 to 500 nm in length Crystallinity 60%–90% Few fold increase in tensile strength and modulus when incorporated in hydrophilic polymers Thermally stable up to 200–230 °C Tensile modulus ~150 GPa Tensile strength ~77 GPa	Drug delivery vehicle Rheology modifier Support for catalysts and sensors Diaphragms in earphones Tissue engineering, scaffolds Biomimetic foams Optical application Additive to drilling fluids Additive to cement-based materials Water pollutant remediation Toughened paper Flexible panels for flat panel displays Polymer nanocomposites for developing membranes, fibres, textiles, batteries, supercapacitors, electroactive polymers, sensors, and actuators

Data collected from George J, Sabapathi SN. Cellulose nanocrystals: synthesis, functional properties, and applications. Nanotechnol Sci Appl 2015;8:45–54, Kalia S, Dufresne A, Cherian BM, Kaith BS, Avérous L, Njuguna J, et al. Cellulose-based bio- and nanocomposites: a review. Int J Polym Sci 2011;2011, Wei H, Rodriguez K, Renneckar S, Vikesland PJ. Environmental science and engineering applications of nanocellulose-based nanocomposites. Environ Sci Nano 2014;1(4):302–16, Kaushik M, Moores A. Review: nanocelluloses as versatile supports for metal nanoparticles and their applications in catalysis. Green Chem 2016;18(3):622–37, Hurley BRA, Ouzts A, Fischer J, Gomes T. Paper presented at IAPRI world conference 2012: effects of private and public label packaging on consumer purchase patterns. Packag Technol Sci 2013;29(January):399–412, Hubbe MA, Ferrer A, Tyagi P, Yin Y, Salas C, Pal L, et al. Nanocellulose in thin films, coatings, and plies for packaging applications: a review. Bioresources 2017;12(1):2143–233, Moon RJ, Schueneman GT, Simonsen J. Overview of cellulose nanomaterials, their capabilities and applications, Jom 2016;68(9):2383–94, Kafy A, Kim HC, Zhai L, Kim JW, Hai L Van, Kang TJ, et al. Cellulose long fibers fabricated from cellulose nanofibers and its strong and tough characteristics. Sci Rep 2017;7(1):1–8.

solvents, but also enhances their compatibility and dispersibility inside polymer matrices [46]. Table 1.2 presents selected examples of the main CNC surface functionalization.

Depending on the origin of the starting cellulose, experimental technique used, and hydrolysis conditions, several CNC geometries, dimensions, and crystallinities can be obtained [51–53]. Typical CNC geometries originating from different cellulose sources are shown in Table 1.3. For instance, CNC extracted from wood (100-300 nm), have smaller length than CNC isolated from tunicates (>1000 nm) [51, 54]. A key parameter for CNC is the aspect ratio which is defined as the ratio between the length and the diameter (width). The aspect ratio can vary between 10 and 30 for cotton, up to 100 for tunicates.

TABLE 1.2 Examples of CNC surface functionalization.

Reaction pathway	Functionalizing agent	Research focus	Reference
Esterification	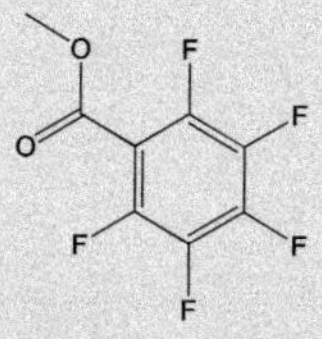 Methyl 2,3,4,5,6-pentafluorobenzoate	Hydrophobic and oleophobic materials	[47]
	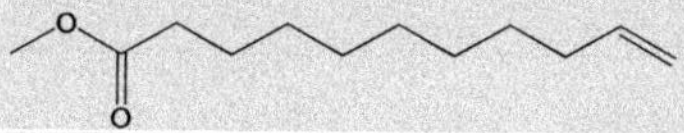 Methyl undec-10-enoate	Poly(butadiene) rubber reinforcement via cross-linking by thiol-ene click reaction	[44]
	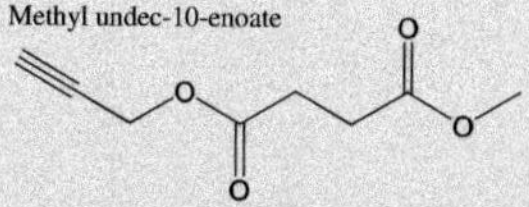 Methyl prop-2-yn-1-yl succinate	Reinforcement in GAP/PTPB polymer matrix	[48]
Silylation	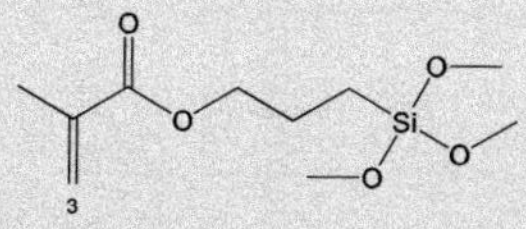 3-(trimethoxysilyl)propyl methacrylate	Carbon-carbon double bond for free radical polymerization	[43]
	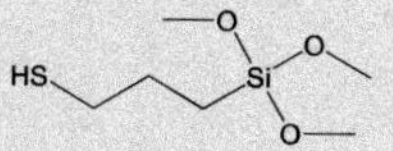 3-(trimethoxysilyl)propane-1-thiol	Allows facile functionalization of cellulose under mild conditions	[45]

Cationization	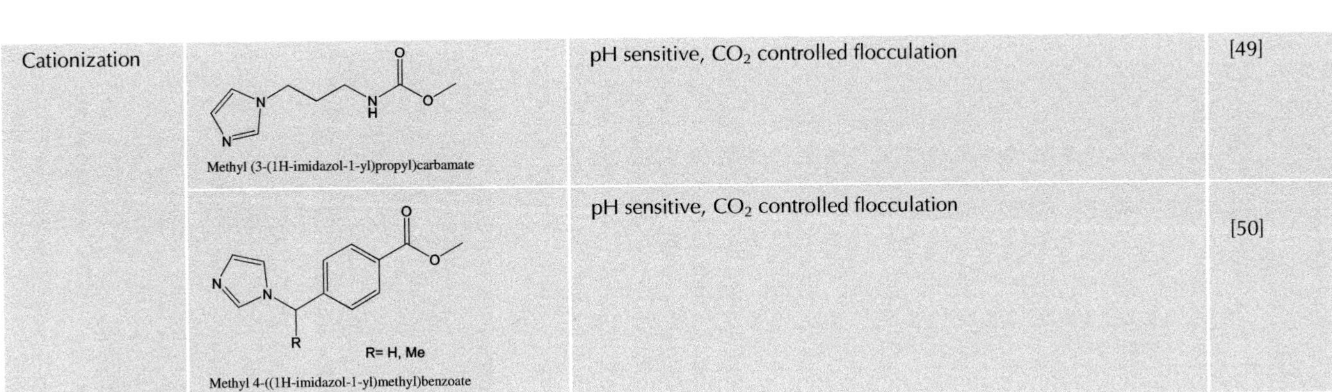 Methyl (3-(1H-imidazol-1-yl)propyl)carbamate	pH sensitive, CO_2 controlled flocculation	[49]
	R= H, Me Methyl 4-((1H-imidazol-1-yl)methyl)benzoate	pH sensitive, CO_2 controlled flocculation	[50]

TABLE 1.3 Characteristics (length, width, and aspect ratio) of some CNC from various sources.

Source	Length (nm)	Width (nm)	Aspect ratio	References
Wood	100–300	3–5	20–100	[53, 54]
Cotton	100–150	5–10	10–30	[55]
Ramie	70–200	5–15	~12	[22]
Banana pseudostems	123–147	5.3–9.1	21.2–2.8	[56]
Sisal	100–300	3–5	~60	[57]
Rice straw	117–270	11.2–30.7	8.8–10.5	[58]
Wastepaper	100–300	3–10	30–70	[59]
Tunicates	>1000	10–20	~100	[60]
Bacteria	100–1000	10–25	2–100	[61]

Numerous characterization techniques have been used to evaluate the morphology of different CNC including transmission electron microscopy (TEM), scanning electron microscopy (SEM), field-emission scanning electron microscopy (FE-SEM), and atomic force microscopy (AFM). The most conventional and common one is TEM. Several researchers isolated CNC from various sources and investigated the size and morphology of the obtained CNC. Fig. 1.3 presents typical transmission electron micrographs (TEM) of selected

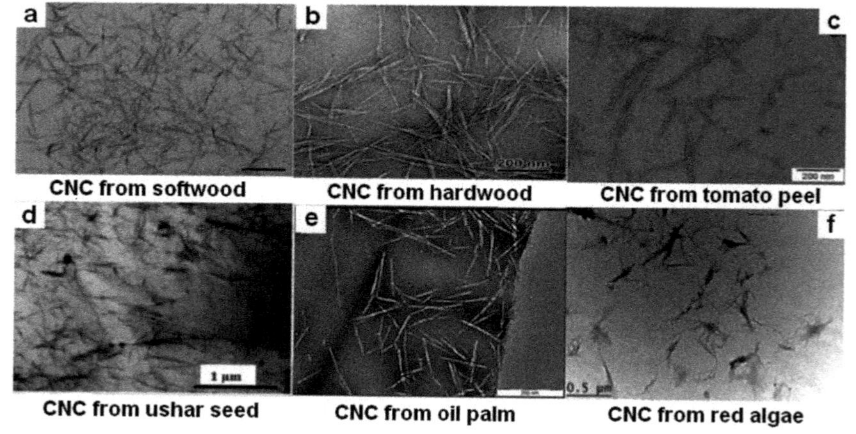

FIG. 1.3 Transmission electron microscope (TEM) images of cellulose nanocrystals derived from: (A) softwood, (B) hardwood, (C) tomato peel, (D) calotropis procera, (E) oil palm, and (F) red algae [62–67].

cellulose nanocrystals obtained from different cellulosic sources. As seen in Fig. 1.3, each rod or needle can be considered as a cellulose crystal with no apparent defect.

Hence, the extraction and analysis of the CNC characteristics from various abundant and renewable cellulosic resources is necessary and particularly relevant for the efficient comparison and application of these green resources [68–70]. In 1953, CNC were initially obtained from natural fibers by Mukherjee et al. [71]. Today, they can be isolated from several renewable bioresources including vegetable and animal sources [72, 73]. The main sources used for the production of CNC are cotton and wood pulp owing to their high cellulose content (more than 60%) and simple pretreatment method on their lignocellulosic structure [74]. Nevertheless, environmental concerns and decreasing forest resources caused by the increased demand for wood products led to increased interest towards non-wooden cellulosic materials such as [75, 76]: non-wood plants, tunicates, certain species of bacteria, and algae [77, 78]. So far, several studies have been conducted to successfully isolate CNC from numerous agricultural wastes or by-products: coconut fibers, maize cobs, peanut shells, stalks of cereal crops, sugar palm tree, soy hulls, rice husk, barley husk, tomato peels, walnut shell, vine shoots, pineapple crown waste, grape pomace, pistachio shells, grain straw, wheat straw, sugarcane bagasse, kenaf, waste cotton, kiwi pruning wastes, rice straw, and sesame husk [25, 62, 69, 75, 79–93]. These agricultural residues are of particular interest because of their abundance, low cost, renewability, and biodegradability [94]. Nevertheless, the production and yield efficiency of these extraction sources are low, making it essential to increase the scope of the biomass sources. Consequently, seeking of high-quality cellulose resources is highly desirable and strongly needed.

1.3 Production of cellulose nanocrystals

Generally, the production of CNC involves three stages: (i) mechanical size reduction and washing treatment, (ii) Purification by alkali and bleaching treatments, and (iii) liberation of cellulose nanocrystal. These general steps are briefly described in Fig. 1.4.

Firstly, the primary raw fibers (Fig. 1.5A) are broken down by milled/grinding/cutting with a mill by a mechanical treatment using high shear, high energy transfer, and high impact as seen in Fig. 1.5B [96], and then ground as well as repeatedly sieved with ~40–80 mesh screen to achieve a uniform size of fine particles/fibers [97, 98]. These steps increase the contact surface area and the swelling capacity of fiber in water, allowing them to be more efficiently treated by chemical methods [99]. Afterward, the resulting ground powder is washed with deionized water or some dewaxing solvents namely ethanol, hexane, methanol, benzene, and toluene [100, 101], to remove dirt and some other impurities including waxes, pectins, pigments, oils, fatty acids, phenolic substances, and chlorophyll [39]. The second stage consists of an alkalization treatment with sodium hydroxide (NaOH) or potassium hydroxide (KOH) under variable conditions (Fig. 1.5C) [102], followed

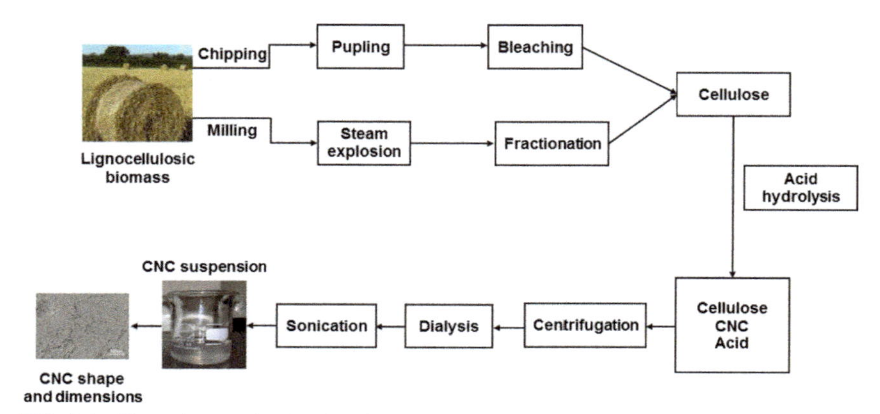

FIG. 1.4 The main steps involved in the preparation of cellulose nanocrystals [95].

FIG. 1.5 Photographs of vine shoots (VS) raw fiber leading to different materials: (A) collected raw vine shoots, (B) grinded VS fibers, (C) alkali-treated VS, (D) bleached VS fibers, (E) extracted cellulose nanocrystals (CNC), and (F) lyophilized CNC [82].

by a conventional bleaching treatment also known as delignification through acidified sodium chlorite ($NaClO_2$), or hydrogen peroxide (H_2O_2) with varying parameters (Fig. 1.5D) [103]. The purpose of the alkali-bleaching treatment is to remove or eliminate the non-cellulosic components in the raw fibers including pectins, lignins, hemicelluloses and other extractive substances [104], which are strongly linked with cellulose [105], thereby leading to the production of purified cellulose fibers with higher level of brightness [106]. In addition, these processes can modify the surface of natural fibers, improve the thermal stability, as well as the adhesion with the polymer matrix. However, there are various drawbacks of these treatments including their significant negative impact on the environment owing to the emission of toxic chlorinated organic compounds from bleached fibers, as well as the high amount of water consumed during the alkali process [107]. A detailed description of these treatments is available for the following cellulosic sources: plant, wood, algae, tunicate, and bacteria [108–112].

The last stage involves the conversion of the obtained purified cellulose into its nanodimension structure with high crystallinity, high aspect ratio, and variable geometries/dimensions [113]. Historically, the preparation of elongated rod-like CNC was firstly reported by Rånby et al. [114], who produced colloidal suspensions of CNC via sulfuric acid hydrolysis. Several processes have been used to isolate CNC including ultrasonic technique, enzymatic hydrolysis, TEMPO-mediated oxidation, acid hydrolysis, as well as mechanical disintegration [115, 116]. These treatments can be used alone, or in combination, to achieve the targeted CNC particle morphology. Among these methods, sulfuric acid hydrolysis treatment is by far the most popular and effective method since it produces sulfated CNC with high crystallinity and small particle size, improving the tensile strength, flexibility, and dispersibility of CNC in aqueous media [117]. Under strictly controlled hydrolysis conditions of agitation, duration, temperature, and sulfuric acid concentration, the less organized and more accessible fraction of cellulose, also called the amorphous or disordered regions, were preferentially hydrolyzed or attacked by hydronium ions to form glucose molecules, while the crystalline regions remained intact due to their high resistance to acid attack [118]. Overall, a prolonged reaction time in H_2SO_4 increases the CNC crystallinity to 70%–80%, while both CNC aspect ratio and length decrease with increasing temperature or hydrolysis time [119]. In addition, sulfuric acid concentration significantly affects the CNC in yield, shape, size, and crystalline structure [120]. Unfortunately, sulfuric acid hydrolysis has some drawbacks including large water consumption, severe operating conditions, equipment corrosion, high costs, and generation of high amount of waste [121]. As an alternative to sulfuric acid hydrolysis, other acids (hydrochloric, nitric, formic, citric, phosphoric, and hydrobromic) or some organic acids (oxalic, *p*-toluenesulfonic, maleic, and benzenesulphonic) are also applied to hydrolyze cellulose into CNC [122, 123]. In this sense, each acid treatment confers specific functional groups on the surface of the isolated CNC having a direct effect on their colloidal stability. For instance, CNC produced from HCl exhibit poor colloidal stability and their aqueous suspensions tend to flocculate [124]. Some acid hydrolysis processes of CNC obtained from various cellulosic sources are reported in Table 1.4. Moreover, the CNC aqueous suspensions are in the form of a white gel as shown in Fig. 1.5E. After hydrolysis, the resulting CNC suspension is washed with cold water and then centrifuged until constant pH (neutral) [132, 133]. A supplementary neutralization step of the CNC suspension is conducted using dialysis tubing with ultra-pure water, deionized water, as well as distilled water using regenerated cellulose membranes to remove any free acid

TABLE 1.4 Various acid hydrolysis procedures to extract CNC from various cellulosic sources.

Cellulosic source	Process	References
Waste paper	H_2SO_4 hydrolysis	[59]
Rice husk	H_2SO_4 hydrolysis	[25]
Wood cotton cloth	HCl, H_2SO_4 hydrolysis	[125]
Kenaf fiber	HCl hydrolysis	[126]
Oil palm trunk	H_2SO_4 hydrolysis	[127]
Posidonia oceanica leaves	H_2SO_4 hydrolysis	[128]
Cotton	H_2SO_4 hydrolysis	[129]
Tomato peels	H_2SO_4 hydrolysis	[62]
Corn cob	H_2SO_4 hydrolysis	[68]
Coconut husk	HNO_3, H_2SO_4 hydrolysis	[96]
Hibiscus leaf	Oxalic acid hydrolysis	[130]
Cotton	H_3PO_4 hydrolysis	[131]

molecules [134]. Finally, a dry CNC powder is obtained through freeze-drying for easier use in industrial applications as illustrated in Fig. 1.5F.

1.4 Cellulose nanocrystal/nanoparticles hybrid nanocomposites

Recently, the use of CNC as a new class of reinforcing agents for polymer nanocomposite development has attracted considerable interest in the materials community, owing to its nanoscale dimensions, large specific surface area, unique morphology, low density, biocompatibility, availability, surface functionality, as well as unique optical, rheological and mechanical properties, and most importantly biodegradable in nature [135, 136]. In 1995, Favier et al. [137] reported the first research study related to the use of CNC as fillers in the field of nanocomposites. A literature survey reported that CNC have the potential to significantly reinforce polymers at 1 wt%. Their excellent properties make CNC a promising class of materials for the production of cost-effective, highly durable polymer nanocomposites with high mechanical, thermal, and barrier properties than neat polymers or conventional composites [138, 139]. However, the main issue of CNC nanoparticles is their homogeneous dispersion inside a polymeric matrix. To solve this issue, modification of the CNC surface to enhance compatibility with hydrophobic matrices in necessary. On the other hand, CNC have abundant hydroxyl groups on their surfaces which facilitates their dispersions with water-soluble polymers [140]. Furthermore, CNC-based polymer nanocomposites have various potential applications in automotive, construction, and aerospace industries, as well as electronics, cosmetics, packaging, and biomedicine purposes [141].

Several hydrophilic and hydrophobic polymeric matrices have been used: starch, chitosan, polyvinyl alcohol (PVA), polylactic acid (PLA), polycaprolactone (PCL), epoxy, polypropylene (PP), polystyrene (PS), and polyethylene (PE) [142, 143]. The most common uses for the preparation of CNC-reinforced nanocomposites are shown in Fig. 1.6. Generally, CNC have been applied to reinforce polymers by conventional processes including solvent casting, melt compounding, electrospinning, freeze-drying, as well as hot-pressed processing techniques. The CNC-based nanocomposites are produced by mixing a stable CNC aqueous suspension within water soluble or water dispersible polymers, and the resulting mixture is usually cast with solvent evaporation to form a solid nanocomposite film.

More recently, hybrid inorganic/organic polymer nanocomposites represent an emerging new class of nanomaterials exhibiting enhanced thermal, optical, and mechanical properties due to the synergistic effect resulting from chemical or physical interactions between inorganic and organic nanofillers [144]. In addition, these hybrid polymer nanocomposites present advantages of light weight, flexibility, low cost, as well as high impact resistance. Several studies have been published on the combination of different reinforcements in a polymer matrix to obtain multifunctional polymeric systems with high performance and enhanced properties [145, 146]. In particularly, CNC/nanoparticles hybrid nanocomposites have recently attracted much attention both from academia and industry. In these materials, CNC and numerous inorganic nanoparticles such as clays (montmorillonite and organoclays), have been combined in a polymer matrix to obtain unusual hybrid nanocomposites. Arjmandi et al. [147] investigated the effect of CNC on the biodegradability, thermal, tensile, morphological, water absorption, and transparency properties of PLA/MMT/CNW hybrid nanocomposites prepared by solution casting. The results showed that the replacement of 1 wt% of nanoclay with 1 wt% of CNC in a PLA/MMT nanocomposite containing 5 wt% nanoclay led to an eight-fold increase (10% to 80%) of strain to failure for the hybrid nanocomposites. The highest tensile strength for the hybrid nanocomposites (~36 MPa) was obtained by incorporating 1 wt% of CNC into PLA/MMT nanocomposite. The PLA/MMT/CNW hybrid nanocomposites were highly transparent and exhibited similar performance as PLA/MMT nanocomposite and neat PLA. More recently, Vaezi et al. [148] synthesized cationic starch/montmorillonite/nanocrystal cellulose bionanocomposites by solution casting. They found that hybrid bionanocomposites with 5 wt% of CNC and MMT showed the best improvement in barrier properties against oxygen and water vapor, as well as mechanical properties: the tensile strength and tensile modulus increased by up to 61% and 73%

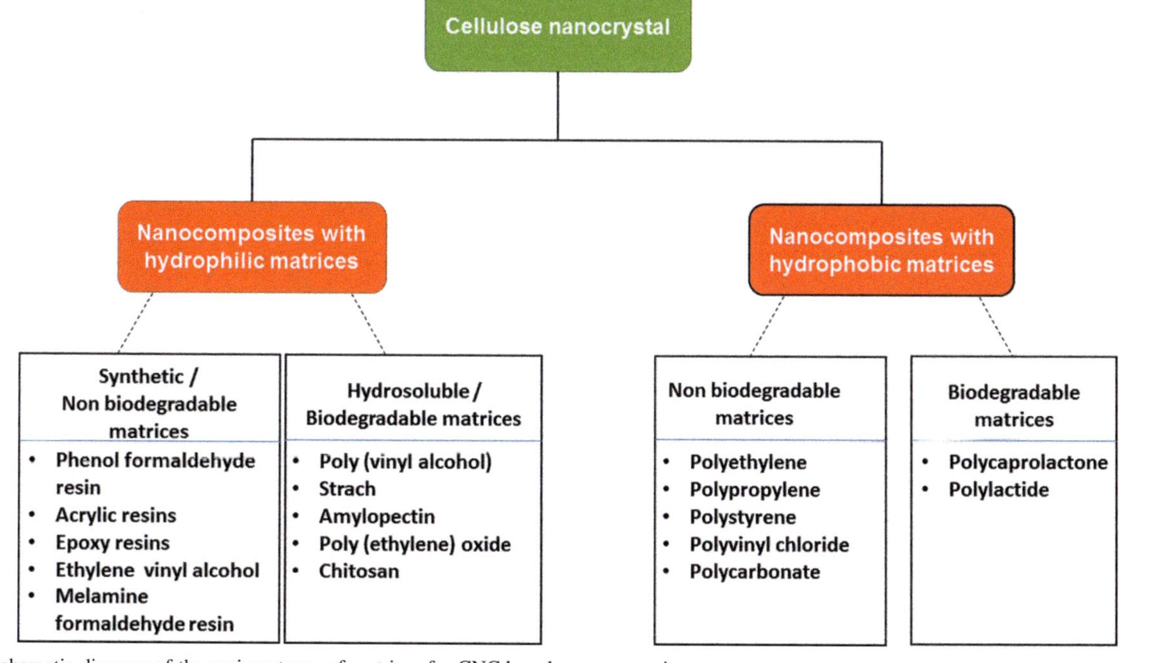

FIG. 1.6 Schematic diagram of the various types of matrices for CNC based nanocomposites.

respectively, compared to the neat chitosan film. Similarly, Naderizadeh et al. [149] blended starch with poly(vinyl alcohol) (PVA) to enhance its film formability and then added MMT and starch nanocrystals (SNC) to the blend to enhance the mechanical properties films produced by solvent casting. The tensile tests and thermogravimetric analysis (TGA) showed that the presence of hybrid nanofillers SNC and Na-MMT at 3 wt% of the total amount of nanofillers (50:50 (w/w)) led to improved thermal stability (an increase in onset decomposition temperature and residual mass), and mechanical properties (tensile strength and Young's modulus improved by 113% and 917%, respectively).

Hybrid polymer nanocomposites based on CNC and other organic nanoparticles, such as graphene and reduced graphene oxide, have also been extensively studied. Using a simple and environmentally friendly process, cellulose nanocrystal-stabilized graphene (GR-CNC) was incorporated into poly(vinyl alcohol) (PVA) aqueous solutions to obtain PVA-based nanocomposites (GR-CNC/PVA) by a casting method. At 1 wt% nanofiller loading, GR-CNC/PVA exhibited improvements in tensile strength and Young's modulus of about 20% and 50% respectively, compared with the neat PVA, CNC/PVA and GR-T/PVA. This substantial reinforcement found for GR-CNC/PVA is ascribed to the strong interaction between PVA and GR-CNC, suggesting the synergistic reinforcing effect of CNC and graphene (GR) on the properties of the hybrid nanocomposites [150]. In another interesting study, Pal et al. [151] developed a novel PLA/CNC/reduced graphene oxide (rGO) nanocomposite via solution casting. They incorporated cellulose nanocrystals (CNC) and rGO into polylactic acid (PLA) matrix using different concentrations of CNC and rGO. The effect of CNC and rGO addition on the crystallinity, morphology, thermal, mechanical, and bio-medical properties were investigated. The results showed a good dispersion of CNC and rGO within the PLA matrix leading to higher tensile strength (23%) and Young's modulus (29%), as well as higher onset decomposition temperature (from 340 °C for PLA to 350 °C for PLA/CNC/rGO). In addition, better biocompatibility and improved antibacterial response were also reported. The resulting hybrid nanocomposite film can be a promising alternative for biomedical and packaging applications.

In their recent work, Pal et al. [152] reported a novel PLA based biodegradable nanocomposite film reinforced with modified CNC and rGO through solvent casting. To improve the uniform dispersion of CNC and rGO in PLA, the CNC surface was modified using TEMPO oxidation followed by surface grafting of TEMPO-oxidized CNC (TOCNC) performed with polyethylene glycol (PEG). According to the results, the surface modification of CNC and addition of rGO at the optimum content (94.5% PLA, 5% PEG-TOCCNC and 0.5% rGO) enhanced the mechanical properties (tensile strength and Young's modulus were 47.54 MPa and 4.305 GPa, respectively) of the film compared to the PLA matrix. In addition, the PEG-TOCNC/rGO/PLA nanocomposite film exhibited improvement in stiffness, strength, and toughness (modulus ranging

2000–4300 GPa and tensile strengths of 30–47 MPa) with increasing blend ratio of PEG-TOCNC in PLA/rGO matrix. As a result, the reinforcement or the combined effect of CNC and rGO resulted into outstanding increase in PLA properties. These PLA/PEG-TOCNC/rGO nanocomposites can be used as scaffolds in tissue engineering and in material packaging.

Recently, researchers have shown a great interest in the combination of CNC and carbon nanotubes (CNT) as reinforcements to enhance the mechanical properties, as well as the electrical conductivity, of polymer matrices. In light of these facts, Yu et al. [153] proposed an environmentally friendly strategy to fabricate PLA/CNC/CNT composites with excellent electrical conductivity, as well as good mechanical properties (tensile strength=45 MPa, Young's modulus=3152 MPa) through a Pickering emulsion method. In addition, the resulting composites were shown to have an outstanding electromagnetic interference (EMI) shielding of 41.8 dB with only 4.3 wt% CNT. Zhu et al. [154] reported the preparation of flexible TPU/CNT−CNC composites strain sensor in one step. In their method, carbon nanotube (CNT) suspensions with cellulose nanocrystals (CNC) were directly pumped into porous electrospun thermoplastic polyurethane (TPU) membranes through a simple filtration process. The authors showed that the strain sensor can be used in visual control and detection of the full range of human body motions, as well as large-scale human body motion measurements.

Metal nanoparticles are important types of nanofillers that have been successfully used in the functionalization of polymer materials. CNC and metal nanoparticles were also combined in a polymer matrix. Liu et al. [155] synthesized nanocomposites composed on bifunctional nanofillers, like carboxylated cellulose nanocrystals (CCN) and silver nanoparticles (AgNP), to enhance the mechanical and antimicrobial properties of waterborne polyurethane (WPU). The morphology, mechanical behavior, and antibacterial activity of WPU-based composites and neat WPU were analyzed. It was found that the WPU-based composite films were homogeneous and highly reinforced. The WPU/CCN/AgNP composite films also showed a strong antibacterial activity against *E. coli* and *S. aureus*. Therefore, the CCN/AgNP nanocomposites can be used as bifunctional nanofillers within WPU. A ternary bionanocomposite film based on PLA with modified cellulose nanocrystals (s-CNC) and synthesized silver nanoparticles (Ag) was produced via solvent casting and further characterized. It was shown that a homogeneous particle dispersion in the polymer matrix did not affect the original PLA transparency. But reduction in PLA barrier properties was detected for the ternary systems loaded with 1 wt% Ag, suggesting the positive effect of silver and its combination with cellulose nanostructures. In addition, PLA bionanocomposites showed a significant antibacterial activity influenced by the Ag content. Thus, these new biodegradable PLA/CCN/AgNP nanocomposites can be used for fresh food packaging applications [156]. Furthermore, Rescignano et al. [157] reported the production of a new generation of hybrid bionanocomposites based on a PVA matrix reinforced with a combination of cellulose nanocrystals and poly(D,L-lactide-coglycolide) (PLGA) nanoparticles (NP) loaded with bovine serum albumin (BSA) fluorescein isothiocynate conjugate (FITC-BSA). The results showed that the bionanocomposite films are suitable to successfully transport biopolymeric nanoparticles to adult bone marrow mesenchymal stem cells, representing a new tool for drug delivery strategies. These PVA ternary bio-nanocomposites can be used for therapeutic applications.

1.5 Conclusion

In summary, cellulose nanocrystal reinforcements have attracted substantial attention from scientific and industrial areas owing to their high surface functionality and reactivity, renewable nature, availability throughout the world, low density, and low cost. These cellulosic nanofillers can be applied as eco-friendly reinforcing agents for various polymers even at low filler loadings (1 wt%). Cellulose nanocrystal/nanoparticles hybrid nanocomposites, in which a cellulosic nanoreinforcement is combined with another nanoparticle, offer the possibility of broadening the area of applications of polymer nanocomposites. In this chapter, the structure, properties, sources, as well as extraction techniques of CNC were discussed. Some recent investigations related to the hybridization of CNC with numerous types of inorganic/organic nanoparticles such as clay, graphene, carbon nanotube, and metal nanoparticles into polymers were briefly reviewed for a wide variety of functional applications such as biomedical, packaging, electronics, therapeutic, and tissue engineering.

References

[1] Thakur VK, Voicu SI. Recent advances in cellulose and chitosan based membranes for water purification: a concise review. Carbohydr Polym 2016;146:148–65.

[2] Voicu SI, Condruz RM, Mitran V, Cimpean A, Miculescu F, Andronescu C, et al. Sericin covalent immobilization onto cellulose acetate membrane for biomedical applications. ACS Sustain Chem Eng 2016;4(3):1765–74.

[3] Park NM, Choi S, Oh JE, Hwang DY. Facile extraction of cellulose nanocrystals. Carbohydr Polym 2019;223(June):115114.

[4] Griffith JD, Willcox S, Powers DW, Nelson R, Baxter BK. Discovery of abundant cellulose microfibers encased in 250 Ma permian halite: a macromolecular target in the search for life on other planets. Astrobiology 2008;8(2):215–28.

[5] Bhat AH, Khan I, Usmani MA, Umapathi R, Al-Kindy SMZ. Cellulose an ageless renewable green nanomaterial for medical applications: an overview of ionic liquids in extraction, separation and dissolution of cellulose. Int J Biol Macromol 2019;129:750–77.

[6] Mokhena TC, John MJ. Cellulose nanomaterials: new generation materials for solving global issues. In: Cellulose, Vol. 27. Netherlands: Springer; 2020. p. 1149–94.

[7] Belouadah Z, Ati A, Rokbi M. Characterization of new natural cellulosic fiber from Lygeum spartum L. Carbohydr Polym 2015;134:429–37.

[8] Thomas MG, Abraham E, Jyotishkumar P, Maria HJ, Pothen LA, Thomas S. Nanocelluloses from jute fibers and their nanocomposites with natural rubber: preparation and characterization. Int J Biol Macromol 2015;81:768–77.
[9] Gazzotti S, Rampazzo R, Hakkarainen M, Bussini D, Ortenzi MA, Farina H, et al. Cellulose nanofibrils as reinforcing agents for PLA-based nanocomposites: an in situ approach. Compos Sci Technol 2019;171(October 2018):94–102.
[10] Lee KY, Tang M, Williams CK, Bismarck A. Carbohydrate derived copoly(lactide) as the compatibilizer for bacterial cellulose reinforced polylactide nanocomposites. Compos Sci Technol 2012;72(14):1646–50.
[11] Abbasian M, Hasanzadeh P, Mahmoodzadeh F, Salehi R. Novel cationic cellulose-based nanocomposites for targeted delivery of methotrexate to breast cancer cells. J Macromol Sci Part A Pure Appl Chem 2020;57(2):99–115.
[12] Abitbol T, Rivkin A, Cao Y, Nevo Y, Abraham E, Ben-Shalom T, et al. Nanocellulose, a tiny fiber with huge applications. Curr Opin Biotechnol 2016;39(I):76–88.
[13] Nasir M, Hashim R, Sulaiman O, Asim M. Nanocellulose: preparation methods and applications. In: Cellulose-reinforced nanofibre composites: production, properties and applications. Elsevier Ltd; 2017. p. 261–76.
[14] Farooq A, Patoary MK, Zhang M, Mussana H, Li M, Naeem MA, et al. Cellulose from sources to nanocellulose and an overview of synthesis and properties of nanocellulose/zinc oxide nanocomposite materials. Int J Biol Macromol 2020;154:1050–73.
[15] Charreau H, Cavallo E, Foresti ML. Patents involving nanocellulose: analysis of their evolution since 2010. Carbohydr Polym 2020;237(February):116039.
[16] Mariano M, El Kissi N, Dufresne A. Cellulose nanocrystals and related nanocomposites: review of some properties and challenges. J Polym Sci Part B Polym Phys 2014;52(12): 791–806.
[17] Moon RJ, Martini A, Nairn J, Simonsen J, Youngblood J. Cellulose nanomaterials review: structure, properties and nanocomposites. Chem Soc Rev 2011;40:3941–94.
[18] Nechyporchuk O, Belgacem MN, Bras J. Production of cellulose nanofibrils: a review of recent advances. Ind Crop Prod 2016;93:2–25.
[19] Hokkanen S, Bhatnagar A, Sillanpää M. A review on modification methods to cellulose-based adsorbents to improve adsorption capacity. Water Res 2016;91:156–73.
[20] Perumal AB, Sellamuthu PS, Nambiar RB, Sadiku ER. Development of polyvinyl alcohol/chitosan bio-nanocomposite films reinforced with cellulose nanocrystals isolated from rice straw. Appl Surf Sci 2018;449:591–602.
[21] Azizi Samir MAS, Alloin F, Dufresne A. Review of recent research into cellulosic whiskers, their properties and their application in nanocomposite field. Biomacromolecules 2005;6(2): 612–26.
[22] Junior de Menezes A, Siqueira G, Curvelo AAS, Dufresne A. Extrusion and characterization of functionalized cellulose whiskers reinforced polyethylene nanocomposites. Polymer (Guildf) 2009;50(19):4552–63.
[23] Chen W, Yu H, Liu Y, Chen P, Zhang M, Hai Y. Individualization of cellulose nanofibers from wood using high-intensity ultrasonication combined with chemical pretreatments. Carbohydr Polym 2011;83(4):1804–11.
[24] Hsieh YC, Yano H, Nogi M, Eichhorn SJ. An estimation of the Young's modulus of bacterial cellulose filaments. Cellulose 2008;15(4):507–13.
[25] Johar N, Ahmad I, Dufresne A. Extraction, preparation and characterization of cellulose fibres and nanocrystals from rice husk. Ind Crop Prod 2012;37(1):93–9.
[26] Tshikovhi A, Mishra SB, Mishra AK. Nanocellulose-based composites for the removal of contaminants from wastewater. Int J Biol Macromol 2020;152:616–32.

[27] Choudhary P, Jaiswal A, Singh S, Gupta SK. Bacterial cellulose based composites: preparation and characterization. Mater Sci Forum 2020;978 MSF:183–90.
[28] Zhao Y, Zhang Y, Lindström ME, Li J. Tunicate cellulose nanocrystals: preparation, neat films and nanocomposite films with glucomannans. Carbohydr Polym 2015;117:286–96.
[29] Kumbhar JV, Jadhav SH, Bodas DS, Barhanpurkar-Naik A, Wani MR, Paknikar KM, et al. In vitro and in vivo studies of a novel bacterial cellulose-based acellular bilayer nanocomposite scaffold for the repair of osteochondral defects. Int J Nanomedicine 2017;12:6437–59.
[30] Flauzino Neto WP, Mariano M, da Silva ISV, Silvério HA, Putaux JL, Otaguro H, et al. Mechanical properties of natural rubber nanocomposites reinforced with high aspect ratio cellulose nanocrystals isolated from soy hulls. Carbohydr Polym 2016;153:143–52.
[31] Thwe MM, Liao K. Durability of bamboo-glass fiber reinforced polymer matrix hybrid composites. Compos Sci Technol 2003;63(3–4):375–87.
[32] Fortunati E, Puglia D, Monti M, Peponi L, Santulli C, Kenny JM, et al. Extraction of cellulose nanocrystals from Phormium tenax fibres. J Polym Environ 2013;21(2):319–28.
[33] Siqueira G, Tapin-Lingua S, Bras J, da Silva PD, Dufresne A. Mechanical properties of natural rubber nanocomposites reinforced with cellulosic nanoparticles obtained from combined mechanical shearing, and enzymatic and acid hydrolysis of sisal fibers. Cellulose 2011;18(1):57–65.
[34] Kamel S. Nanotechnology and its applications in lignocellulosic composites, a mini review. Express Polym Lett 2007;1(9):546–75.
[35] Ludueña LN, Vecchio A, Stefani PM, Alvarez VA. Extraction of cellulose nanowhiskers from natural fibers and agricultural byproducts. Fibers Polym 2013;14(7):1118–27.
[36] Nascimento DM, Almeida JS, Dias AF, Figueirêdo MCB, Morais JPS, Feitosa JPA, et al. A novel green approach for the preparation of cellulose nanowhiskers from white coir. Carbohydr Polym 2014;110:456–63.
[37] Salas C, Nypelö T, Rodriguez-Abreu C, Carrillo C, Rojas OJ. Nanocellulose properties and applications in colloids and interfaces. Curr Opin Colloid Interface Sci 2014;19(5): 383–96.
[38] Rojas J, Bedoya M, Ciro Y. Current trends in the production of cellulose nanoparticles and nanocomposites for biomedical applications. In: Cellulose—fundamental aspects of current trends. InTech; 2015.
[39] Malucelli LC, Lacerda LG, Dziedzic M, da Silva Carvalho Filho MA. Preparation, properties and future perspectives of nanocrystals from agro-industrial residues: a review of recent research. Rev Environ Sci Biotechnol 2017;16(1):131–45.
[40] Lagerwall JPF, Schütz C, Salajkova M, Noh J, Park JH, Scalia G, et al. Cellulose nanocrystal-based materials: from liquid crystal self-assembly and glass formation to multifunctional thin films. NPG Asia Mater 2014;6(1):1–12.
[41] Dufresne A. Nanocellulose: a new ageless bionanomaterial. Mater Today 2013;16(6):220–7.
[42] Geyer U, Heinze T, Stein A, Klemm D, Marsch S, Schumann D, et al. Formation, derivatization and applications of bacterial cellulose. Int J Biol Macromol 1994;16(6):343–7.
[43] Yang J, Han CR, Duan JF, Ma MG, Zhang XM, Xu F, et al. Studies on the properties and formation mechanism of flexible nanocomposite hydrogels from cellulose nanocrystals and poly(acrylic acid). J Mater Chem 2012;22(42):22467–80.
[44] Rosilo H, Kontturi E, Seitsonen J, Kolehmainen E, Ikkala O. Transition to reinforced state by percolating domains of intercalated brush-modified cellulose nanocrystals and poly(butadiene) in cross-linked composites based on thiol-ene click chemistry. Biomacromolecules 2013; 14:1547–54.
[45] Huang J-L, Li C-J, Gray DG. Functionalization of cellulose nanocrystal films via "thiol-ene" click reaction. RSC Adv 2014;4(14):6965–9.

[46] Tingaut P, Zimmermann T, Sèbe G. Cellulose nanocrystals and microfibrillated cellulose as building blocks for the design of hierarchical functional materials. J Mater Chem 2012;22(38): 20105.

[47] Salam A, Lucia LA, Jameel H. Fluorine-based surface decorated cellulose nanocrystals as potential hydrophobic and oleophobic materials. Cellulose 2015;22(1):397–406.

[48] Chen J, Lin N, Huang J, Dufresne A. Highly alkynyl-functionalization of cellulose nanocrystals and advanced nanocomposites thereof via click chemistry. Polym Chem 2015;6(24): 4385–95.

[49] Wang HD, Jessop PG, Bouchard J, Champagne P, Cunningham MF. Cellulose nanocrystals with CO2-switchable aggregation and redispersion properties. Cellulose 2015;22(5):3105–16.

[50] Eyley S, Vandamme D, Lama S, Van Den Mooter G, Muylaert K, Thielemans W. CO2 controlled flocculation of microalgae using pH responsive cellulose nanocrystals. Nanoscale 2015;7(34):14413–21.

[51] Elazzouzi-hafraoui S, Nishiyama Y, Putaux J, Heux L, Dubreuil F, Rochas C. The shape and size distribution of crystalline nanoparticles prepared by acid hydrolysis of native cellulose. Biomacromolecules 2008;9:57–65.

[52] Martins DF, de Souza AB, Henrique MA, Silvério HA, Flauzino Neto WP, Pasquini D. The influence of the cellulose hydrolysis process on the structure of cellulose nanocrystals extracted from capim mombaça (Panicum maximum). Ind Crop Prod 2015;65:496–505.

[53] Beck-Candanedo S, Roman M, Gray DG. Effect of reaction conditions on the properties and behavior of wood cellulose nanocrystal suspensions. Biomacromolecules 2005;6(2): 1048–54.

[54] Peng BL, Dhar N, Liu HL, Tam KC. Chemistry and applications of nanocrystalline cellulose and its derivatives: a nanotechnology perspective. Can J Chem Eng 2011 Oct;89(5): 1191–206.

[55] Araki J, Wada M, Kuga S. Steric stabilization of a cellulose microcrystal suspension by poly (ethylene glycol) grafting. Langmuir 2001;17(1):21–7.

[56] Pereira ALS, Nascimento DMD, Souza Filho MDSM, Morais JPS, Vasconcelos NF, Feitosa JPA, et al. Improvement of polyvinyl alcohol properties by adding nanocrystalline cellulose isolated from banana pseudostems. Carbohydr Polym 2014;112:165–72.

[57] Garcia de Rodriguez NL, Thielemans W, Dufresne A. Sisal cellulose whiskers reinforced polyvinyl acetate nanocomposites. Cellulose 2006;13(3):261–70.

[58] Lo HY. Cellulose nanocrystals and self-assembled nanostructures from cotton, rice straw and grape skin: a source perspective. J Mater Sci 2013;48(22):7837–46.

[59] Danial WH, Abdul Majid Z, Mohd Muhid MN, Triwahyono S, Bakar MB, Ramli Z. The reuse of wastepaper for the extraction of cellulose nanocrystals. Carbohydr Polym 2015; 118:165–9.

[60] Kimura F, Kimura T, Tamura M, Hirai A, Ikuno M, Horii F. Magnetic alignment of the chiral nematic phase of a cellulose microfibril suspension. Langmuir 2005;21(5):2034–7.

[61] George J, Bawa AS, Siddaramaiah. Synthesis and characterization of bacterial cellulose nanocrystals and their PVA nanocomposites. Adv Mater Res 2010;123–125:383–6.

[62] Jiang F, Lo HY. Cellulose nanocrystal isolation from tomato peels and assembled nanofibers. Carbohydr Polym 2015;122:60–8.

[63] Chen YW, Lee HV, Juan JC, Phang SM. Production of new cellulose nanomaterial from red algae marine biomass Gelidium elegans. Carbohydr Polym 2016;151:1210–9.

[64] Rohaizu R, Wanrosli WD. Sono-assisted TEMPO oxidation of oil palm lignocellulosic biomass for isolation of nanocrystalline cellulose. Ultrason Sonochem 2017;34:631–9.

[65] Oun AA, Rhim JW. Characterization of nanocelluloses isolated from Ushar (Calotropis procera) seed fiber: effect of isolation method. Mater Lett 2016;168:146–50.

[66] Flauzino Neto WP, Putaux JL, Mariano M, Ogawa Y, Otaguro H, Pasquini D, et al. Comprehensive morphological and structural investigation of cellulose I and II nanocrystals prepared by sulphuric acid hydrolysis. RSC Adv 2016;6(79):76017–27.

[67] Salajková M, Berglund LA, Zhou Q. Hydrophobic cellulose nanocrystals modified with quaternary ammonium salts. J Mater Chem 2012;22(37):19798–805.

[68] Silvério HA, Flauzino Neto WP, Dantas NO, Pasquini D. Extraction and characterization of cellulose nanocrystals from corncob for application as reinforcing agent in nanocomposites. Ind Crop Prod 2013;44:427–36.

[69] Pires W, Neto F, Alves H, Oliveira N, Pasquini D. Extraction and characterization of cellulose nanocrystals from agro-industrial residue—soy hulls. Ind Crop Prod 2013;42:480–8.

[70] Chen W, Yu H, Liu Y, Hai Y, Zhang M, Chen P. Isolation and characterization of cellulose nanofibers from four plant cellulose fibers using a chemical-ultrasonic process. Cellulose 2011;18(2):433–42.

[71] Mukherjee SM, Woods HJ. X-ray and electron microscope studies of the degradation of cellulose by sulphuric acid. Biochim Biophys Acta 1953;10(C):499–511.

[72] da Silva ISV, Neto WPF, Silvério HA, Pasquini D, Zeni Andrade M, Otaguro H. Mechanical, thermal and barrier properties of pectin/cellulose nanocrystal nanocomposite films and their effect on the storability of strawberries (Fragaria ananassa). Polym Adv Technol 2017; 28(8):1005–12.

[73] Tang J, Sisler J, Grishkewich N, Tam KC. Functionalization of cellulose nanocrystals for advanced applications. J Colloid Interface Sci 2017;494:397–409.

[74] Fan JS, Li YH. Maximizing the yield of nanocrystalline cellulose from cotton pulp fiber. Carbohydr Polym 2012;88(4):1184–8.

[75] Ashori A, Nourbakhsh A. Bio-based composites from waste agricultural residues. Waste Manag 2010;30(4):680–4.

[76] Tang XZ, Kumar P, Alavi S, Sandeep KP. Recent advances in biopolymers and biopolymer-based nanocomposites for food packaging materials. Crit Rev Food Sci Nutr 2012;52(5): 426–42.

[77] Heux L, Chauve G, Bonini C. Nonflocculating and chiral-nematic self-ordering of cellulose microcrystals suspensions in nonpolar solvents. Langmuir 2000;16(21):8210–2.

[78] Roman M, Winter WT. Effect of sulfate groups from sulfuric acid hydrolysis on the thermal degradation behavior of bacterial cellulose. Biomacromolecules 2004;5:1671–7.

[79] Zainuddin SYZ, Ahmad I, Kargarzadeh H, Abdullah I, Dufresne A. Potential of using multi-scale kenaf fibers as reinforcing filler in cassava starch-kenaf biocomposites. Carbohydr Polym 2013;92(2):2299–305.

[80] Prado KS, Spinacé MAS. Isolation and characterization of cellulose nanocrystals from pineapple crown waste and their potential uses. Int J Biol Macromol 2019;122:410–6.

[81] Rahimi M, Behrooz R. Effect of cellulose characteristic and hydrolyze conditions on morphology and size of nanocrystal cellulose extracted from wheat straw. Int J Polym Mater Polym Biomater 2011;60(8):529–41.

[82] El Achaby M, El Miri N, Hannache H, Gmouh S, Ben Youcef H, Aboulkas A. Production of cellulose nanocrystals from vine shoots and their use for the development of nanocomposite materials. Int J Biol Macromol 2018;117:592–600.

[83] Hemmati F, Jafari SM, Kashaninejad M, Barani MM. Synthesis and characterization of cellulose nanocrystals derived from walnut shell agricultural residues. Int J Biol Macromol 2018;120:1216–24.

[84] Fortunati E, Benincasa P, Balestra GM, Luzi F, Mazzaglia A, Del Buono D, et al. Revalorization of barley straw and husk as precursors for cellulose nanocrystals extraction and their effect on PVA_CH nanocomposites. Ind Crop Prod 2016;92:201–17.

[85] Rosa SML, Rehman N, De Miranda MIG, Nachtigall SMB, Bica CID. Chlorine-free extraction of cellulose from rice husk and whisker isolation. Carbohydr Polym 2012;87(2):1131–8.
[86] Sahari J, Sapuan SM, Zainudin ES, Maleque MA. Mechanical and thermal properties of environmentally friendly composites derived from sugar palm tree. Mater Des 2013;49:285–9.
[87] Panthapulakkal S, Zereshkian A, Sain M. Preparation and characterization of wheat straw fibers for reinforcing application in injection molded thermoplastic composites. Bioresour Technol 2006;97(2):265–72.
[88] Yang Z, Peng H, Wang W, Liu T. Crystallization behavior of poly(ε-caprolactone)/layered double hydroxide nanocomposites. J Appl Polym Sci 2010;116(5):2658–67.
[89] Kasiri N, Fathi M. Production of cellulose nanocrystals from pistachio shells and their application for stabilizing Pickering emulsions. Int J Biol Macromol 2018;106:1023–31.
[90] Lu P, Hsieh Y. Lo. Preparation and characterization of cellulose nanocrystals from rice straw. Carbohydr Polym 2012;87(1):564–73.
[91] Mandal A, Chakrabarty D. Isolation of nanocellulose from waste sugarcane bagasse (SCB) and its characterization. Carbohydr Polym 2011;86(3):1291–9.
[92] Purkait BS, Ray D, Sengupta S, Kar T, Mohanty A, Misra M. Isolation of cellulose nanoparticles from sesame husk. Ind Eng Chem Res 2011;50(2):871–6.
[93] Luzi F, Fortunati E, Giovanale G, Mazzaglia A, Torre L, Balestra GM. Cellulose nanocrystals from Actinidia deliciosa pruning residues combined with carvacrol in PVA_CH films with antioxidant/antimicrobial properties for packaging applications. Int J Biol Macromol 2017; 104:43–55.
[94] Julie Chandra CS, George N, Narayanankutty SK. Isolation and characterization of cellulose nanofibrils from arecanut husk fibre. Carbohydr Polym 2016;142:158–66.
[95] Brinchi L, Cotana F, Fortunati E, Kenny JM. Production of nanocrystalline cellulose from lignocellulosic biomass: technology and applications. Carbohydr Polym 2013;94(1): 154–69.
[96] Rosa MF, Medeiros ES, Malmonge JA, Gregorski KS, Wood DF, Mattoso LHC, et al. Cellulose nanowhiskers from coconut husk fibers: effect of preparation conditions on their thermal and morphological behavior. Carbohydr Polym 2010;81(1):83–92.
[97] Follain N, Belbekhouche S, Bras J, Siqueira G, Marais S, Dufresne A. Water transport properties of bio-nanocomposites reinforced by Luffa cylindrica cellulose nanocrystals. J Membr Sci 2013;427:218–29.
[98] Bano S, Negi YS. Studies on cellulose nanocrystals isolated from groundnut shells. Carbohydr Polym 2017;157:1041–9.
[99] Zimmermann T, Bordeanu N, Strub E. Properties of nanofibrillated cellulose from different raw materials and its reinforcement potential. Carbohydr Polym 2010;79(4):1086–93.
[100] Jiang Y, Zhou J, Zhang Q, Zhao G, Heng L, Chen D, et al. Preparation of cellulose nanocrystals from Humulus japonicus stem and the influence of high temperature pretreatment. Carbohydr Polym 2017;164:284–93.
[101] Taflick T, Schwendler LA, Rosa SML, Bica CID, Nachtigall SMB. Cellulose nanocrystals from acacia bark—influence of solvent extraction. Int J Biol Macromol 2017;101:553–61.
[102] Ilyas RA, Sapuan SM, Ishak MR. Isolation and characterization of nanocrystalline cellulose from sugar palm fibres (Arenga Pinnata). Carbohydr Polym 2018;181:1038–51.
[103] Hernandez CC, Ferreira FF, Rosa DS. X-ray powder diffraction and other analyses of cellulose nanocrystals obtained from corn straw by chemical treatments. Carbohydr Polym 2018;193:39–44.
[104] Mohamed MA, Salleh WNW, Jaafar J, Ismail AF, Abd Mutalib M, Mohamad AB, et al. Physicochemical characterization of cellulose nanocrystal and nanoporous self-assembled CNC membrane derived from Ceiba pentandra. Carbohydr Polym 2017;157:1892–902.

[105] Pakarinen A, Zhang J, Brock T, Maijala P, Viikari L. Enzymatic accessibility of fiber hemp is enhanced by enzymatic or chemical removal of pectin. Bioresour Technol 2012;107: 275–81.
[106] Frone AN, Chiulan I, Panaitescu DM, Nicolae CA, Ghiurea M, Galan AM. Isolation of cellulose nanocrystals from plum seed shells, structural and morphological characterization. Mater Lett 2017;194:160–3.
[107] El Oudiani A, Ben Sghaier R, Chaabouni Y, Msahli S, Sakli F. Physico-chemical and mechanical characterization of alkali-treated Agave americana L. fiber. J Text Inst 2012;103(4): 349–55.
[108] Persson J, Chanzy H, Sugiyama J. Combined infrared and electron diffraction study of the polymorphism of native celluloses. Macromolecules 1991;24(9):2461–6.
[109] Hanley SJ, Revol JF, Godbout L, Gray DG. Atomic force microscopy and transmission electron microscopy of cellulose from Micrasterias denticulata; evidence for a chiral helical microfibril twist. Cellulose 1997;4(3):209–20.
[110] Hubbe M, Rojas OJ, Lucia L, Sain M. Cellulosic nanocomposites: a review. Bioresources 2008;3(3):929–80.
[111] Alemdar A, Sain M. Isolation and characterization of nanofibers from agricultural residues—wheat straw and soy hulls. Bioresour Technol 2008;99(6):1664–71.
[112] Iguchi M, Yamanaka S, Budhiono A. Bacterial cellulose—a masterpiece of nature's arts. J Mater Sci 2000;35(2):261–70.
[113] García A, Gandini A, Labidi J, Belgacem N, Bras J. Industrial and crop wastes: a new source for nanocellulose biorefinery. Ind Crop Prod 2016;93:26–38.
[114] Rånby BG. III. Fibrous macromolecular systems. Cellulose and muscle. The colloidal properties of cellulose micelles. Discuss Faraday Soc 1951;11(111):158–64.
[115] Siró I, Plackett D. Microfibrillated cellulose and new nanocomposite materials: a review. Cellulose 2010;17:459–94.
[116] Trache D, Hussin MH, Haafiz MKM, Thakur VK. Recent progress in cellulose nanocrystals: sources and production. Nanoscale 2017;9:1763–86.
[117] El Achaby M, Kassab Z, Barakat A, Aboulkas A. Alfa fibers as viable sustainable source for cellulose nanocrystals extraction: application for improving the tensile properties of biopolymer nanocomposite films. Ind Crop Prod 2018;112(July 2017):499–510.
[118] Miao X, Lin J, Tian F, Li X, Bian F, Wang J. Cellulose nanofibrils extracted from the byproduct of cotton plant. Carbohydr Polym 2016;136:841–50.
[119] Dong XM, Revol JF, Gray DG. Effect of microcrystallite preparation conditions on the formation of colloid crystals of cellulose. Cellulose 1998;5(1):19–32.
[120] Sèbe G, Ham-Pichavant F, Ibarboure E, Koffi ALC, Tingaut P. Supramolecular structure characterization of cellulose II nanowhiskers produced by acid hydrolysis of cellulose i substrates. Biomacromolecules 2012;13(2):570–8.
[121] Nascimento DM, Almeida JS, Vale MS, Leitão RC, Muniz CR, Figueirêdo MCB, et al. A comprehensive approach for obtaining cellulose nanocrystal from coconut fiber. Part I: proposition of technological pathways. Ind Crop Prod 2016;93:66–75.
[122] Sirviö JA, Visanko M, Liimatainen H. Acidic deep eutectic solvents as hydrolytic media for cellulose nanocrystal production. Biomacromolecules 2016;17(9):3025–32.
[123] Chen L, Zhu JY, Baez C, Kitin P, Elder T. Highly thermal-stable and functional cellulose nanocrystals and nanofibrils produced using fully recyclable organic acids. Green Chem 2016;18(13):3835–43.
[124] Araki J, Wada M, Kuga S, Okano T. Flow properties of microcrystalline cellulose suspension prepared by acid treatment of native cellulose. Colloids Surfaces A Physicochem Eng Asp 1998;142(1):75–82.

[125] Wang Z, Yao ZJ, Zhou J, Zhang Y. Reuse of waste cotton cloth for the extraction of cellulose nanocrystals. Carbohydr Polym 2017;157:945–52.
[126] Zaini LH, Jonoobi M, Tahir PM, Karimi S. Isolation and characterization of cellulose whiskers from kenaf (Hibiscus cannabinus L.) bast fibers. J Biomater Nanobiotechnol 2013; 04(01):37–44.
[127] Lamaming J, Hashim R, Sulaiman O, Leh CP, Sugimoto T, Nordin NA. Cellulose nanocrystals isolated from oil palm trunk. Carbohydr Polym 2015;127:202–8.
[128] Bettaieb F, Khiari R, Hassan ML, Belgacem MN, Bras J, Dufresne A, et al. Preparation and characterization of new cellulose nanocrystals from marine biomass Posidonia oceanica. Ind Crop Prod 2015;72:175–82.
[129] Maiti S, Jayaramudu J, Das K, Reddy SM, Sadiku R, Ray SS, et al. Preparation and characterization of nano-cellulose with new shape from different precursor. Carbohydr Polym 2013;98(1):562–7.
[130] Sonia A, Priya DK. Chemical, morphology and thermal evaluation of cellulose microfibers obtained from Hibiscus sabdariffa. Carbohydr Polym 2013;92(1):668–74.
[131] Espinosa SC, Kuhnt T, Foster EJ, Weder C. Isolation of thermally stable cellulose nanocrystals from spent coffee grounds via phosphoric acid hydrolysis. Biomacromolecules 2013; 14(4):1223–30.
[132] Mendes CADC, Ferreira NMS, Furtado CRG, De Sousa AMF. Isolation and characterization of nanocrystalline cellulose from corn husk. Mater Lett 2015;148:26–9.
[133] Mo Y, Guo R, Liu J, Lan Y, Zhang Y, Xue W, et al. Preparation and properties of PLGA nanofiber membranes reinforced with cellulose nanocrystals. Colloids Surf B: Biointerfaces 2015;132:177–84.
[134] Prathapan R, Thapa R, Garnier G, Tabor RF. Modulating the zeta potential of cellulose nanocrystals using salts and surfactants. Colloids Surfaces A Physicochem Eng Asp 2016;509:11–8.
[135] Zhou Y, Fan M, Chen L, Zhuang J. Lignocellulosic fibre mediated rubber composites: an overview. Compos Part B 2015;76:180–91.
[136] Huang J, Rodrigue D. Comparison of the mechanical properties between carbon nanotube and nanocrystalline cellulose polypropylene based nano-composites. Mater Des 2015;65:974–82.
[137] Favier V, Chanzy H, Cavaille JY. Polymer nanocomposites reinforced by cellulose whiskers. Macromolecules 1995;28:6365–7.
[138] Moriana R, Vilaplana F, Ek M. Cellulose nanocrystals from forest residues as reinforcing agents for composites: a study from macro- to nano-dimensions. Carbohydr Polym 2016; 139:139–49.
[139] Song T, Tanpichai S, Oksman K. Cross-linked polyvinyl alcohol (PVA) foams reinforced with cellulose nanocrystals (CNCs). Cellulose 2016;23(3):1925–38.
[140] Grishkewich N, Mohammed N, Tang J, Tam KC. Recent advances in the application of cellulose nanocrystals. Curr Opin Colloid Interface Sci 2017;29:32–45.
[141] Tang Y, Du Y, Li Y, Wang X, Hu X. A thermosensitive chitosan/poly(vinyl alcohol) hydrogel containing hydroxyapatite for protein delivery. J Biomed Mater Res A 2009;91(4):953–63.
[142] Xu S, Girouard N, Schueneman G, Shofner ML, Meredith JC. Mechanical and thermal properties of waterborne epoxy composites containing cellulose nanocrystals. Polymer (Guildf) 2013;54(24):6589–98.
[143] Oun AA, Rhim JW. Isolation of cellulose nanocrystals from grain straws and their use for the preparation of carboxymethyl cellulose-based nanocomposite films. Carbohydr Polym 2016;150:187–200.
[144] Marques PAAP, Trindade T, Neto CP. Titanium dioxide/cellulose nanocomposites prepared by a controlled hydrolysis method. Compos Sci Technol 2006;66(7–8):1038–44.

[145] Xiao W, Wu T, Peng J, Bai Y, Li J, Lai G, et al. Preparation, structure, and properties of chitosan/cellulose/multiwalled carbon nanotube composite membranes and fibers. J Appl Polym Sci 2013;128(2):1193–9.

[146] Malho J, Walther A, Ikkala O, Linder MB. Facile method for stiff, tough, and strong nanocomposites by direct exfoliation of multilayered graphene into native nanocellulose matrix. Biomacromolecules 2012;13:1093–9.

[147] Arjmandi R, Hassan A, Haafiz MKM, Zakaria Z, Islam MS. Effect of hydrolysed cellulose nanowhiskers on properties of montmorillonite/polylactic acid nanocomposites. Int J Biol Macromol 2016;82:998–1010.

[148] Vaezi K, Asadpour G, Sharifi SH. Bio nanocomposites based on cationic starch reinforced with montmorillonite and cellulose nanocrystals: fundamental properties and biodegradability study. Int J Biol Macromol 2020;146:374–86.

[149] Naderizadeh S, Shakeri A, Mahdavi H, Nikfarjam N, Taheri QN. Hybrid Nanocomposite films of starch, poly(vinyl alcohol) (PVA), starch nanocrystals (SNCs), and montmorillonite (Na-MMT): structure–properties relationship. Starch/Staerke 2019;71(1–2).

[150] Montes S, Carrasco PM, Ruiz V, Cabañero G, Grande HJ, Labidi J, et al. Synergistic reinforcement of poly(vinyl alcohol) nanocomposites with cellulose nanocrystal-stabilized graphene. Compos Sci Technol 2015;117:26–31.

[151] Pal N, Dubey P, Gopinath P, Pal K. Combined effect of cellulose nanocrystal and reduced graphene oxide into poly-lactic acid matrix nanocomposite as a scaffold and its anti-bacterial activity. Int J Biol Macromol 2017;95:94–105.

[152] Pal N, Banerjee S, Roy P, Pal K. Reduced graphene oxide and PEG-grafted TEMPO-oxidized cellulose nanocrystal reinforced poly-lactic acid nanocomposite film for biomedical application. Mater Sci Eng C 2019;104:109956.

[153] Yu B, Zhao Z, Fu S, Meng L, Liu Y, Chen F, et al. Fabrication of PLA/CNC/CNT conductive composites for high electromagnetic interference shielding based on Pickering emulsions method. Compos Part A Appl Sci Manuf 2019;125(January):105558.

[154] Zhu L, Zhou X, Liu Y, Fu Q. Highly sensitive, ultrastretchable strain sensors prepared by pumping hybrid fillers of carbon nanotubes/cellulose nanocrystal into electrospun polyurethane membranes. ACS Appl Mater Interfaces 2019;11(13):12968–77.

[155] Liu H, Song J, Shang S, Song Z, Wang D. Cellulose nanocrystal/silver nanoparticle composites as bifunctional nanofillers within waterborne polyurethane. ACS Appl Mater Interfaces 2012;4(5):2413–9.

[156] Fortunati E, Rinaldi S, Peltzer M, Bloise N, Visai L, Armentano I, et al. Nano-biocomposite films with modified cellulose nanocrystals and synthesized silver nanoparticles. Carbohydr Polym 2014;101(1):1122–33.

[157] Rescignano N, Fortunati E, Montesano S, Emiliani C, Kenny JM, Martino S, et al. PVA bionanocomposites: a new take-off using cellulose nanocrystals and PLGA nanoparticles. Carbohydr Polym 2014;99:47–58.

Chapter 2

Characterization techniques for hybrid nanocomposites based on cellulose nanocrystals/nanofibrils and nanoparticles

Khadija El Bourakadi, Rachid Bouhfid, and Abou el Kacem Qaiss
Moroccan Foundation for Advanced Science, Innovation and Research (MAScIR), Composites and Nanocomposites Center, Rabat Design Center, Rabat, Morocco

2.1 Introduction

Currently, the use of nanocellulose-based products is becoming a good alternative to deal with environmental problems. Their use is also important to counter the future depletion of fossil resources such as oil resources, greenhouse gas emissions, and waste recycling problems. The development of new bio-sourced materials with a low environmental impact and properties similar to synthetic materials is one of the challenges. Thus, the most promising approach is to include some dispersed micro and nanoparticles such as nanocellulose and its derivatives during polymerization or as additives to develop hybrid nanocomposites for functional applications [1]. In recent years, there has been growing interest in the use of natural fibers as reinforcing elements in composite materials based on macromolecular matrices. The driving force behind this research activity lies in the fact that cellulose fiber based composites are more recyclable, unlike their glass fiber counterparts [2]. In addition, cellulose fibers have many other advantages including low density, obtained from a renewable source [3], and available worldwide in a variety of forms at low cost [4, 5]. Continuous increases in oil prices are only encouraging research efforts on this renewable material. These cellulosic nanoparticles are mainly used as an alternative to conventional composites, since less than 5% of the mass of reinforcement is required to obtain materials with better properties [6].

Hybrid nanocomposites based on cellulose are generally formulated by incorporating nanoscale cellulose fillers with other nanofillers to a variety of polymer matrices, leading to mechanical strength and improved properties [7].

Cellulose Nanocrystal/Nanoparticles Hybrid Nanocomposites. https://doi.org/10.1016/B978-0-12-822906-4.00010-4

Recent studies revealed that nanoscale celluloses are being intensively researched for potential use in a range of applications such as the development of new hybrid nanocomposites in different fields [8, 9]. The preparation of these nanocomposites may be done by using several techniques including cross-linking graft reaction, organo-micas, solvent-casting at room temperature, and sol-gel processing. Once cellulosic hybrid materials are produced, several characterization tools must be selected to determine their chemical structure and morphological properties, as well as thermal and dynamic mechanical stability. In this context, several studies focused their effort in this field. For example, the micro Fourier transform infrared spectroscopy (FT-IR), silicon element analysis, X-ray diffraction, scanning electron microscopy (SEM), atomic force microscopic (AFM,) and differential scanning calorimetry (DSC) were used to identify the cellulose hybrids based on polyhedral oligomeric silsesquioxane [10]. According to several investigations, the addition of cellulose nanoparticles, even a small content into the composite (5, 10, 25 wt%), has a positive effect on improving the hybrid nanocomposite mechanical properties [11]. In this chapter, a brief overview of the latest research in the field of cellulose nanocrystals and nanofibers, as well as hybrid nanocomposites, is presented. To start, the chemical characteristics, properties, and application of cellulose fibers and cellulose particles, especially as reinforcing agents for several polymers, is discussed. Then, a focus is made on specific characterization techniques to examine these specific materials. Finally, a conclusion is given on the methods with respect to the materials investigated.

2.2 Cellulose: Chemical structure, properties, and application

Cellulose is the most abundant organic biopolymer on the planet. It is assumed to be an inexhaustible and biodegradable resource [11]. It is estimated that more than 10^{11} tons of cellulose can be harvested on earth, 80% of which is located in forests which represents an annual production of over 7.5×10^{10} tons [12]. This biopolymer is a long linear and homogeneous chain made up of macromolecular groups of the polysaccharide type. Cellulose is a polymer linked by β-glycosides bonds made up of glucopyranose units linked together by β-(1,4) bonds. This linear polysaccharide is a homopolymer whose repeating unit is the cellobiose dimer as shown in Fig. 2.1. Cellulose is the main structure in a wide variety of species including plants such as wood, cotton, and wheat straw, as well as algae like valonia and marine animals [13].

Cellobioses form microfibrils linked together mainly by hydrogen bridges due to the presence of numerous hydroxyl groups along the chains. The polarity of the cellulose chains allows a parallel or antiparallel arrangement of the chains inside the cellulose crystal. In general, in its solid-state, cellulose is considered as a semi-crystalline polymer since it is composed of crystalline regions where the cellulosic molecules present an almost perfect parallelism, alternating with

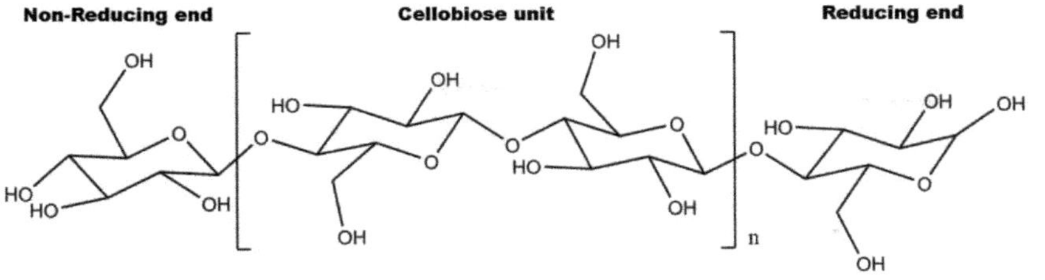

FIG. 2.1 Structure of the cellulose chain.

amorphous zones where the degree of organization is weak [14]. Also, due to the presence of numerous hydroxyl groups (-OH), the cellulose chain has a great tendency to form intramolecular and intermolecular hydrogen bonds. The intermolecular bonds make it possible to give a strong cohesion between the various cellulose chains which are most of the time oriented in the same direction [15].

Cellulose, which has a polar character due to the presence of numerous hydroxyl groups, will have affinity only for polar solvents or liquids. Cellulose reactions are essentially similar to those characteristics of alcohols. On the one hand, cellulose being a macromolecule, there is often steric hinderance between the hydroxyl functions and the reagents. On the other hand, the presence of amorphous zones which are easily accessible and crystalline zones that are not very reactive greatly affects reactivity [11]. Cellulose chains can also undergo acid hydrolysis in the presence of acids such as HCl or H_2SO_4. This process results in a reduction of the chain length and the formation of reducing functions.

Based on these structural properties, cellulose and its derivatives have been characterized by several specific properties including high Young's modulus, dimensional stability, low thermal expansion coefficient, exceptional reinforcing potential, and transparency. It is extremely crystalline with a very high Young's modulus in the range of 150 GPa [16]. Additionally, cellulose derivatives have good thermal properties in terms of thermo-chemical degradation which is depending on the heating rate, particle type, and type of surface modification [17]. On the other hand, the limited thermal stability of natural fibers is one of the main disadvantages, mainly in the case of their use in composite materials. Numerous studies have shown that the chemical modification of cellulosic fibers can improve their thermal stability. The thermal degradation of modified and unmodified plant fibers has been widely studied by several researchers [18, 19]. Cellulose is considered a relatively thermally stable polymer because it retains its structure and mechanical properties up to 200 °C. Cellulose does not have a melting temperature (it is not fusible). On the other hand, it is the site of significant decomposition reactions for temperatures between 250 °C and 350 °C.

Another attractive advantage attributed to cellulose derivatives is their low density, nonabrasive nature, nontoxic behavior, biocompatibility, biodegradability, and relatively high tensile/flexural modulus. Additionally, since this bio-molecule is made from renewable natural sources which are very abundant and therefore low-cost, it is not necessary to synthesize them leading to the manufacture of composite materials with excellent performances [20].

Owing to their remarkable advantages, cellulose and its derivatives have attracted a great deal of interest in the development and the research of new composites and nanocomposites hybrid materials for numerous applications. Several studies have shown the great role of this bio-polymer in the manufacture of different materials. In this context, different fields of application for celluloses are:

✓ as a reinforcing agent for composites [21],
✓ for its conductive properties allowing the formation of conductive materials [22],
✓ as an oxygen barrier [23],
✓ in the medical field [24],
✓ for memory storage and its magnetic capacity [25].

But the most studied application of cellulose and its derivatives is as a reinforcement in composite materials. For instance, cellulose nanocrystals (extracted from kenaf fibers) were used as reinforcing agents for polylactic acid (PLA) to elaborate a new bio-degradable nanocomposites with excellent mechanical improvement (2% in modulus, 15% in tensile stress, and 27% in tensile strain), thermal (up 300 °C), and morphological properties [26]. In addition, several scientific works reported on the huge potential of cellulose derivatives in different areas such as pharmaceutical, bio-medical implants, drugs delivery, separation membrane, electroactive polymers, and hygiene products [11, 27, 28].

Regarding the use of nanocellulose as a reinforcing agent, Ahola et al. [21] showed that the addition of poly(amideamine) and epichlorohydrin to cellulose nanofibrils prepared by TEMPO oxidation improved the strength of paper. These results showed that nanocelluloses have the potential for reinforcement in a composite material. Another important application of this polysaccharide is to manufacture a fully bio-based conductive separator prepared from cellulose polysaccharide for a potential application as a separator in electrolyte polymers fuel cells [29].

Nanocelluloses cannot directly conduct current. However, researchers have shown that it is possible to obtain a semi-conductive material from cellulose nanoparticles [30]. This property is of particular interesting in the manufacture of electronic papers [25].

It was also reported that nanocelluloses can be used in the medical field. Dong and Roman [24] showed that it is possible to use fluorescent agents on cellulose nanocrystals. This application allows fluorescent nanocelluloses to be attached to target cells in the body in order to identify them. Tissue growth would also be possible by using nanocelluloses as anchors.

Currently, one of the objectives of scientific research in the field of composite materials and plastics is to replace petroleum products with biodegradable ones. In this context, cellulose and its derivatives are playing a vital role to reduce the chemical products used, especially in the field of composite and nanocomposite materials.

The incorporation of nanocelluloses in a polar matrix showed that mechanical properties were significantly improved. Made from a renewable raw material, they are advantageous in contrast to petroleum derivatives. Another positive point is the biodegradability of nanocellulose [25].

The fibrous material of wood is the main source used to manufacture nanocelluloses on a large scale, environmental standards associated with wood uses must be respected. While nanocellulose has many applications, its use will have an impact on forest resources since the production of nanocellulose is generally derived from bleached Kraft pulp with a yield of 50% [31]. On the other hand, if the wood use increases, this leads to a decrease in petroleum products use and this type of fiber will become a benefit to the environment. The use of cellulose nanoparticles in bio or petroleum matrices, such as polyvinyl alcohol, polylactic acid, polyolefins, polyesters, and regenerated celluloses, has been reported [28]. These new elaborated hybrid materials based on cellulose derivatives have been characterized using different tools such as: Fourier transform infrared spectroscopy, UV-Vis, nuclear magnetic resonance (NMR), Raman, X-ray diffraction, scanning electron microscopy (SEM), transmission electron microscopy (TEM), atomic force microscopy (AFM), dynamic mechanical thermal analysis (DMA), thermal and mechanical properties among others.

In the next section, the main characterization used to identify the properties of hybrid materials based on cellulose is presented.

2.3 Characterization of cellulose-based hybrid nanocomposites

In recent years, research and development in the field of composite and nanocomposite hybrid materials based on cellulose derivatives as a reinforcing agent has attracted a great deal of attention. The preparation of these materials can be done by several routes under different conditions. The most important part of cellulose-based hybrid materials is their characterization in terms of structure, morphology, thermal, as well as mechanical behavior. Fig. 2.2 presents an overview these techniques.

2.3.1 Structural characterization

In order to identify the chemical structure of cellulose alone or mixed with other nanofillers into polymeric matrices, several techniques are combined. The principles and operation of each method is presented, as well as their advantages to be discussed and compared.

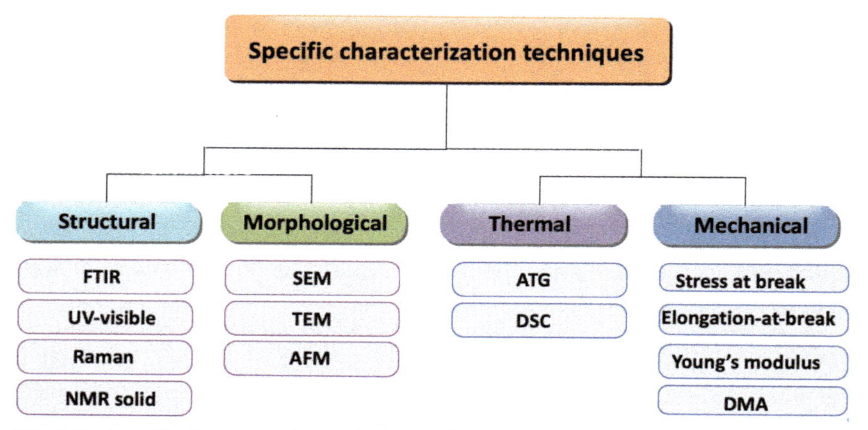

FIG. 2.2 Specific characterization tools frequently used for cellulose hybrid materials.

2.3.1.1 Fourier transform infrared (FTIR)

Fourier transform infrared (FTIR) spectroscopy is the main characterization technique based on measuring the infrared (IR) absorption, transmission or emission spectrum of most existing materials. The measurements can be performed over a wide range of frequencies (600–4000 cm^{-1}) with several advantages such as: resolution, speed, signal-to-noise ratio, and detection limits. An important and strong feature of this method compared to others is that almost all compounds have a characteristic absorption/emission in the infrared region, allowing both qualitative and quantitative analysis of the compounds in different forms like organic and inorganic molecules, composite-nanocomposites materials, and many others [32]. The characterization of cellulose-based hybrid materials via FTIR is a fundamental method for studying the functional groups of these materials. This is especially important to confirm successful surface modification and their interaction when added into a polymeric matrix. There is also the possibility to determine interaction with other nanofillers inside hybrid composites.

The use of this technique for identification and interpretation of cellulose derivative materials has been extensively reported in the literature. For instance, Ashori et al. [33] used FTIR to confirm chemical interactions between cellulose and silica. The FTIR spectra for untreated bacterial cellulose (BC), amorphous silica, and BC/silica composites is illustrated in Fig. 2.3. This figure shows the presence of several characteristic peaks such as 3200–3700 cm^{-1}, 1110–830 cm^{-1}, and 460 cm^{-1} attributed to Si-OH, Si-O, and Si-O-Si groups, respectively.

In another work, FTIR was used to get more information about chemical interactions and the behavior of organic and inorganic components into the same material. In this context, Angelova et al. [34] aimed to characterize

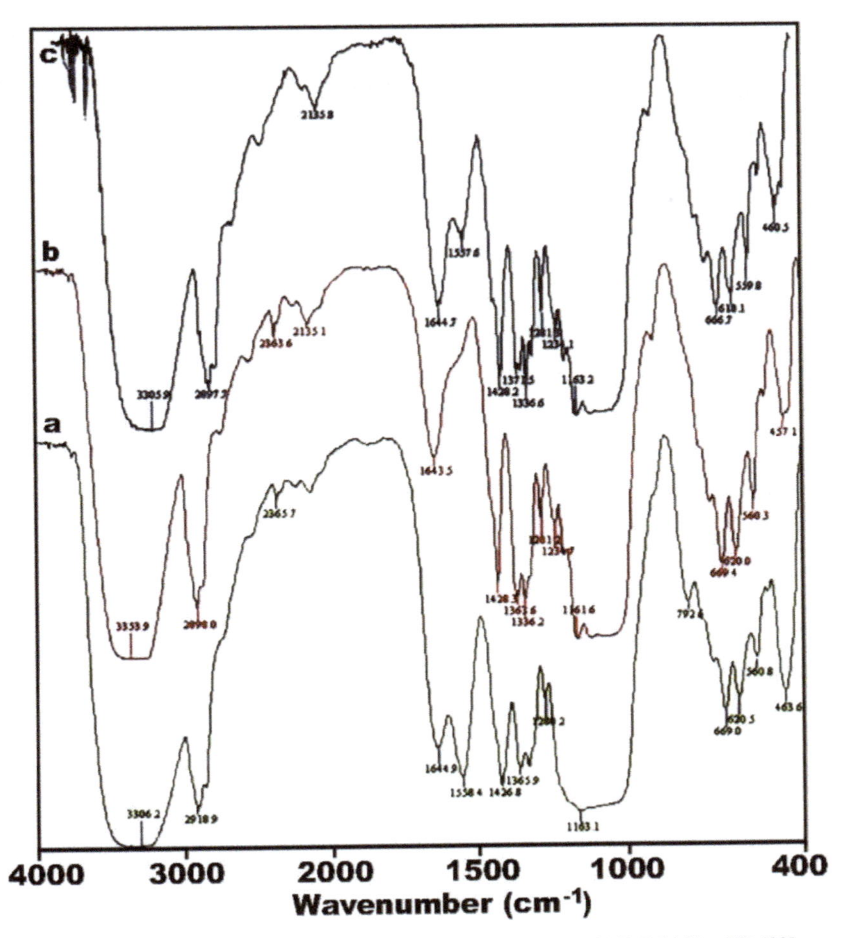

FIG. 2.3 FTIR spectra of: (A) BC/silica 3%, (B) BC/silica 5%, and (C) BC/silica 7% [33].

SiO_2 hybrid materials based on tetraethyl orthosilicate (TEOS), hydroxypropylmethylcellulose (HPMC), and silver. The FITR spectra reported in Fig. 2.4 shows the presence of the silanol (Si-O) group stretch and absorbed water on the surface which can be confirmed by two bands between 3200 and 3600 cm^{-1}, while the more intense bands at 3500 cm^{-1} is due to the vibration of the hydrogen bonds between the organic and inorganic components of the hybrid materials. The characteristic bands of the silica network are between 1300 and 600 cm^{-1}: asymmetric stretch group (Si-O-Si) in the 1000–1200 cm^{-1} region. From these data, it seems that FTIR spectra of the hybrid composites exhibit characteristic peaks for a SiO_2 network.

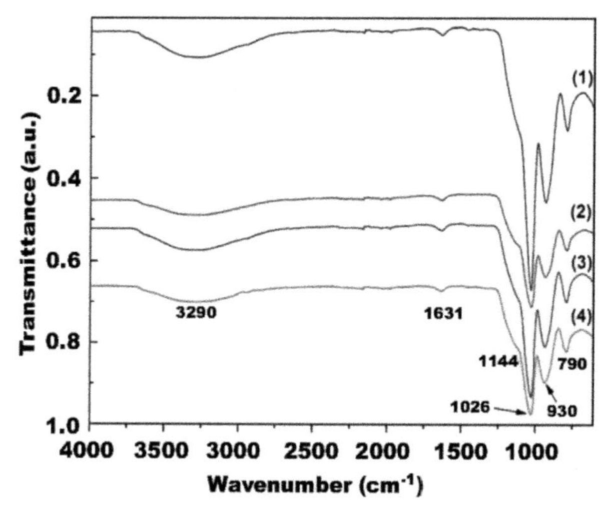

FIG. 2.4 Infrared spectra of hybrid composites with different silver content: 0.0 wt% Ag (1), 0.5 wt% Ag (2), 1.5 wt% Ag (3) and 2.5 wt% Ag (4) [34].

Recently, a group of researchers from Argentina reported the changes in chemical structure of cellulose, which was recognized by FTIR [35]. This technique was used to characterize poly(vinyl alcohol)/cellulose nanowhiskers (CNW) nanocomposite hydrogels. The samples were ground with KBr and the results are presented in Fig. 2.5. From the FTIR spectra, it seems that some peaks are present for both the CNW nanocomposite and commercial crystalline micro-cellulose (MCC). For example, the peaks in the range of 3060 and 3640 cm^{-1} in the MCC and CNW spectra are typical of C-H and O-H groups, while around 1635 cm^{-1} the O-H bending of the absorbed water is detected. The position and intensity of these bands did not change, showing that the cellulose constituents were neither removed nor degraded after acid hydrolysis. As a result of this investigation, the MCC and CNW spectra confirm the high purity of both products as they do not contain lignin, hemicellulose or wax residues (the characteristics bands are absent). In addition, the changes in the chemical structure of the cellulose channels can be monitored by FTIR. For instance, the peak at 1102 cm^{-1} was not significantly altered after acid hydrolysis, indicating that polymorph I of cellulose is present in both MCC and CNW. Furthermore, the acid hydrolysis was able to remove the amorphous cellulose which is visible in the CNW spectrum with a finer peak at 900 cm^{-1} and a less intense band at 2900 cm^{-1}.

Similarly, Xie et al. [36] developed a new generation of hybrid biomaterials based on cellulose/silica using a sol-gel covalent crosslinking process. The chemical structure of these biomaterials was investigated by micro-FTIR

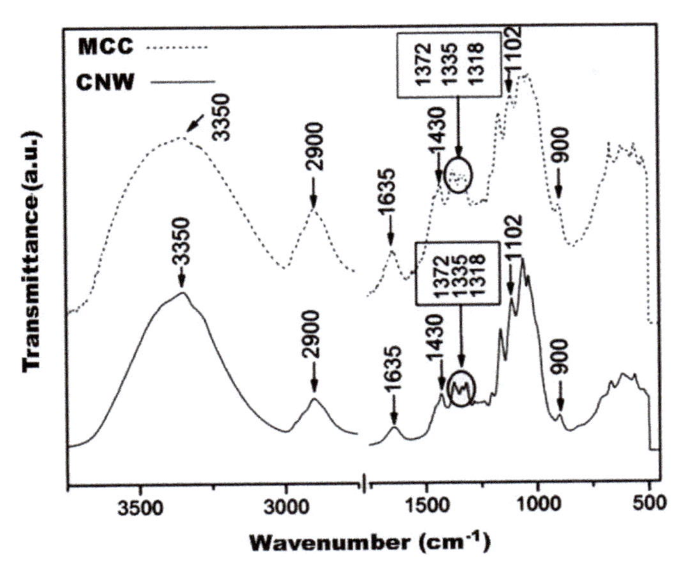

FIG. 2.5 FTIR spectra of MCC (commercial crystalline micro-cellulose) and CNW (cellulose nanowhiskers) nanocomposites [35].

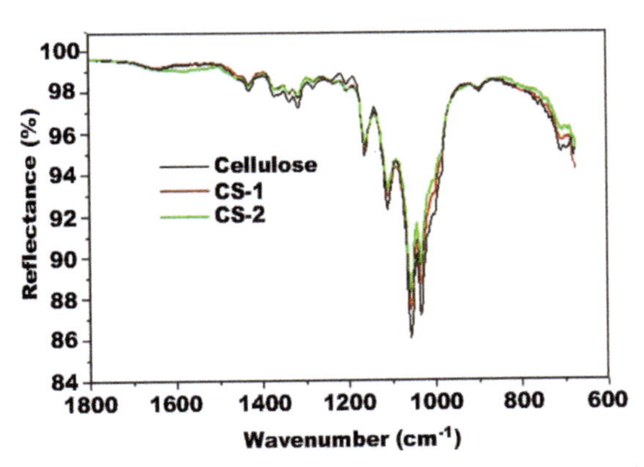

FIG. 2.6 Micro-FTIR spectra of cellulose and cellulose/silica hybrids composites [36].

spectroscopy. The results are reported in Fig. 2.6. The presence of —N=C— of 1,3,5-triazine cycle was confirmed by the presence of two bands in the range of 1458 and 1442 cm^{-1}. Also, the bands at 1079 and 710 cm^{-1} are attributed to —Si—O—Si— vibrations, while the band at 1079 cm^{-1} can be assigned to the presence of —Si—O—C—, —C—O—C—, and Si—C bonds.

2.3.1.2 Raman spectroscopy

According to the literature, Raman spectroscopy was found to be the main technique used to pick up rotation, vibration, as well as other states of the molecular system while being able to study their chemical composition [37]. Composites and nanocomposite hybrid materials based on cellulose derivatives represent a real meeting point for Raman spectroscopic studies as they are considered as innovative materials, but also poses a real challenge for this technique [38]. Raman analysis provides a comprehensive overview of the various properties of these hybrid materials, especially because they are often prepared in aqueous suspensions at extremely low concentrations. But this technique allows to determine the characteristics of hydrated materials (presence of water). For a long time, Raman spectroscopy was used to determine the elastic modulus of the native cellulose crystals of tunic and cotton, leading to values of 143 GPa [39] and 105 GPa [40], respectively. In the discussion below, the role of this technique in the characterization of hybrid materials based on cellulose and its derivatives is presented.

Chen et al. [41] reported the fabrication and characterization of a pH sensor using titanium dioxide (TiO_2)/multiwall carbon nanotube/cellulose hybrid nanocomposite. From the Raman spectra, it was revealed that high anatase crystalline TiO_2 nanoparticles were well formed on the surface of the hybrid nanocomposite (MWCNT), as presented in Fig. 2.7. After coating by TiO_2 nanoparticle, new peaks were observed in the TiO_2/MWCNT Raman spectra. The main anatase Raman-active modes appeared in the region of 143, 395, 515, and 638 cm^{-1}, which confirmed that anatase was found to be the most important element for the TiO_2-coated MWCNT.

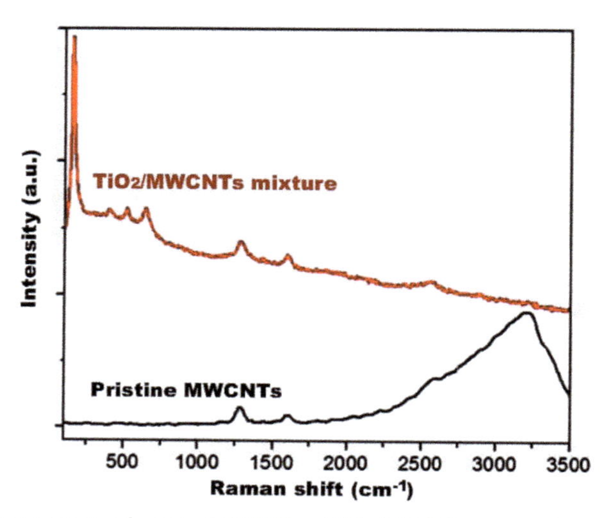

FIG. 2.7 Raman spectra of pristine MWCNT and TiO_2/MWCNT mixture [41].

In another interesting study, Roy and Guthrie [42] reported the synthesis and characterization of a new generation of novel natural-synthetic hybrid materials based on cellulose by using a reversible addition-fragmentation chain-transfer process. Raman spectroscopy was used to investigate the interaction between the components, as well as the changes in the chemical structure inside the hybrid materials. The high intensities of peaks related to monosubstituted aromatic ring of polystyrene was easily detected.

2.3.1.3 X-ray photoelectron spectroscopy (XPS)

X-ray photoelectron spectroscopy (XPS) is a quantitative method of photoelectron spectrometry based on measuring the spectra of photoelectrons induced by X-ray photons. It is mostly used to provide qualitative and quantitative information on the elementary compositions of matter, mainly solid surfaces. But this technique also provides information on the structure. XPS is well suited to study lignocellulosic fibers [43] and their derivatives [44]. In recent years, much attention has been given to this technique to investigate the elemental composition of composites and nanocomposites surface. In this section, a focus is made on its use in the field of hybrid materials based on cellulose derivatives.

Mahadeva and Kim [45] used XPS to characterize the chemical state of SnO_2 coating. This technique is used to investigate the effect of SnO_2 deposition on the electrical conductivity of the cellulose films as determined by measuring their current-voltage (I-V) characteristics. Fig. 2.8 reports

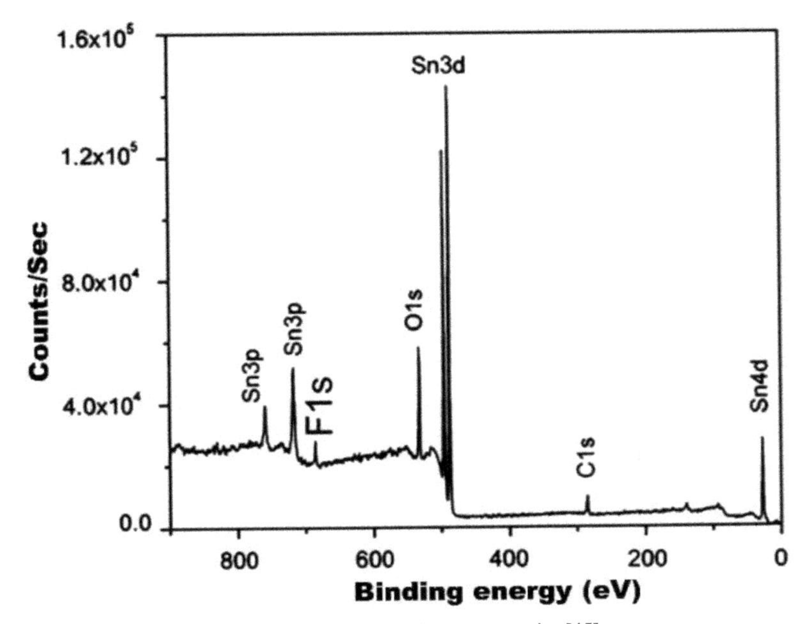

FIG. 2.8 XPS survey of a cellulose-SnO_2 hybrid nanocomposite [45].

the XPS spectrum of the cellulose-SnO_2 nanocomposite after 24 hours of deposition. According to this figure, all the cellulose films coated with SnO_2 show traces of tin, oxygen, and fluorine. The peaks appeared at 487.1 eV for $Sn3d_{5/2}$ and at 495.4 eV for $Sn3d_{3/2}$. Consequently, the chemical composition of the coatings deposited for 12, 18, and 24 hours were estimated as Sn:O:F:C = 13.7:16.0:2.1:36.0, 24.7:29.5:4.5:22.1, and 27.7:29.5:5.1:16.6 respectively, indicating that the amount of SnO_2 increased with time.

In an investigation by Freire et al. [46], two powerful methods were combined (XPS and time-of-flight secondary ion mass spectrometry, TOF-SIMS) to analyze the surface of partially esterified cellulose fibers. The XPS analysis revealed that the surface area covered by fatty acid fragments increased with increasing chain length, while the degree of substitution had a moderate effect, especially with higher esterification performance (higher SD value). When dimethylformamide (DMF) was used to swell the cellulose, it was suggested that an important reaction involving OH groups on the fibers surface occurred. The results of this analysis are summarized in Table 2.1 where the nomenclature of PC products is based on reaction conditions: the first number indicates the length of the esterification agent chain, the following letters indicate the agent used, and the last number indicates the reaction time in h. For example, PC6T1h indicates fibers treated with hexanoyl chloride toluene for 1 hour.

TABLE 2.1 XPS results of unmodified (PECF) and modified (PC series) fibers [46].

Sample	Degree of substitution	XPS results					$\%\theta_{fa}$
		O/C	*C1*	*C2*	*C3*	*C4*	
PECF	-	0.75	22.0	61.7	13.0	3.3	–
PC6T1h	0.42	0.46	39.2	45.2	12.9	4.8	17.2
PC6T6h	0.53	0.47	45.9	33.4	12.9	7.8	23.9
PC6DMF6h	1.26	0.40	51.7	34.7	11.0	2.6	29.7
PC12T1h	0.61	0.24	62.6	23.8	6.8	6.7	40.6
PC12T6h	1.17	0.24	65.0	23.0	2.6	6.0	43.0
PC12DMF6h	1.28	0.22	68.0	19.5	7.2	4.9	46.0
PC22T1h	0.008	0.17	77.8	14.2	4.7	3.3	55.8
PC22T6h	0.070	0.14	82.7	11.4	4.0	1.9	60.7
PC22DMF6h	1.01	0.12	83.4	10.0	4.0	2.6	61.4

Interestingly, Maniruzzaman et al. [47] fabricated a novel hybrid nanocomposite using titanium dioxide (TiO_2) and cellulose. This innovative material is expected to work as a potential conductometric glucose biosensor. In this work, XPS confirmed that glucose oxidase was successfully immobilized on TiO_2-cellulose hybrid nanocomposite in which TiO_2 and glucose oxidase (GOx) were covalently bonded. Fig. 2.9 presents the XPS spectrum of the TiO_2-cellulose hybrid nanocomposites (TCHN) with 30% TiO_2 after GOx immobilization.

In general, XPS was used to give an idea about the nature of the chemical bond between the cellulosic fibers and polymer coating deduced via cyclic voltammetry. XPS analysis supported the assumption that the chemical bond is, in fact, a hydrogen bond between the nitrogen single pairs of the polymer coating with the OH groups of the cellulose substrate underneath. Fig. 1.10A and B show the XPS spectra for cellulose and cellulose/polyaniline (PAN) composite. Of particular interest is the appearance of a shoulder directly related to the C—OH bond with N in the O1s spectrum for the cellulose/PAN composite (Fig. 1.10B). The shift in the primary O1s peak with respect to C—OH bonds, from 533.3 to 530.6 eV for cellulose and the PAN-coated analogue respectively, indicates the presence of a chemical interaction between OH and N. The bond of the hydroxyl groups of cellulose with nitrogen single pairs of PAN will lead to a decrease in the binding energy of oxygen, from which the attached hydrogen atoms are localized [48].

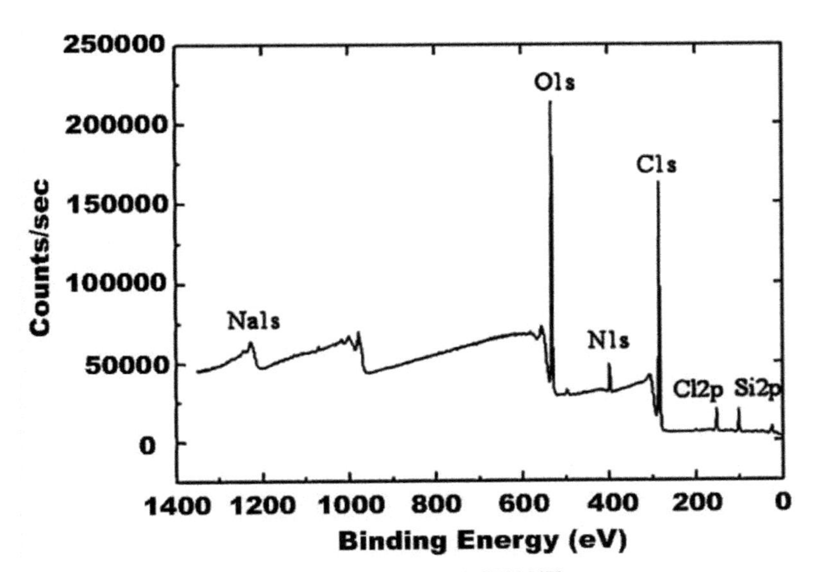

FIG. 2.9 XPS survey spectrum of GOx immobilized TCHN [47].

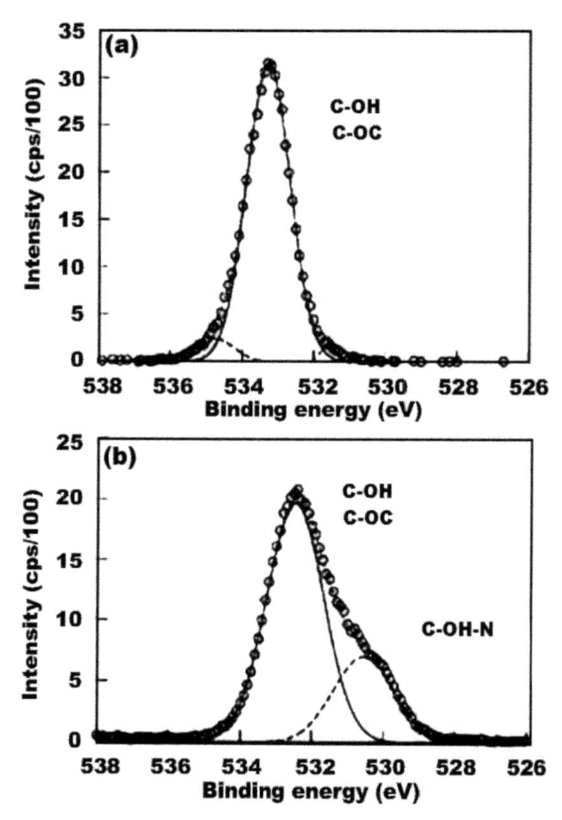

FIG. 1.10 XPS spectra for: (A) O1s for cellulose and (B) O1s for cellulose/polyaniline composite [48].

2.3.1.4 UV-Vis spectroscopy

Ultraviolet-visible spectroscopy (UV-Vis) is a spectroscopic technique involving photons whose wavelengths are in the ultraviolet (100–400 nm), visible (400–750 nm) or near infrared (750–1400 nm) range. This technique is another type of electron spectroscopy method. The substrates analyzed are most often in solution, but can also be in the gas phase and less frequently in the solid state. Due to its great importance in scientific research as a practical characterization tool, UV-Vis spectroscopy has been used in several field such as: analytical food chemistry [49], measurement of proteins and other large molecules [50], bioprocess and fermentation monitoring [51], and surface modification of cellulose nanocrystals [52]. Recently, this spectroscopic method has gained a high interest in the characterization of cellulose derivatives and hybrid nanocomposites.

To study the optical behavior of cellulose-ZnO nanocomposites and regenerated cellulose, Bagheri and Rabieh [53] used the technique, in which an amount of each sample was dispersed in ethanol under ultrasonic radiation for 5 min before measuring the sample's absorption spectrum. Fig. 1.11 shows that the composite has a peak at 358 nm which is due to the ZnO nanoparticles, while the regenerated cellulose does not have a significant absorption peak in this spectral range. As a result, the absorption spectrum analysis is considered as a useful tool to understand the band gap energy of metal oxides.

Similarly, a novel hybrid nanocomposite has been prepared by combined the synergetic effects of ZnO functionality and the renewability of cellulose. The optical properties were investigated by measuring the photoluminescence and UV-Vis absorption spectra. Fig. 1.12 clearly shows that the cellulose film exhibits excellent transparency (over 85%), compared with the cellulose film in the visible light region, while the cellulose film with ZnO reveals a transparency of less than 85% below 380 nm. More absorption was observed from cellulose and ZnO hybrid nanocomposite (CEZOHN), which is associated to the ZnO seed layer [54].

2.3.1.5 Nuclear magnetic resonance (NMR)

NMR spectroscopy is a technique exploiting the magnetic properties of specific atomic nuclei. It is based on the phenomenon of nuclear magnetic resonance (NMR). The most important applications for organic chemistry are proton and carbon-13 NMR performed in liquid solutions. But NMR is also applicable to any nucleus with a nonzero spin, whether in liquid solutions or in solids. Some gases, such as xenon, can also be measured when they are adsorbed in

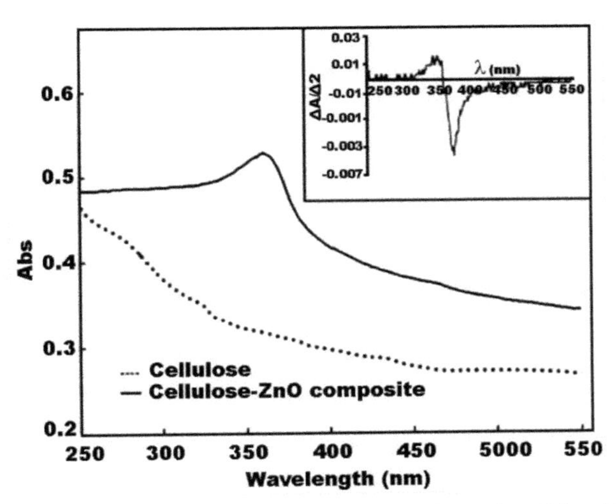

FIG. 1.11 UV-Visible absorbance spectra of regenerated cellulose and a cellulose-ZnO nanocomposite [53].

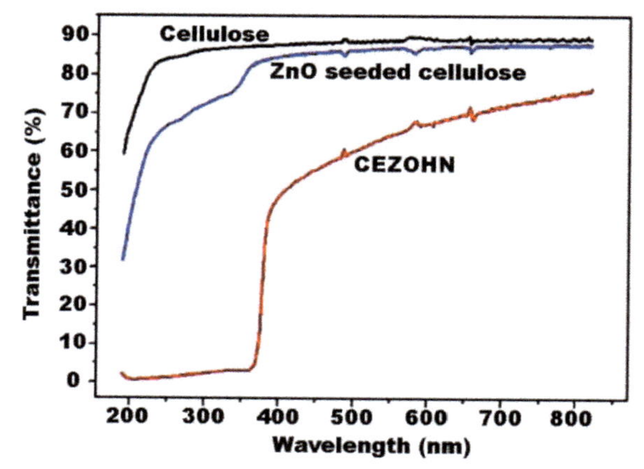

FIG. 1.12 UV-visible transmittance spectra of cellulose film and CEZOHN [54].

porous materials for example. The use of this method has long been an essential tool in the study of cellulose and its nanocomposites, mainly due to the limited number of active solvent systems and the importance of the plant's microstructured hierarchical cell wall. Solid-state NMR spectroscopy not only provides chemical information, but also information on the chemical environment and ultrastructural details, normally unattainable by other nondestructive spectral methods [55].

In this context, several studies are reported in the literature showing the major role of this spectroscopy technique in chemical structure identification of cellulose alone [56] or combined with other polymeric matrix or fillers [38]. For example, it was established that the orientation of the liquid crystalline cellulose nanocrystals suspension in the magnetic field of an NMR spectrometer was helpful in interpreting the NMR spectra of proteins added to a suspension [57]. Fortunati et al. [58] developed an eco-friendly nanocomposites based on poly (butylene succinate) containing ether-oxygen sequences based on a polymer backbone and modified cellulose nanocrystals. The chemical and degradation properties of the films was studied, which correlated with the cellulose nanocrystals content and the microstructure of the polymer chain. Proton nuclear magnetic resonance (^{1}H NMR) has been used in this case to determine the polymer's structure and composition. It was found that the chemical structure was as expected and the composition of the random copolymer (P(BSxTESy)) was close to that of the feed. For instance, the ^{1}H NMR spectrum of P(BS70TES30) is reported in Fig. 1.13, along with the chemical shift assignments.

In another work, a simple approach was presented to control the grafting of poly(2-(dimethylamino) ethyl methacrylate) from the surface of cellulosic fibers using reversible chain-transfer fragmentation (RAFT). To measure the monomer conversion percentage, ^{1}H NMR spectroscopy was used. In this case,

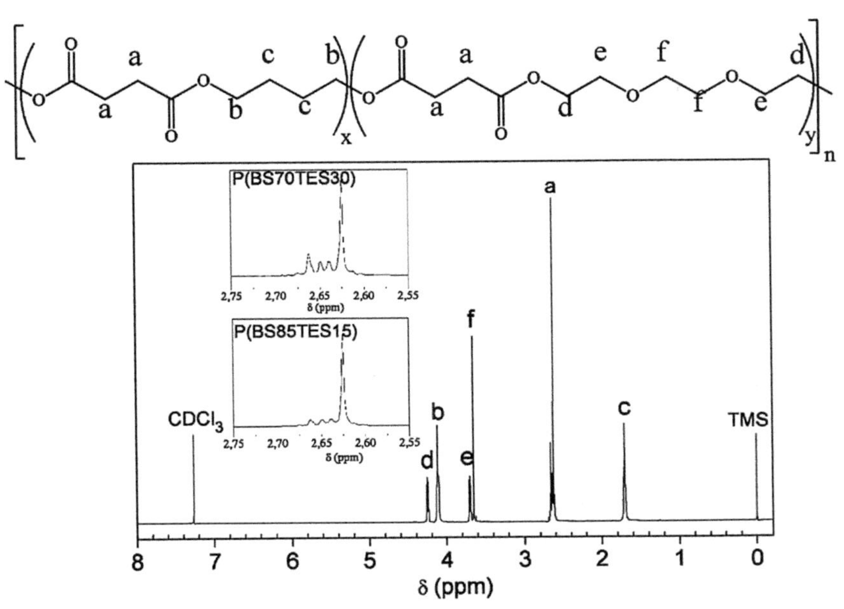

FIG. 1.13 Chemical structure and ^{1}H NMR of different polyesters with cellulose nanocrystals [58].

the polymeric solution was precipitated in a large amount of cold hexane. Then, the supernatant was removed and the polymer was again dissolved in dichloromethane and re-precipitated in cold hexane. As a conclusion of this work, the aminolysis reaction showed effective elimination of the red color characteristic of the cellulose graft copolymer formed by the RAFT polymerization [42].

Cunha et al. [59] fabricated a new organic-inorganic hybrid material by using chemically modified cellulose fibers. The chemical modification was carried out in two steps: with (3-isocyanatopropyl)triethoxysilane followed by acid hydrolysis. To characterize these hybrid materials, solid-state ^{29}Si NMR spectroscopy was used which is considered as a suitable technique for the examination of the structure of silicon-type hybrid materials [60]. The results, which obviously confirmed the enhance in the Si—O—Si bridges content combined with a decrease in Si–OEt linkages, are illustrated in Fig. 1.14. After the cellulose fibers modification with different proportions of reinforcing agent, the spectra of the subsequent derivatives showed signals in four different areas: approx. 45 (only for sample B1), 49, 57, and 67 ppm (only for sample C1), attributed to the media T^0, T^1, T^2, and T^3, respectively. Therefore, ^{29}Si NMR showed that, during the hydrolysis of the ethoxy groups, condensation reactions occurred mainly leading to linear Si-O-Si sequences, followed by the minor contribution of the ramifications from more branched units, and that a considerable amount of OH groups was also formed resulting in hydrophilic surfaces.

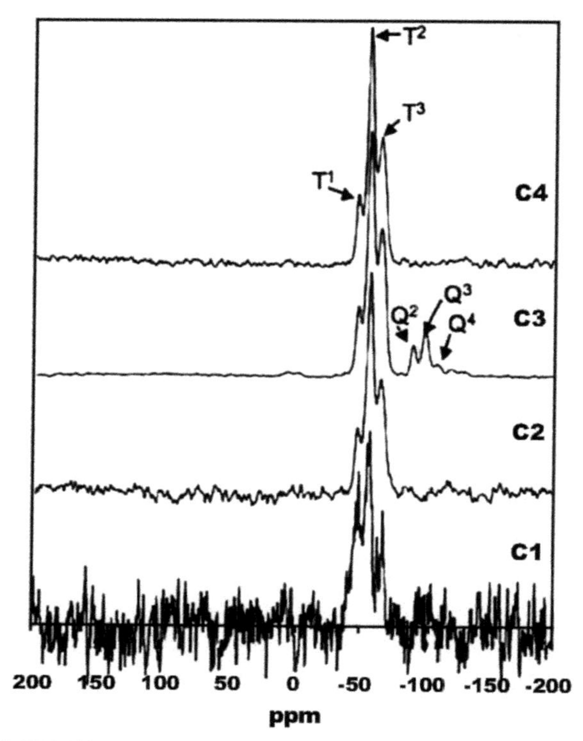

FIG. 1.14 ^{29}Si CP-MAS NMR spectra of cellulose fibers after modification [60].

2.3.1.6 *X-ray diffraction (XRD)*

X-ray diffraction (XRD) is an analytical technique based on the diffraction of X-rays by matter, especially for crystalline materials. X-ray diffraction is an elastic scattering (without loss of photon energy) giving rise to increasing interference as a more ordered materials is analyzed. For noncrystalline materials, the term "diffusion" is more appropriate. This is a powerful nondestructive technique to characterize crystalline materials as it provides information on structures, phases, preferred crystal orientations (texture), and other structural parameters such as average grain size, crystallinity, tension, and crystal defects. This method has a huge importance in science, engineering, and technology [61]. Interestingly, X-ray diffraction has been commonly used in the characterization of cellulose and its derivatives, as well as the characterization of nanocomposite-based cellulose [62]. The structural properties of nanocomposites (size and crystallinity index of cellulose crystallite) can be characterized by X-ray diffraction. XRD is also used to determine possible changes in the crystal structure of the matrix after adding nanoparticles.

Nypelo et al. [63] fabricated several self-standing hybrid films using cellulose nanocrystals (CNC) and electrospun composite fibers using CNC and polyvinyl alcohol with magnetic properties resulting from cobalt iron oxide nanoparticles in a CNC matrix. The characterization of these materials was carried out by various tools such as scanning electron microscopy, atomic force microscopy and X-ray diffraction. Of the properties that affected the magnetic behavior of the inorganic ferric particles, crystallinity and size are the most important. XRD was used to estimate the crystalline structure and the particles size. Fig. 2.15A shows that diffraction peaks originating from CNC dominated in inorganic CNC material. On top of these diffraction peaks, the XRD results of the CNC-inorganic mixture showed strong peaks at 30, 34, 57, and 62 degree. Bragg diffraction angles are characteristic of cobalt ferrite particles. So they were removed from the XRD pattern of neat CNC to that of CNC-inorganic material and the resulting pattern was compared with the pattern of the magnetic reference particles precipitated in the absence of CNC (Fig. 2.15B). From the XRD measurements (Fig. 2.15), the main difference in the magnetic properties

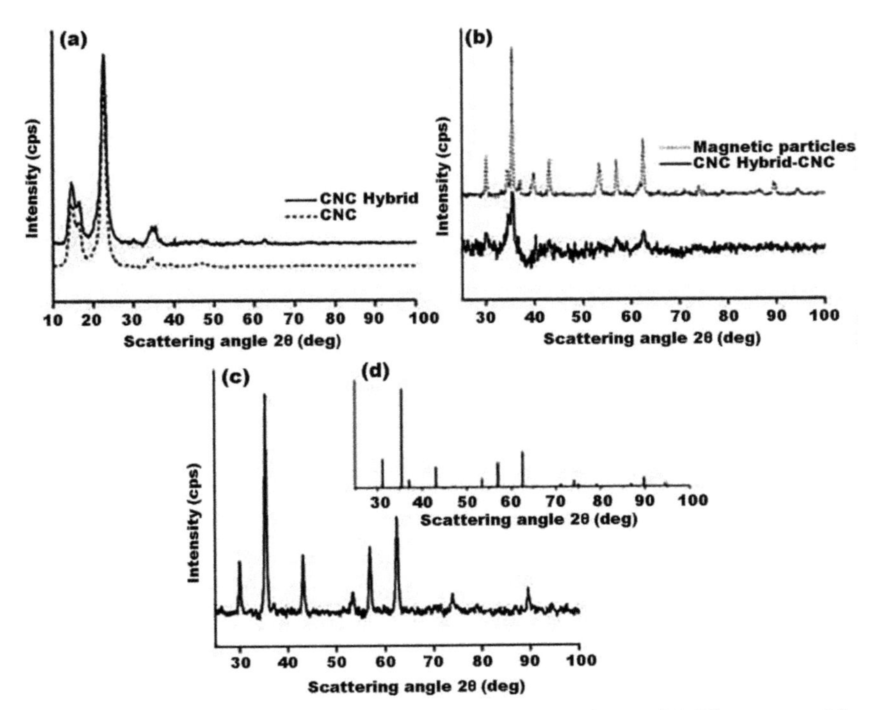

FIG. 2.15 (A) XRD profiles of the neat CNC and CNC-inorganic material, (B) spectrum of the inorganic particle contribution after extracting the data corresponding to CNC from the spectrum of magnetic CNC-inorganic material, (C) spectrum of the magnetically purified CNC-inorganic material, and (D) a reference cobalt ferrite spectrum [63].

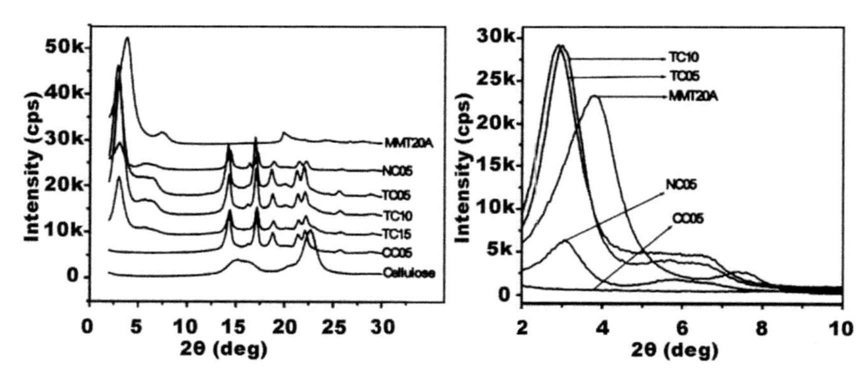

FIG. 2.16 XRD patterns of cellulose nanocomposites for the range of: (A) 2°–30° and (B) 2°–10° [64].

of pure cobalt ferrite particles and CNC inorganic systems are related to size and structure.

Pratheep and Singh [64] aimed to prepare three novel hybrid materials by adding layered silicate and microcrystalline cellulose to a thermoplastic polymer using ethylene-propylene (EP) copolymer as a thermoplastic polymer matrix and maleated EP (MEP) copolymer as a compatibility agent. These three types of nanocomposites were characterized via X-ray diffraction, differential scanning calorimetry, thermogravimetric analysis, and Fourier transform infrared spectroscopy. The intercalation of the layered silicate into the nanocomposite was confirmed by XRD. The X-ray diffraction patters of the hybrid materials are reported in Fig. 2.16 in which the shift peaks in the 001 basal space moves from a larger angle (2θ =4°–6°) to a smaller angle (2θ=1.5°–3°) which generally represents the spacing (d) between the silicate layers. Higher d-spacing are associated to macromolecular chains entering the gallery of layered silicates, which is normally referred to as "interleaving," while the delamination of these silicate layers from the order of the gallery is called "exfoliation," for which the peaks will not be seen at lower angles. In the case of pure crystalline cellulose fibers, peaks around 16 and 23 degree appeared. These peaks in other samples were difficult to detect, which might be attributed to the destruction of the cellulose crystalline region during processing or as a result of reaction with the compatibilizer. The d-spacing of the interlayer in clay (MMT20A) was showed around 232.3 nm (2θ=3.8 degree).

2.3.2 Morphological characterization

The morphological properties of cellulose whiskers are frequently examined via microscopy such as scanning electron microscopy, atomic force microscopy, and transmission electron microscopy, as well as light scattering techniques such as small angle neutron scattering [65] and polarized-depolarized dynamic light scattering [66]. In the discussion below, the use of these microscopic techniques in the characterization of cellulose-based hybrid nanocomposites is reported.

2.3.2.1 Scanning electron microscopy (SEM)

Scanning electron microscopy (SEM) is an electron microscopy technique able of producing high-resolution images of the surface of a sample using the principle of electron-matter interactions. Currently, SEM is used in fields ranging from biology to materials science, and several manufacturers offer devices with secondary electron detectors as standard equipment with a resolution between 0.4 and 20 nm. SEM can also be used to analyze the structure and morphology of cellulose nanocomposites. SEM enables the examination of the homogeneity of the nanocomposite-based cellulose, the presence of voids, the degree of dispersion of the cellulose nanocrystals inside the continuous matrix, the presence of aggregates, and the possible orientation of the cellulose nanocrystals through sedimentation [38].

Small and Johnston [67] reported the elaboration of novel hybrid materials based on photoluminescent cellulose fibers by coating bleached Kraft fibers with ZnS nanocrystals doped with Mn^{2+} and Cu^{2+} metals ions through a wet chemical synthesis process in which zinc salts and transition metal ions were precipitated with sodium sulfide in homogenous solutions. The obtained materials were characterized using several tools such as photoluminescence spectroscopy, X-ray diffraction, energy dispersive spectroscopy, X-ray photoelectron spectroscopy, as well as scanning electron microscopy. SEM was carried out to confirm the presence of metal ions on the cellulose fibers surface and the results are presented in Fig. 2.17. It can be seen that the doped ZnS nanocrystals are uniformly covering the surface and maintain the photoluminescent properties of their independent

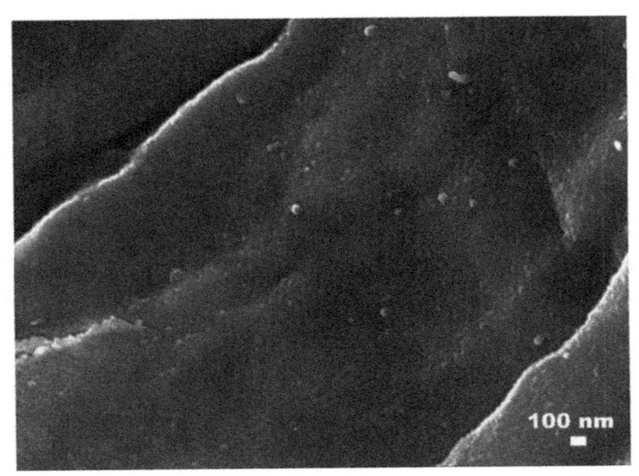

FIG. 2.17 SEM image of the surface of a cellulose fiber coated with Mn^{2+} doped ZnS nanocrystals [67].

powders. In addition, it was observed from SEM images that spherical nanocrystals of doped ZnS have been formed on the surface. The metal salts are around 30 nm in diameter and show a consistent coating.

In the development of hybrid nanocomposite materials based on cellulose derivatives, two researchers from the Modern Textile Institute at Donghua University aimed at developing a novel cellulose hybrid material using polyhedral oligomeric silsesquioxane (POSS) via a cross-linking grafting reaction [10]. The chemical and surface morphological structures of the elaborated nanocomposites were investigated by micro-FTIR, silicon element analysis, X-ray diffraction, atomic force microscopy, differential scanning calorimetry, and SEM. SEM analysis was carried out to investigate any change in relation to the surface morphology of the amino-POSS grafted cellulose hybrids. As shown in Fig. 2.18, the SEM images confirmed that the grafted cellulose hybrids have POSS nano-silica particles. From their SEM data, the authors concluded that:

- the surface of cellulose hybrids grafted with POSS have a high number of nano-sized particles.
- POSS nanoparticles are well dispersed in the cellulose matrix.
- the grafted cellulose hybrid has a well-miniaturized nano-sized heterogeneous structure.
- the particle surface of CP-2 is larger than that of CP-1.
- Increasing the POSS concentration increases the nanoparticles size.

More recently, Yao et al. [68] explored the mechanism of ultrasound on the preparation of cellulose/$Cu(OH)_2$/CuO hybrid materials. Several characterization techniques were used to study the effect of different reaction conditions, as well as the volume of H_2O_2, heating technique, pulse mode of ultrasound irradiation, time of sonication, and power density on the cellulose/$Cu(OH)_2$/CuO hybrid materials. Among these characterization tools, SEM provided various information of the surface morphology of the hybrid materials synthesized by diverse reaction with different parameters.

The SEM images showed that the addition of H_2O_2 changed the shape of the cellulose hybrids. In addition, these different results indicate that the synthesis process has a major effect on the shape and size of the cellulose hybrids.

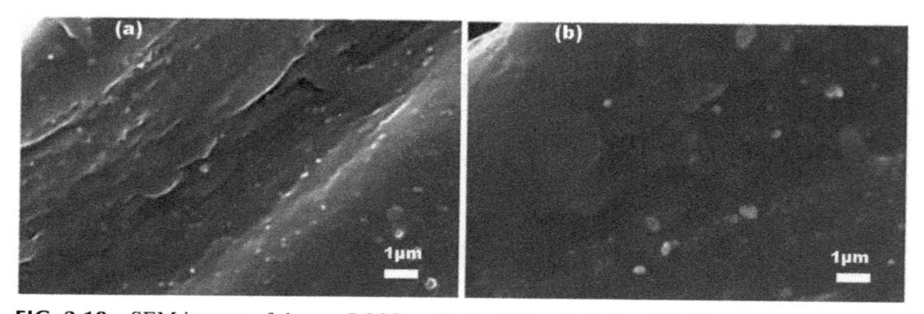

FIG. 2.18 SEM images of the am-POSS grafted cellulose hybrids: (A) CP-1 and (B) CP-2 [10].

Apparently, the ultrasonic radiation method can produce a cellulose hybrid having a more uniform shape and size than the oil heating method.

2.3.2.2 Atomic force microscopy (AFM)

Among the microscopic methods frequently used for the morphological characterization of different materials is atomic force microscopy (AFM). This technique allows to gather information about the surface morphology and topography of nanocomposite materials [69, 70]. Interestingly, AFM has long been a major tool in the study of cellulose and its derivatives morphologies [71]. Characterization of cellulose derivatives via AFM is an important method to investigate the surface modification of these important materials and their application to nanocomposites.

Bonardd et al. [72] combined two of the most abundant biodegradable materials on earth which are chitosan and cellulose nanocrystals. This combination was carried out to elaborate a novel biocomposite with improved dielectric constant. The characterization of the starting and final samples was done by numerous methods such as X-ray diffraction, infrared spectroscopy, solid-state NMR, thermogravimetric analysis, attenuated total reflectance spectroscopy, and atomic force microscopy (AFM) to determine the morphological properties of the bio-based films. The results of this characterization are reported in Fig. 2.19 which is showing AFM images for chitosan and the nanocomposites at three different wt% of CN-CNC (10, 30, and 50 wt%) The main conclusion from this study are:

- ✓ cyanoethylated CNC were well dispersed into the film for all compositions.
- ✓ The presence of a higher amount of CN-CNC is obviously visible in the nanocomposites with 30 and 50 wt%, with a good filler dispersion even

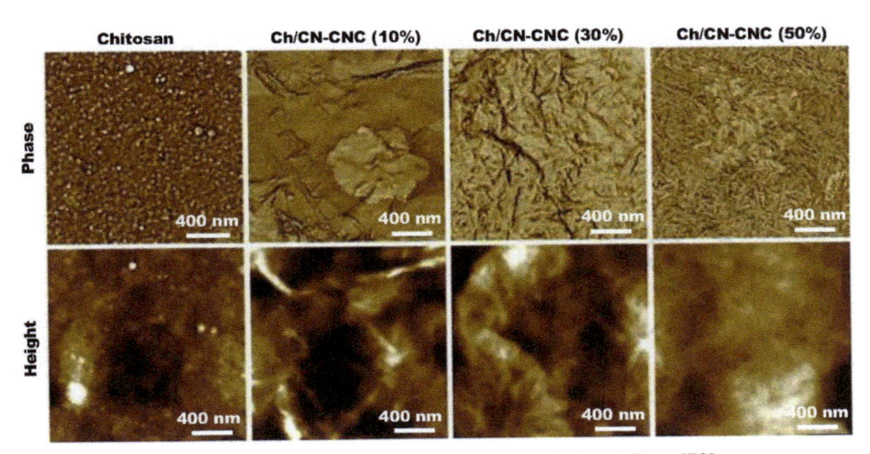

FIG. 2.19 AFM images of chitosan-cellulose nanocrystals bio-based films [72].

at the maximum concentration without the presence of significant agglomerates.

✓ all CNC are presenting the same nanometric size in all the films elaborated.

Owning to its advantages as a powerful tool in the characterization of cellulose whiskers, AFM was commonly used to analyze cellulose nanocomposites. In this case, AFM was especially used to determine the structural properties of cellulose whiskers and their nanocomposites using poly(lactic acid) as a matrix. According to the results on cellulose whiskers, AFM turned out to be an excellent alternative to SEM without restrictions in terms of contrast and resolution. However, the shape of the whiskers appeared different from that observed via TEM. Generally, AFM tips are limited in size and shape [73].

In another appealing work, Silvério et al. [74] aimed to evaluate corn cob as a bio-material to isolate nanocrystals using acid hydrolysis. The hydrolysis was done at 45 °C for 30, 60 or 90 min. The obtained cellulose was found to be an appropriate candidate to reinforce polyvinyl alcohol composites. The morphological properties in terms of shape and size were investigated through AFM. Fig. 2.20 presents the AFM images of the resulting CNC nanocomposites. The AFM images showed needle-like nanoparticles which confirmed the success of the cellulose nanocrystal extraction from the corn cobs. In this case, AFM topography measurements were carried out instead of transmission electron microscope to more accurately characterize the size of individual crystallites. Due to the preparation conditions used, an increase in the hydrolysis time led to a decrease in the nanocrystal aspect ratio (L/D). But the aspect ratio was above 10, which is considered as a minimum for a good stress transfer from the matrix to the fibers leading to improved mechanical properties [74].

2.3.2.3 Transmission electron microscopy (TEM)

Transmission electron microscopy (TEM) is a microscopy technique where an electron beam is "transmitted" through a very thin sample. The interactions between the electrons and the sample result in an image with a resolution down to 0.08 nm. The images obtained are generally not explicit and must be

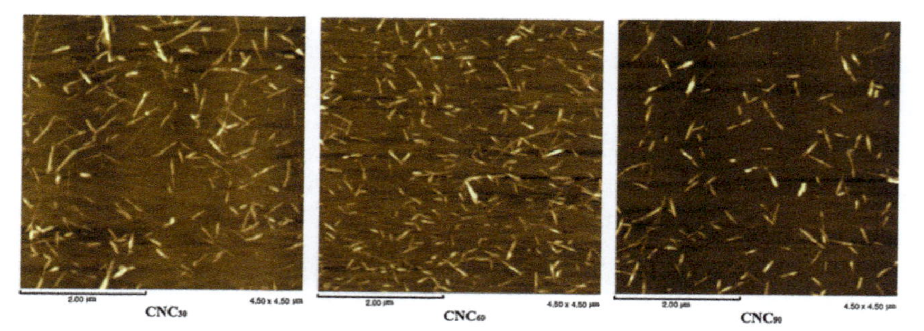

FIG. 2.20 AFM images of nanocomposite based on cellulose nanocrystals [74].

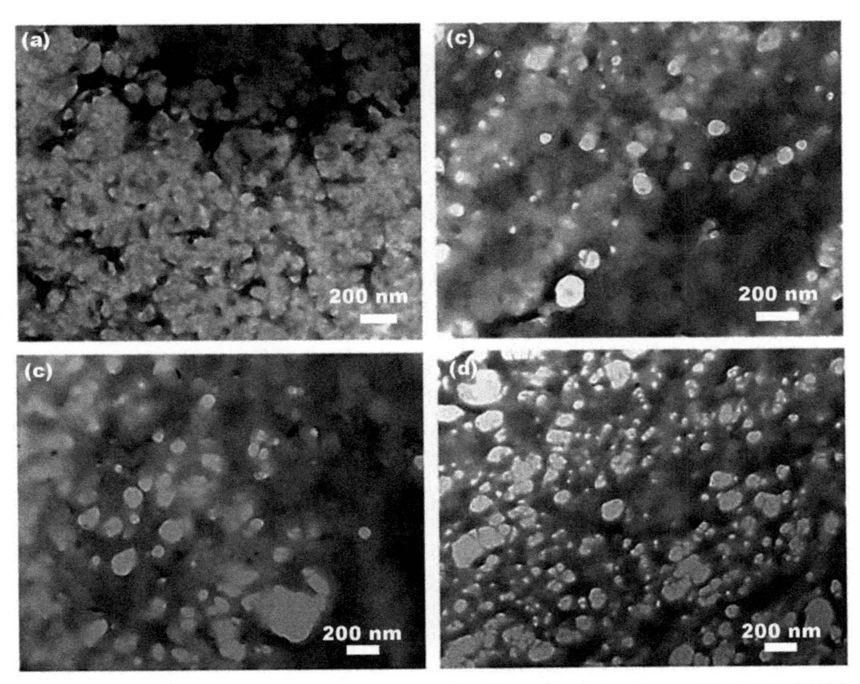

FIG. 2.21 TEM images of GO-cellulose composite films after being freeze-dried: (A) CG-02, (B) CG-04, (C) CG-05, and (D) CG-06 [76].

interpreted with the help of theoretical support. The main interest of this microscopy technique is to combine high resolution with Fourier space information; i.e., the diffraction. It is also possible to identify the chemical composition of the sample by studying the X-radiation generated by the electron beam. This method attracted the interest of the several scientists, especially for the characterization of cellulose nanoparticles and their nanocomposites [75].

Wu et al. [76] tried to elaborate new nanocomposite films by introducing fully exfoliated graphene oxide nanosheets into a cellulose matrix through a simple process. These nanosheets acted as a barrier against the fracture propagation of the cellulose composite films. The microstructures of the prepared cellulose materials before and after graphene oxide nanoparticles (GO) addition were investigated by SEM and TEM. Fig. 2.21 illustrates the TEM images of the GO-cellulose composite films. The pore environment was not composed of cellulose nanofibrils, but of a film, and the GO nanosheets were barely distinguishable. As GO content increased, the composite films had a larger pore size in the internal part.

More recently, Van Rie and Thielemans [77] highlighted the great advantage of combining cellulose with gold nanoparticles. Due to the unique properties of these two materials, this combination can lead to potential practical nanocomposites with excellent performances. In this work the researchers reviewed recent development in the field of cellulose-gold nanoparticle composites, followed by the preparation methods of cellulose-gold hybrid materials. According to the authors, the characterization of these hybrid composite may be performed via several tools including SEM, XPS, Raman, and UV-visible. The morphological properties can also be investigated by TEM. TEM images showed a nano-network morphology consisting of gold nanoparticles linked by cellulose nanofiber threads. It was assumed that this precise arrangement and entanglement of the cellulose chains is the consequence of a limiting terminal immobilization of sulfur-gold bonds on the surface of the gold nanoparticles.

Similarly, TEM was used to examine the morphological characteristics of hybrid materials based on cellulose nanofibrils (CNF) and gold nanoparticles (AuNP) [78]. The gold nanoparticles were fixed on the functionalized surface of the cellulose nanofibrils by electrostatic interactions or click reactions. The confirmation of this reaction was carried out using TEM and the results are reported in Fig. 2.22 illustrating that the AuNP were attached to the functionalized CNF surfaces. All TEM images evidently show the presence of AuNP on the functionalized CNF surfaces. In addition, TEM images revealed that AuNP

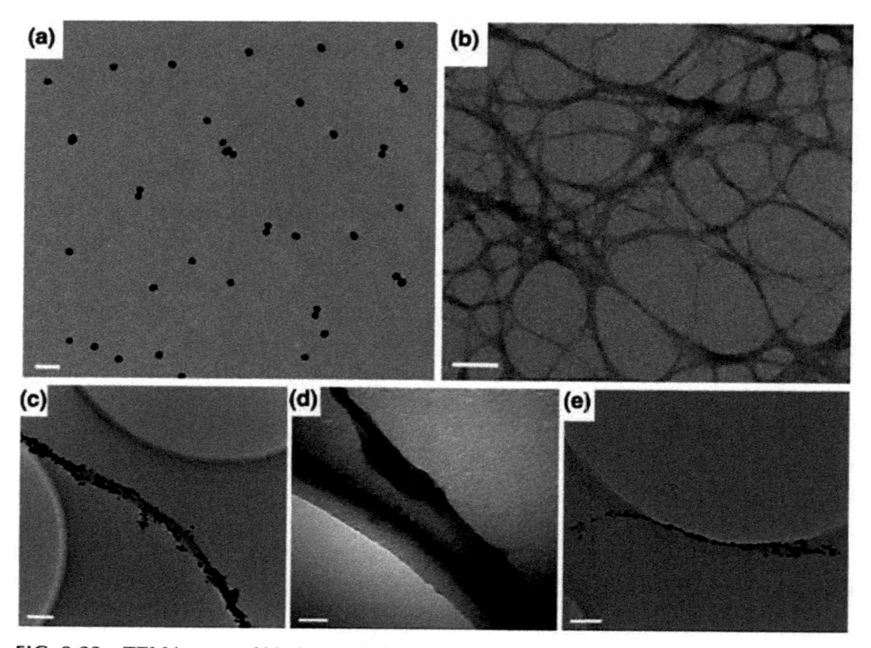

FIG. 2.22 TEM images of bio-inorganic hybrid materials: (A) as-synthesized AuNP with citrate reduction, (B) unmodified CNF, (C) AuNP on amino-CNF, (D) AuNP on propargyl-CNF, and (E) AuNP on EPTMAC-CNF. The scale bar in (A,D) is 50 nm and (B,C,E) is 200 nm [78].

have been effectively linked to the CNF surface. This indicates the resistance to the "click" reaction and its potential as a process of CNF functionalization with AuNP, as well as other nanoparticles.

2.3.3 Thermal properties

Generally, the industrial application of composite materials regardless of their composition is related to their thermal behavior. The two main analyses are thermogravimetric analysis (TGA) and differential scanning calorimetry (DSC) which are performed to evaluate the thermal character of a large number of composite materials. Here, a focus on hybrid materials based on cellulose is presented.

2.3.3.1 Thermogravimetric analysis (TGA)

Thermogravimetric analysis (TGA) is a thermal analysis technique that consists of measuring the change in mass of a sample as a function of time, for a given temperature or temperature profile. Such an analysis requires good accuracy for all three measurements: mass, time, and temperature. As the curves of mass variations are often similar, it is often necessary to perform data treatments for better interpretation. The derivation of these curves (DTG) shows at which points these variations are the most important. TGA is often used in research and testing to determine the characteristics of materials such as polymers, to estimate oxidation kinetics at high-temperature, to determine the degradation temperatures (stability), the moisture absorption, the amount of organic and inorganic compounds, the decomposition point of an explosive, and solvent residues. The most important information from TGA analysis of cellulose nanocomposites is to study their thermal behavior [79].

In this context, several works have been reported describing the major role of this technique to evaluate the thermal properties of cellulosic composites [80]. A series of bio-based nanocomposites using cellulose nanoparticles were elaborated and characterized in terms of mechanical and thermal properties. To evaluate these nanocomposites, about 15 mg of samples was placed in a platinum pan and heated from 25 °C to 900 °C at a speed of 10 °C/min under a nitrogen atmosphere with a flow rate of 20 mL/min. The test was repeated at least twice. Thermal degradation of pure samples and bleached fibers derived from *posidonia oceanica* were studied and the TGA curves are illustrated in Fig. 2.23. A slight weight loss appeared around 100 °C. As with any hydrophilic material, it is clearly due to moisture evaporation. Hemicelluloses, which are considered as the least thermally unstable products, begin to degrade around 250 °C. Around 350 °C, the slope of the curve for the raw materials and the cleaned cellulose fibers is completely different, which indicates a change in the chemical kinetics. The decreased in the thermal degradation curve of the pure materials is attributed to lignins.

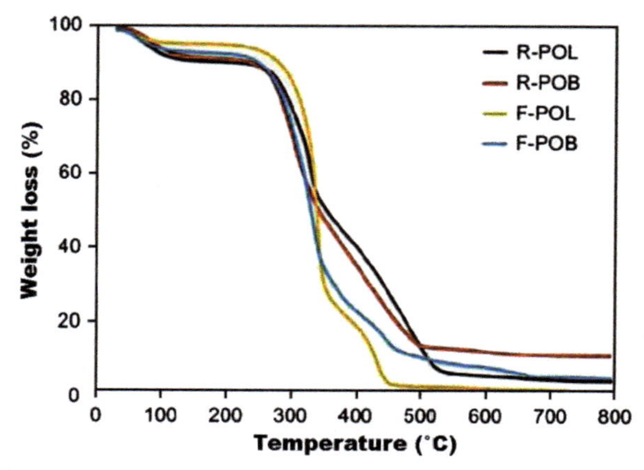

FIG. 2.23 TGA curves of the different materials: raw leaves (R-POL) and balls (R-POB), as well as cellulose fibers purified from the leaves (F-POL) and balls (F-POB) [79].

A novel multifunctional systems was developed by using cellulose nanocrystals (CNC) isolated from microcrystalline cellulose via acid hydrolysis combined with a blend of poly(lactic acid)-poly(hydroxybutyrate) (PLA-PHB) [81]. The morphological and structural properties, as well as the thermal stability of the nanocomposites were studied. The isothermal test was conducted at 200 °C for 40 min in air, while the dynamic measurements were made in a nitrogen atmosphere between 30 °C and 900 °C at 10 °C/min. The thermal properties of PLA and PLA-PHB nanocomposite films are summarized in Table 2.2. It was noted that CNC increased the thermal stability of the neat PLA, while

TABLE 2.2 Thermal properties of samples based on PLA, PHB, and CNC [81].

Formulation	TGA parameters	
	T_{max} (°C)	T_0 (°C)
PLA	357	320
PLA-CNC	343	293
PLA-CNCs	267	261
PLA-PHB	278	266
PLA-PHB-CNC	280	278
PLA-PHB-CNCs	280	278

CNC: Sulphuric acid hydrolysis. CNCs: Cellulose nanocrystals were also modified by adding STEFAC TM 8170 (Stepan Company Norhfield) surfactant, an acid phosphate ester of ethoxylated nonylphenol.

losing less than 1% of the original weight at the processing temperature. On the other hand, PLA-PHB blends based on nanocrystals were found to have poor thermal stability compared with the neat PLA-PHB mixture. It should be noted that each sample was melt mixed (extrusion) using a temperature profile of 180–190–200 °C for 2 min and then immediately transformed into a film for a maximum of 5 min below 200 °C. In contrast, adding CNC shifted the beginning of the degradation process of the PLA-PHB mixture from 266 °C to 278 °C, thus improving the thermal stability of the ternary nanocomposite.

2.3.3.2 Differential scanning calorimetry (DSC)

Differential scanning calorimetry (DSC) is a thermal analysis technique. It measures the differences in heat exchange between a sample to be analyzed and a reference.

DSC is used to determine phase transitions:

✓ The glass transition temperature (T_g) of amorphous materials: polymers, glasses (inorganic, organic or metallic) and ionic liquids;
✓ Melting and crystallization temperatures;
✓ Reaction enthalpies, to know the cross-linking rates of certain polymers.

Among its practical application, DSC can also be used to determine the thermal stability of certain composite and nanocomposite materials, especially the ones based on cellulose and its derivatives [82].

In this purpose, cellulose nanocrystals as a green nanomaterials was used by a group of researchers from India [83]. In their investigation, a top-down approach to the production of bacterial cellulose nanocrystals was based on a commercially available cellulose enzyme to preserve the native properties of the bacterial cellulose in its nanodimension. These bacterial cellulose nanocrystals (BCNC) were combined with polyvinyl alcohol (PVA) to elaborate a new nanocomposite film. The thermal stability of these films was carried out via DSC. The thermal behavior of PVA and its nanocomposites is showed in Table 2.3. The glass transition temperature (T_g) of PVA increased with BCNC

TABLE 2.3 DSC results of PVA and PVA/BCNC nanocomposites [83].

BCNC content in PVA (wt%)	T_g (°C)	T_m (°C)	ΔH_m (J/g)
0	66.7	203.3	83.2
1	67.8	208.0	84.9
2	68.3	208.1	87.0
3	69.3	210.1	87.1
4	69.7	212.7	88.3

addition. The T_g of polymeric nanocomposites generally depends on the degree of interaction between the nanoparticles and the polymer chains. In addition, adding BCNC also increased the melting temperature (T_m) and the enthalpy of fusion (ΔH_m) compared to the neat PVA. Finally, BCNC addition improved the crystalline behavior of PVA as the nanoparticles can act as heterogeneous nucleating agents.

Nanocrystalline cellulose (CNC) was also used as a reinforcing agent for alginate to produce bio-based nanocomposite films [84]. The NCC was incorporated into the matrix in different content (1 to 8%w/w). Their incorporation increased the thermal stability of the alginate films up to 227 °C.

2.3.4 Mechanical properties

A mechanical property is a characteristic describing the behavior of a material subjected to one or more mechanical stresses. The mechanical properties depend on the temperature, surface state, applied forces, and the speed of deformation. The main mechanical properties are:

- Young's modulus
- Elongation at break
- Stress at break

In some cases, to broaden the scope of certain composite materials, researchers aimed to increase the mechanical performances of these materials by adding some specific nanoparticles, also called reinforcing agents, to these composites. Among the nanoparticles which have a real positive impact on improving all the properties of the resulted materials, cellulose is considered as one of the greenest materials frequently used in this case. Taking all this into account, several authors found that the mechanical properties of cellulose-containing materials increases with cellulose concentration [85].

According to several studies, the most important parameters influencing the mechanical properties of composite materials are: fiber length, weight content, fiber orientation, fiber-matrix interfacial adhesion, type of matrix [86]. But other parameters like humidity may also affect the mechanical performances of hybrid materials.

The effect of water absorption on the mechanical properties of hybrid epoxy composites reinforced with cellulose and cellulose fibers was studied by Maslinda et al. [87]. To investigate the effect of water permeation at the fiber/matrix interface, the fractured samples were studied via field emission scanning electron microscopy. It was shown that the incorporation of kenaf, jute, and hemp fibers into epoxy improved the mechanical properties and water resistance. Furthermore, the tensile and bending strength of interlaced hybrid composites was higher than that of individual woven composites, which may be attributed to the different properties of the load distribution between longitudinal and transverse directional fibers in the woven structure. A longer immersion time reduced the strength and modulus of the composites due to water absorption. Nevertheless, with increasing immersion time, the fracture deformation caused by the destruction of the cellulose structure increased.

In recent years, several works in the literature studied the incorporation of various cellulosic fibers including abaca, jute, banana, hemp, and paper with glass fibers, leading to excellent results indicating that the mechanical properties of the hybrid composites were improved compared to monofilament-reinforced composites [88, 89]. There are also published works about hybrid nanocomposites based on poly-lactic acid (PLA) reinforced by inorganic nanofillers and cellulose fibers. Piekarska et al. [90] prepared and characterized novel hybrid nanocomposites reinforced by 3 wt% of organo-modified montmorillonite, 5 wt% of stearic acid-modified calcium carbonate nanoparticles, and 15 wt% of cellulose fibers using PLA as a matrix. The elaborated hybrid materials (PLA/MMT, PLA/NCC, and PLA/CF) were examined to evaluated their mechanical behavior after reinforcement. As a result, the tensile behavior of the hybrid composites was substantially affected by the fibers and comparable to that of PLA/CF. All the fillers increased the storage modulus below the glass transition temperature (T_g); that of PLA/MMT/CF and PLA/NCC/CF has been enhanced compared to polylactide by 50% and 45% respectively. Yield stress (σ_y), stress at break (σ_b), and strain at break (ε_b) of the neat PLA and PLA composites under uniaxial elongation at 25 °C and 35 °C are summarized in Table 2.4.

In another investigation by Lee et al. [91], it was noted that the mechanical properties of hybrid nanocomposite based on caprolactone mixed with polypropylene reinforced by 1% cellulose nanofiber were strongly increased compared with those without reinforcement. Also, the thermal properties of the PCL in the composite (NanoCom-2) increased from 32.4 °C to 35.9 °C with the CNF addition.

2.3.5 Dynamic mechanical analysis (DMA)

Dynamic mechanical analysis (DMA) is a method to determine the viscoelastic behavior of materials. This thermal analysis method allows to study and characterize the viscous and elastic mechanical properties of materials such as polymers. A DMA instrument can be used to determine the following intrinsic physical quantities:

- Complex Young's modulus (E^*), Coulomb shear modulus (G^*), and the complex viscosity (η^*);
- the damping factor, also known as the loss factor or tangent delta (tan δ);
- the frequency-dependent glass transition temperature (T_v). DMA is more sensitive than other thermal analysis techniques for the determination of T_v and the detection of transitions in composites.

TABLE 2.4 Mechanical properties in tension for different nanocomposites [90].

Samples	T (°C)	σ_y (MPa)	σ_b (MPa)	ε_b (%)
PLA	25	49.3±014	47.3±0.1	29.0±3.8
PLA/CF	25	–	44.8±1.2	7.1±0.3
PLA/NCC	25	44.1±2.4	42.2±2.2	31.7±0.7
PLA/NCC/CF	25	–	41.5±0.1	5.8±0.5
PLA/MMT	25	49.9±1.6	45.1±0.7	7.8±0.6
PLA/MMT/CF	25	–	42.1±2.1	5.1±0.3
PLA	35	47.5±0.5	42.8±0.6	37.4±6.4
PLA/CF	35	33.4±3.0	32.5±1.4	6.2±0.1
PLA/NCC	35	43.0±0.7	32.7±1.8	47.4±4.0
PLA/NCC/CF	35	35.8±2.9	29.8±1.0	6.2±0.1
PLA/MMT	35	43.8±0.6	37.2±0.2	9.4±1.3
PLA/MMTPLA/MMT/CF	35	28.7±4.0	27.6±4.1	4.4±0.3

DMA is frequently used to evaluate the thermomechanical characteristics of films elaborated using cellulose and cellulose nanocrystals [80]. For example, it was found that the addition of cellulose nanofillers into epoxy composites exhibit enhanced properties such as improved and dynamic mechanical properties [92]. The better properties are most probably caused by the synergistic effect of the cellulose fibers structure and the maximized interaction. Accordingly, dynamic mechanical analysis (storage modulus, loss modulus, tan delta and Cole-Cole plots) was also investigated. The data clearly showed that the incorporation of cellulose fillers improved the thermal stability (up to 300 °C) and has significantly enhanced the dynamic properties (E' and E'') of all the epoxy nanocomposites compared to neat matrix.

The effect of cellulose nanofibers (CNF) on the dynamic mechanical properties of polyester resin composites was also studied by Lavoratti et al. [93]. The essential parameters commonly used to evaluate the dynamic properties of composite and nanocomposite materials are the modulus in the glassy (E'g) and in the rubbery (E'r) regions at 30 °C and 175 °C respectively, the loss modulus and the peak height were calculated from the tan delta curves and for all the composites. Generally, the addition of fillers in the matrix can result in a higher elastic modulus of the rubbery region as a result of limited free movement of the polymer chains as compared to neat polymers. The dynamic mechanical properties of reinforced materials can be affected by several parameters such

TABLE 2.5 Dynamic mechanical properties of epoxy, neat EFB, neat woven jute and their hybrid composites [95].

Samples	Coefficient	Peak height of tan δ curve	T_g from tan δ_{max} (°C)	T_g from E''_{max} (°C)
Epoxy matrix	–	0.36	81.6	66.7
Neat EFB	0.55	0.29	85.0	74.8
Neat J_w	0.47	0.22	91.2	85.7
EFB/J_wEFB	0.52	0.28	81.2	75.3
J_w EFB/J_w	0.51	0.27	73.7	72.9

Coefficient: The effectiveness of fibres on the moduli of the composites.

as: nature of nanofillers, dispersion state, and interfacial interactions between the reinforcing agent (CNF) and the polymeric matrix [94].

On the other hand, researchers from Malaysia tried to develop a new epoxy hybrid composites using cellulose nanoparticles extracted from oil palm empty fruit bunch (EFB) [95]. Their dynamic mechanical properties were evaluated in terms of storage modulus (E_0), glass transition temperature (T_g), and peak height of tan δ. The results in Table 2.5 show that the T_g obtained from the loss modulus was inferior than the T_g obtained from the tan δ curve. The shift in T_g of the polymer matrix with fiber addition shows that their presence plays a vital role on T_g. The highest tan δ of the epoxy matrix indicates that it has higher mobility and better damping behavior.

2.4 Conclusion

The incorporation of cellulose nanocrystals and cellulose nanofibrils in different polymeric matrices is a promising alternative solution to conventional composites as its offers good general properties for standard composites, nanocomposites, and hybrid materials. Improvement in the overall properties can be directly related to the excellent properties of the cellulose (nano)particles which are isolated from different sources. Nevertheless, several other parameters must be taken into account such as: fiber size, surface modification, fiber content, coupling additives, choice of matrix, etc. Good compatibility and high adhesion between cellulose nanofibers with polymers can also affect other characteristics of the obtained nanocomposite such as structural, morphological, thermal, and mechanical properties.

In this chapter, a review of the different techniques used to characterize these materials was presented. The most important ones are morphological, thermal, and mechanical properties. But to get a clear picture of a material's structure and behavior, a combination of different methods is necessary.

References

[1] Saba N, Jawaid M, Alothman OY, Paridah M, Hassan A. Recent advances in epoxy resin, natural fiber-reinforced epoxy composites and their applications. J Reinf Plast Compos 2015;35 (6):447–70.

[2] Czarnecki L, White JL. Shear flow rheological properties, fiber damage, and mastication characteristics of aramid-, glass-, and cellulose-fiber-reinforced polystyrene melts. J Appl Polym Sci 1980;25:1217–44.

[3] Wang W, Cai Z, Yu J. Study on the chemical modification process of jute fiber. J Eng Fiber Fabr 2008;3(2):1–11.

[4] Joly C, Kofman M, Gauthier R. Polypropylene/cellulosic fiber composites chemical treatment of the cellulose assuming compatibilization between the two materials. J Macromol Sci Part A Pure Appl Chem 1996;33(February 2015):37–41.

[5] Zadorecki P, Michell AJ. Future prospects for wood cellulose as reinforcement in organic polymer composites. Polym Compos 1989;10(2):69–77.

[6] Dufresne A. Polysaccharide nanocrystal reinforced nanocomposites. Can J Chem 2008;494:484–94.

[7] Sequeira S, Evtuguin DV, Portugal I. Preparation and properties of cellulose/silica hybrid composites. Polym Compos 2009;30:1275–82.

[8] Zubik K, Singhsa P, Wang Y, Manuspiya H, Narain R. Thermo-responsive poly(N-Isopropylacrylamide)-cellulose nanocrystals hybrid hydrogels for wound dressing. Polymers (Basel) 2017;9(12):119.

[9] Fragal EH, Cellet TSP, Fragal VH, Companhoni MVP, Ueda-Nakamura T, Muniz EC, Rubira AF. Hybrid materials for bone tissue engineering from biomimetic growth of hydroxiapatite on cellulose nanowhiskers. Carbohydr Polym 2016;152:734–46.

[10] Xie K, Zhang Y, Yu Y. Preparation and characterization of cellulose hybrids grafted with the polyhedral oligomeric silsesquioxanes (POSS). Carbohydr Polym 2009;77(4):858–62.

[11] Habibi Y, Lucia LA, Rojas OJ. Cellulose nanocrystals: chemistry, self-assembly, and applications. Chem Rev 2010;110:3479–500.

[12] Online VA, Eyley S, Thielemans W. Surface modi fi cation of cellulose nanocrystals. Nanoscale 2014;6:7764–79.

[13] Corre AC, de Teixeira EM, Pessan LA, LHC M. Cellulose nanofibers from curaua fibers. Cellulose 2010;17:1183–92.

[14] French AD, Johnson ÆGP. Cellulose and the twofold screw axis: modeling and experimental arguments. Cellulose 2009;16:959–73.

[15] Nsor-atindana J, Chen M, Goff HD, Zhong F. Functionality and nutritional aspects of microcrystalline cellulose in food. Carbohydr Polym 2017;172:159–74.

[16] Kim J, Zhai L, Mun S, Ko H-U, Yun Y-M. Cellulose nanocrystals, nanofibers and their composites as renewable smart materials. Nanosens Biosens Info-Tech Sensors Syst 2015;9434:15–20.

[17] Fukuzumi H, Saito T, Iwata T, Kumamoto Y. Transparent and high gas barrier films of cellulose nanofibers prepared by TEMPO-mediated oxidation. Biomacromolecules 2009;10:162–5.

[18] Nguyen T, Zavarin E, Ii EM, Zavarin E. Thermal analysis of lignocellulosic materials. J Macromol Sci Part C 2017;1797(November):1–65.
[19] Singh MS, Chowdhury S. Recent developments in solvent-free multicomponent reactions: a perfect synergy for eco-compatible organic synthesis. RSC Adv 2012;2(11):4547–92.
[20] Peng BL, Dhar N, Liu HL, Tam KC. Chemistry and applications of nanocrystalline cellulose and its derivatives: a nanotechnology perspective. Can J Chem Eng 2011;89(5):1191–206.
[21] Ahola S, Österberg M, Laine J. Cellulose nanofibrils—adsorption with poly (amideamine) epichlorohydrin studied by QCM-D and application as a paper strength additive. Cellulose 2008;15:303–14.
[22] Agarwal M, Lvov Y, Varahramyan K. Conductive wood microfibres for smart paper through layer-by-layer nanocoating. Nanotechnology 2006;17:5319–25.
[23] Rodionova G, Saito T, Lenes M. Mechanical and oxygen barrier properties of films prepared from fibrillated dispersions of TEMPO-oxidized Norway spruce and Eucalyptus pulps. Cellulose 2012;19:705–11.
[24] Dong S, Roman M. Fluorescently labeled cellulose nanocrystals for bioimaging applications. J Am Chem Soc 2007;129(45):13810–1.
[25] Hubbe M, Rojas OJ, Lucia L, Sain M. Cellulosic nanocomposites: a review. Bioresources 2008;3(3):929–80.
[26] Ketabchi MR, Khalid M, Ratnam CT, Walvekar R. Mechanical and thermal properties of polylactic acid composites reinforced with cellulose nanoparticles extracted from kenaf fi bre. Mater Res Express 2016;3:125301.
[27] Nanomaterials G, Klemm D, Klemm D, Kramer F, Moritz S, Lindström T, et al. Nanocelluloses: a new family of nature-based materials. Angew Chem Int Ed 2011;50:5438–66.
[28] Olivera S, Muralidhara HB, Venkatesh K, Guna VK, Gopalakrishna K. Potential applications of cellulose and chitosan nanoparticles/composites in wastewater treatment: a review. Carbohydr Polym 2016;153:600–18.
[29] Vilela C, Silva ACQ, Domingues EM, Gonçalves G, Martins MA, Figueiredo FML, et al. Conductive polysaccharides-based proton-exchange membranes for fuel cell applications: the case of bacterial cellulose and fucoidan. Carbohydr Polym 2019;115604:100632.
[30] Agarwal M, McDonald J, Lvov Y, Varahramyan K. Controlled conductive polymer coating on wood microfibers via layer-by-layer nanoassembly. Mater Res Soc Symp Proc 2006;920:1–7.
[31] Westbye P, Svanberg C, Gatenholm P. The effect of molecular composition of xylan extracted from birch on its assembly onto bleached softwood kraft pulp. Holzforschung 2006;60 (2):143–8.
[32] Bacsik Z, Mink J, Keresztury G. FTIR spectroscopy of the atmosphere. I. Principles and methods. Appl Spectrosc Rev 2004;39(3):295–363.
[33] Ashori A, Sheykhnazari S, Tabarsa T, Shakeri A, Golalipour M. Bacterial cellulose/silica nanocomposites: preparation and characterization. Carbohydr Polym 2012;90(1):413–8.
[34] Angelova T, Rangelova N, Dineva H, Georgieva N, Müller R. Synthesis, characterization and antibacterial assessment of SiO_2-hydroxypropylmethyl cellulose hybrid materials with embedded silver nanoparticles. Biotechnol Biotechnol Equip 2014;2818(October 2015):747–52.
[35] Gonzalez J, Ludueña LN, Ponce A, Alvarez VA. Poly (vinyl alcohol)/cellulose nanowhiskers nanocomposite hydrogels for potential wound dressings. Mater Sci Eng C 2014;34:54–61.
[36] Xie K, Yu Y, Shi Y. Synthesis and characterization of cellulose/silica hybrid materials with chemical crosslinking. Carbohydr Polym 2009;78(4):799–805.
[37] McCreery RL. In: Winefordner JD, editor. Raman spectroscopy for chemical analysis. John Wiley Sons; 2005. p. 22–437.

[38] Torres FG, Troncoso OP, Torres C, Grande CJ. Cellulose based blends, composites and nanocomposites. In: Thomas S, Visakh P, Mathew AP, editors. Advances in natural polymers. 2013th ed. Berlin Heidelberg: Springer-Verlag; 2012. p. 21–54.
[39] Šturcová A, Davies G, Eichhorn SJ. Elastic modulus and stress-transfer properties of tunicate cellulose whiskers. Biomacromolecules 2005;6:1055–61.
[40] Rusli R, Eichhorn SJ. Determination of the stiffness of cellulose nanowhiskers and the fiber-matrix interface in a nanocomposite using Raman spectroscopy. Appl Phys Lett 2008;93, 033111.
[41] Chen Y, Mun SC, Kim J. A wide range conductometric pH sensor made with titanium dioxide/multiwall carbon nanotube/cellulose hybrid nanocomposite. IEEE Sensors J 2013;13 (11):4157–62.
[42] Roy D, Guthrie JT. Synthesis of natural—synthetic hybrid materials from cellulose via the RAFT process. Soft Matter 2008;4:145–55.
[43] Fardim P, Holmbom B. Origin and surface distribution of anionic groups in different papermaking fibres. Colloids Surfaces A Physicochem Eng Asp 2005;252:237–42.
[44] Zafeiropoulos NE, Vickers PE, Baillie CA. An experimental investigation of modified and unmodified flax fibres with XPS. J Mater Sci 2003;38:3903–14.
[45] Mahadeva SK, Kim J. Hybrid nanocomposite based on cellulose and tin oxide: growth, structure, tensile and electrical characteristics. Sci Technol Adv Mater 2011;12(5), 055006.
[46] Freire CSR, Silvestre AJD, Pascoal C, Gandini A. Surface characterization by XPS, contact angle measurements and ToF-SIMS of cellulose fibers partially esterified with fatty acids. J Colloid Interface Sci 2006;301:205–9.
[47] Maniruzzaman M, Jang S, Kim J. Titanium dioxide—cellulose hybrid nanocomposite and its glucose biosensor application. Mater Sci Eng B 2012;177(11):844–8.
[48] Kelly F, Johnston JH, Borrmann T, Richardson MJ. Functionalised hybrid materials of conducting polymers with individual fibres of cellulose. Eur J Inorg Chem 2007;35:5571–7.
[49] Cozzolino D, Cynkar WU, Shah N, Smith P. Multivariate data analysis applied to spectroscopy: potential application to juice and fruit quality. Food Res Int J 2011;44(7):1888–96.
[50] Schmid F-X. Biological macromolecules: spectrophotometry concentrations. Encycl Life Sci 2001;3:1–4.
[51] Lopes JA, Almeida CF, Pinheiro HM. Bioreactor monitoring with spectroscopy and chemometrics: a review. Anal Bioanal Chem 2012;404:1211–37.
[52] Nielsen LJ, Eyley S, Aylott JW. Dual fluorescent labelling of cellulose nanocrystals for pH sensing. Chem Commun 2010;46:8929–31.
[53] Bagheri M, Rabieh S. Preparation and characterization of cellulose-ZnO nanocomposite based on ionic liquid ([C4 mim]Cl). Cellulose 2013;20:699–705.
[54] Mun S, Kim HC, Ko H-U, Zhai L, Kim JW, Kim J. Flexible cellulose and ZnO hybrid nanocomposite and its UV sensing characteristics. Sci Technol Adv Mater 2017;18(1):1–10.
[55] Foston M. Advances in solid-state NMR of cellulose. Curr Opin Biotechnol 2014;27:176–84.
[56] Atalla RH, Vanderhart DL. The role of solid state 13 C NMR spectroscopy in studies of the nature of native celluloses. Solid State Nucl Magn Reson 1999;15:1–19.
[57] Fleming K, Gray D, Prasannan S, Matthews S. Cellulose crystallites: a new and robust liquid crystalline medium for the measurement of residual dipolar couplings. J Am Chem Soc 2000;10:5224–5.
[58] Fortunati E, Gigli M, Luzi F, Dominici F, Lotti N, Gazzano M, et al. Processing and characterization of nanocomposite based on poly (butylene/triethylene succinate) copolymers and cellulose nanocrystals. Carbohydr Polym 2017;165:51–60.

[59] Cunha AG, Freire CSR, Silvestre AJD, Neto CP, Gandini A. Preparation and characterization of novel highly omniphobic cellulose fibers organic-inorganic hybrid materials. Carbohydr Polym 2010;80(4):1048–56.

[60] Asakura T, Ando I. In: Asakura T, Ando I, editors. Solid state NMR of polymers. Elsevier; 1998. p. 1017.

[61] Tech JAB, Chauhan A, Chauhan P. Powder XRD technique and its applications in science and technology. J Anal Bioanal Tech 2014;5(5).

[62] Rangelova N, Radev L, Nenkova S, Miranda Salvado I, Vas Fernandes M, Herzog M. Methylcellulose/SiO2 hybrids: sol-gel preparation and characterization by XRD, FTIR and AFM. Cent Eur J Chem 2011;9(1):112–8.

[63] Nypelo T, Rodriguez-Abreu C, Jose R, Dickey MD, et al. Magneto-responsive hybrid materials based on cellulose nanocrystals. Cellulose 2014;21:2557–66.

[64] Kumar AP, Singh RP. Novel hybrid of clay, cellulose, and thermoplastics. I. Preparation and characterization of composites of ethylene—propylene copolymer. Appl Polym Sci 2006;104 (4):2672–82.

[65] Terech P, Chazeau L, Cavaille JY. A small-angle scattering study of cellulose whiskers in aqueous suspensions. Macromolecules 1999;32:1872–5.

[66] Lima MMDS, Wong JT, Paillet M, Borsali R, Pecora R. Translational and rotational dynamics of rodlike cellulose whiskers. Langmuir 2003;19(23):24–9.

[67] Small AC, Johnston JH. Novel hybrid materials of cellulose fibres and doped ZnS nanocrystals. Curr Appl Phys 2008;8:512–5.

[68] Yao K, Dong Y, Bian J, Ma M, Li J. Ultrasonics Sonochemistry Understanding the mechanism of ultrasound on the synthesis of cellulose/Cu(OH)2/CuO hybrids. Ultrason Sonochem 2015;24:27–35.

[69] Pankaj SK, Bueno-ferrer C, Misra NN, Neill LO, Jiménez A, Bourke P, et al. Characterization of polylactic acid fi lms for food packaging as affected by dielectric barrier discharge atmospheric plasma. Innov Food Sci Emerg Technol 2013;21.

[70] Kumar SSD, Surianarayanan M, Vijayaraghavan R, Mandal AB, Macfarlane DR. Curcumin loaded poly(2-hydroxyethyl methacrylate) nanoparticles from gelled ionic liquid—in vitro cytotoxicity and anti-cancer activity in SKOV-3 cells. Eur J Pharm Sci 2014;51(1):34–44.

[71] De Morais Teixeira E, Corrêa AC, Manzoli A, de Lima Leite F, de Oliveira CR, Mattoso LHC. Cellulose nanofibers from white and naturally colored cotton fibers. Cellulose 2010;17:595–606.

[72] Bonardd S, Robles E, Barandiaran I, Saldías C, Leiva Á, Kortaberria G. Biocomposites with increased dielectric constant based on Chitosan and nitrile-modified cellulose nanocrystals. Carbohydr Polym 2018;199:20–30.

[73] Kvien I, Tanem BS, Oksman K. Characterization of cellulose whiskers and their nanocomposites by atomic force and electron microscopy. Biomacromolecules 2005;6:3160–5.

[74] Silvério HA, Flauzino Neto WP, Dantas NO, Pasquini D. Extraction and characterization of cellulose nanocrystals from corncob for application as reinforcing agent in nanocomposites. Ind Crop Prod 2013;44:427–36.

[75] Mabrouk AB, Kaddami H, Boufi S, Erchiqui F, Dufresne A. Cellulosic nanoparticles from alfa fibers (Stipa tenacissima): extraction procedures and reinforcement potential in polymer nanocomposites. Cellulose 2012;19:843–53.

[76] Wu Y, Li W, Zhang X, Li B, Luo X, Liu S. Clarification of GO acted as a barrier against the crack propagation of the cellulose composite films. Compos Sci Technol 2014;104:52–8.

[77] Van Rie J, Thielemans W. Cellulose–gold nanoparticle hybrid materials. Nanoscale 2017;9:8525–54.

[78] Guo J, Filpponen I. Attachment of gold nanoparticles on cellulose nanofibrils via click reactions and electrostatic interactions. Cellulose 2016;23(5):3065–75.
[79] Wibowo AC, Misra M, Park H-M, Drzal LT, Schalek R, Mohanty AK. Biodegradable nanocomposites from cellulose acetate: mechanical, morphological, and thermal properties. Compos Part A 2006;37:1428–33.
[80] Bettaieb F, Khiari R, Dufresne A, Mhenni MF, Belgacem MN. Mechanical and thermal properties of Posidonia oceanica cellulose nanocrystal reinforced polymer. Carbohydr Polym 2015;123:99–104.
[81] Arrieta MP, Fortunati E, Dominici F, Rayón E, López J, Kenny JM. Multifunctional PLA-PHB/cellulose nanocrystal films: processing, structural and thermal properties. Carbohydr Polym 2014;107:16–24.
[82] Hou A, Zhou M, Wang X. Preparation and characterization of durable antibacterial cellulose biomaterials modified with triazine derivatives. Carbohydr Polym 2009;75(2):328–32.
[83] George J, Ramana KV, Bawa AS, Siddaramaiah. Bacterial cellulose nanocrystals exhibiting high thermal stability and their polymer nanocomposites. Int J Biol Macromol 2011;48(1):50–7.
[84] Huq T, Salmieri S, Khan A, Khan RA, Le Tien C, Riedl B, et al. Nanocrystalline cellulose (NCC) reinforced alginate based biodegradable nanocomposite film. Carbohydr Polym 2012;90(4):1757–63.
[85] Venkateshwaran N, Elayaperumal A, Sathiya GK. Prediction of tensile properties of hybrid-natural fiber composites. Compos Part B 2012;43(2):793–6.
[86] John K, Naidu SV. Effect of fiber content and fiber treatment on flexural properties of sisal fiber/glass fiber hybrid composites. J Reinf Plast Compos 2004;23(15):1601–5.
[87] Maslinda AB, Abdul Majid MS, Ridzuan MJM, Afendi M, Gibson AG. Effect of water absorption on the mechanical properties of hybrid interwoven cellulosic-cellulosic fibre reinforced epoxy composites. Compos Struct 2017;167:227–37.
[88] Ridzuan M, Abdul Majid MS, Afendi M, Azduwin K, Amin NAM, Zahri JM, Gibson AG. Moisture absorption and mechanical degradation of hybrid Pennisetum purpureum/glass—epoxy composites. Compos Struct 2016;141:110–6.
[89] Raajeshkrishna C, Chandramohan P, Saravanan D. Effect of surface treatment and stacking sequence on mechanical properties of basalt/glass epoxy composites. Polym Polym Compos 2018;27:1–14.
[90] Piekarska K, Sowinski P, Piorkowska E, Haque MM-U, Pracella M. Structure and properties of hybrid PLA nanocomposites with inorganic nanofillers and cellulose fibers. Compos Part A 2016;82:34–41.
[91] Lee S, Teramoto Y, Endo T. Cellulose nanofiber-reinforced polycaprolactone/polypropylene hybrid nanocomposite. Compos Part A 2011;42(2):151–6.
[92] Saba N, Safwan A, Sanyang ML, Mohammad F, Pervaiz M, Jawaid M, Sain M. Thermal and dynamic mechanical properties of cellulose nanofibers reinforced epoxy composites. Int J Biol Macromol 2017;102:822–8.
[93] Lavoratti A, Scienza LC, Zattera AJ. Dynamic-mechanical and thermomechanical properties of cellulose nanofiber/polyester resin composites. Carbohydr Polym 2015;136:955–63.
[94] Builes DH, Labidi J, Eceiza A, Mondragon I, Tercjak A. Unsaturated polyester nanocomposites modified with fibrillated cellulose and PEO-b-PPO-b-PEO block copolymer. Compos Sci Technol 2013;89:120–6.
[95] Jawaid M, Khalil HPSA, Alattas OS. A Woven hybrid biocomposites: dynamic mechanical and thermal properties. Compos Part A 2012;43(2):288–93.

Chapter 3

Hybrid nanocomposites based on cellulose nanocrystals/nanofibrils and carbon nanotubes: From preparation to applications

Farnaz Shahamati Fard, Hossein Kazemi, Frej Mighri, and Denis Rodrigue
Department of Chemical Engineering, Université Laval, Quebec City, QC, Canada

3.1 Introduction

Polymer composites are now playing an essential role in all aspects of human activities through a wide range of applications. However, conventional methods for preparing these materials are using high concentrations (over 40 wt%) of conventional fillers (mm size) limiting their potential in terms of uses, costs, performances, and processability. Today, based on green chemistry and sustainable development concepts, using lower filler concentration is the main target but should provide excellent properties. This is the main reason why nanoparticles, with at least one dimension between 1 and 100 nm, were seen as the best candidates to develop high performance composites. One family of nanoparticles is carbon-based such as fullerenes, carbon nanotubes (CNT), and graphene, having unique structures. These nanoparticles led to a large number of recent publications on several aspects like (1) processing methods, (2) performance analysis, (3) structure-properties relationships/models, and (4) possible applications of these new materials [1]. This work reported on different polymer matrices, including modification/compatibilization methods, to improve dispersion, adhesion, and efficiency.

CNT are carbon allotropes with a tubular structure based on sp^2 hybridization carbon-carbon bonds. This structure was initially reported by Ijima in 1991 [2–4]. The tubes can be composed of one layer and called single-walled carbon

Cellulose Nanocrystal/Nanoparticles Hybrid Nanocomposites. https://doi.org/10.1016/B978-0-12-822906-4.00006-2

nanotube (SWCNT). In this case, the range of outer diameter is between 1 nm and 30 nm [5]. However, CNT can also be composed of several layers and are called multi-walled carbon nanotube (MWCNT). Typical examples are presented in Fig. 3.1.

Furthermore, three carbon nanotube configurations exist: armchair, zigzag, and chiral. The main difference between these configurations depends on their processing conditions having a direct effect on the axis of the graphitic structure (rolling up) [1, 6, 7].

CNT is considered as a one-dimensional structure due to its high aspect ratio of about 1000 [4, 8]. CNT is known for its exceptional structural, electrical, mechanical, electromechanical, and chemical properties. It has a high aspect ratio and modulus [4, 9]. The density and Young's modulus of CNT are 1.3 g/cm^3 (one-sixth of the stainless steel) and higher than 1 TPa (five times greater than steel), respectively [4, 10, 11]. The measured tensile strength for a CNT is 50 times higher than steel (63 GPa) with high thermal conductivity about ~3000 W/m K [4, 12, 13]. CNT also has valuable electrical properties as its electrical conductivity is between 10^6 and 10^7 S/m, which is similar as copper [4, 13]. A combination of properties such as nano-size, high aspect ratio, high specific surface, high strength and modulus, thermal and chemical stability combined with the ability of chemical conversion and high electrical conductivity make CNT an excellent candidate for more investigations. One of the main application fields of CNT is for the production of polymer nanocomposites with improved performances. In this chapter, the effect of CNT addition

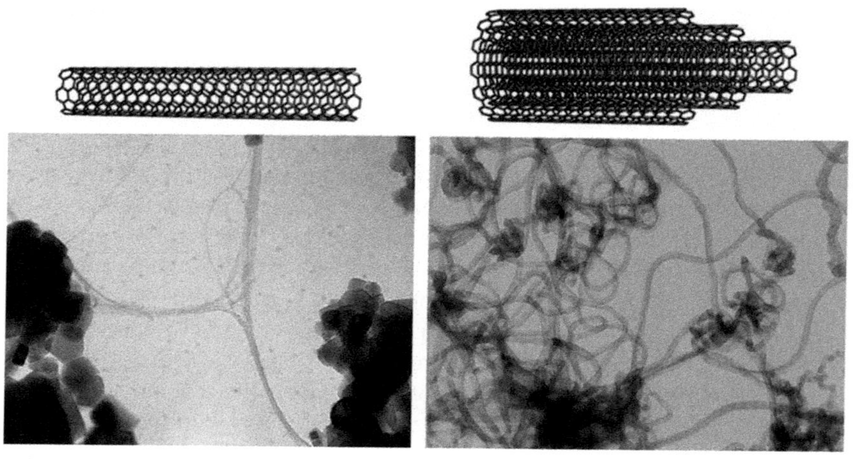

FIG. 3.1 Molecular representations of SWCNT *(top left)* and MWCNT *(top right)* with typical transmission electron micrographs *(below)*. *(Reproduced with permission from Donaldson K, Aitken R, Tran L, Stone V, Duffin R, Forrest G, et al. Carbon nanotubes: a review of their properties in relation to pulmonary toxicology and workplace safety. Toxicol Sci 2006;92(1):5-22. Copyright (2006) Oxford University Press.)*

into cellulose-based materials will be presented with different applications and polymer matrices.

3.2 Thermoplastic polyurethanes

It has confirmed that the addition of a nanofiller into a polymer matrix can improve its mechanical performance, chemical resistance, gas and solvents barrier properties, etc. Furthermore, significant improvements are possible by using relatively low amount of nanofillers (generally less than 5 wt%), avoiding the typical drawbacks associated with traditional micro fillers. In general, the properties of nanocomposites depend on the state of nanofillers dispersion within the matrix and the degree of filler-matrix interactions (interface). Hence, there is an obvious need to develop filler dispersion and strengthen the interfacial interactions to use the full potential of the nanocomposites [14]. Recent studies reported how the incorporation of nanofillers with different geometries and aspect ratios led to substantial improvements in filler dispersion and a more effective use because of synergistic effects.

Thermoplastic polyurethanes (TPU) are an essential class of polymeric materials with remarkable properties such as abrasion resistance, chemical and corrosion resistance, low-temperature flexibility, and a wide range of mechanical strength and toughness. TPU is a block copolymer made from alternating soft and rigid segments, which tend to aggregate into soft domains (SD) and hard domains (HD) due to thermodynamic instability. The final morphology will affect the overall TPU properties. The soft and hard domain size, the nature of the domain interface, as well as the mixing of hard segments into the soft segment phase influence most of the physical properties such as elasticity and toughness [1]. PU is widely used as lacquer and protective coatings, adhesives, sealing components, insulators, gaskets, clothes and footwear, implants, etc. [1].

Significant improvements in TPU properties have been reported through the addition of small amounts of nanostructured materials such as carbon nanotubes (CNT) [1], exfoliated graphite, and graphite oxide with cellulose nanocrystals (CNC). Different procedures have been proposed for their processing such as: in situ polymerization, melt blending, solution mixing techniques, and solvent-exchange processing [3]. Interestingly, some nanofillers such as CNT, CNC, and layered silicates can significantly change the micro-phase domain size and shape of block copolymers, enabling the control of the morphology at the nanoscale and consequently the final properties.

A significant problem appears when nanoparticle-polymer composites, including PU-carbon nanoparticle (CNP) system, are produced to achieve a uniform distribution of nanoparticles in a polymer matrix. When poor uniformity/distribution is obtained, the full potential of the composites cannot be accomplished. The difficulty is the tendency to form stable aggregates. Polymer composites containing CNP can be prepared either by covalent bonding of nanoparticles and their functionalized derivatives with a polymer matrix or by the mechanical introduction of CNP into a polymer when no chemical bonds are formed between the matrix and nanoparticles. The use of functionalized CNP can solve the compatibility problem of nanoparticles leading to a more uniform CNT distribution [1]. Pedrazzoli and Manas-Zloczower [15] investigated the effects of CNT as an active filler in the HD and SD morphology of thermoplastic polyurethanes (TPU). To improve the uniformity of the filler distribution, they examined the combination of CNT with cellulose nanocrystals (CNC). The results showed that nanofiller incorporation can produce different spherulite sizes and morphologies in TPU. Also, the filler surface properties can perform an essential part in filler segregation in the HD regions promoting further phase separation.

Such phase separation behavior is of primary importance to describe a potential route to control the morphology and material properties [15]. TPU-CNT-0.1 (0.1 wt% CNT) is harder than neat TPU indicating that the incorporation of CNT improved the energy dissipation mechanisms during fracture (Fig. 3.2B). The presence of rigid nanotubes limited the crack propagation and their path by acting as deflection sites. The presence of well-distributed CNT aggregates with an average dimension of 200–250 nm is presented in Fig. 3.2B'. The TPU-CNC nanocomposite with 0.1 wt% CNC displays similar aggregates with an average dimension of 250–300 nm (Fig. 3.2C'). On the other hand, the hybrid system CNT0.8-CNC0.2, having the same total filler content, presents smaller aggregates and well-distributed fillers inside the matrix, indicating a synergistic effect between the CNC and the CNT on their dispersion state and aggregation potential (Fig. 3.2D'). The well-dispersed CNC may act as a dispersing agent for CNT improving its dispersion because of their synergistic effect leading to a more homogeneous distribution in TPU nanocomposites [9].

Wu et al. [16] worked on water-responsive shape memory hybrid polymers based on thermoplastic PU crosslinking with CNT and hydroxyethyl cotton cellulose nanofiber (CNF-C) through solution blending. Their study showed that the CNT content has a direct effect on the thermal conductivity of hybrid nanocomposites. On the one hand, they examined the durability of the nanocomposites by mechanical and water-responsive shape memory effect. The results illustrated adequate mechanical and sensing performances for the TPU matrix fully crosslinked with CNF-C and CNT. The TPU-CNF-C-CNT nanocomposites were analyzed under fixed and recovered states under different conditions, such as periodicity and frequency. FTIR analysis was used to investigate the chemical interaction in the nanocomposites for the dry and wet states (Fig. 3.3).

The -NH peaks at 3335 and 2953 cm^{-1} were slightly altered by the presence of a polar group of the nanofillers, as well as hydrogen bonding interaction between carboxylic and hydroxyl groups (at 1640 cm^{-1}) from the TPU with CNF-C. On the other hand, the wet state of the nanocomposites showed hydrogen bonding break-up leading to the elimination of the peak at 1641 cm^{-1}. The drying process of the modified TPU illustrates the reappearing of hydrogen

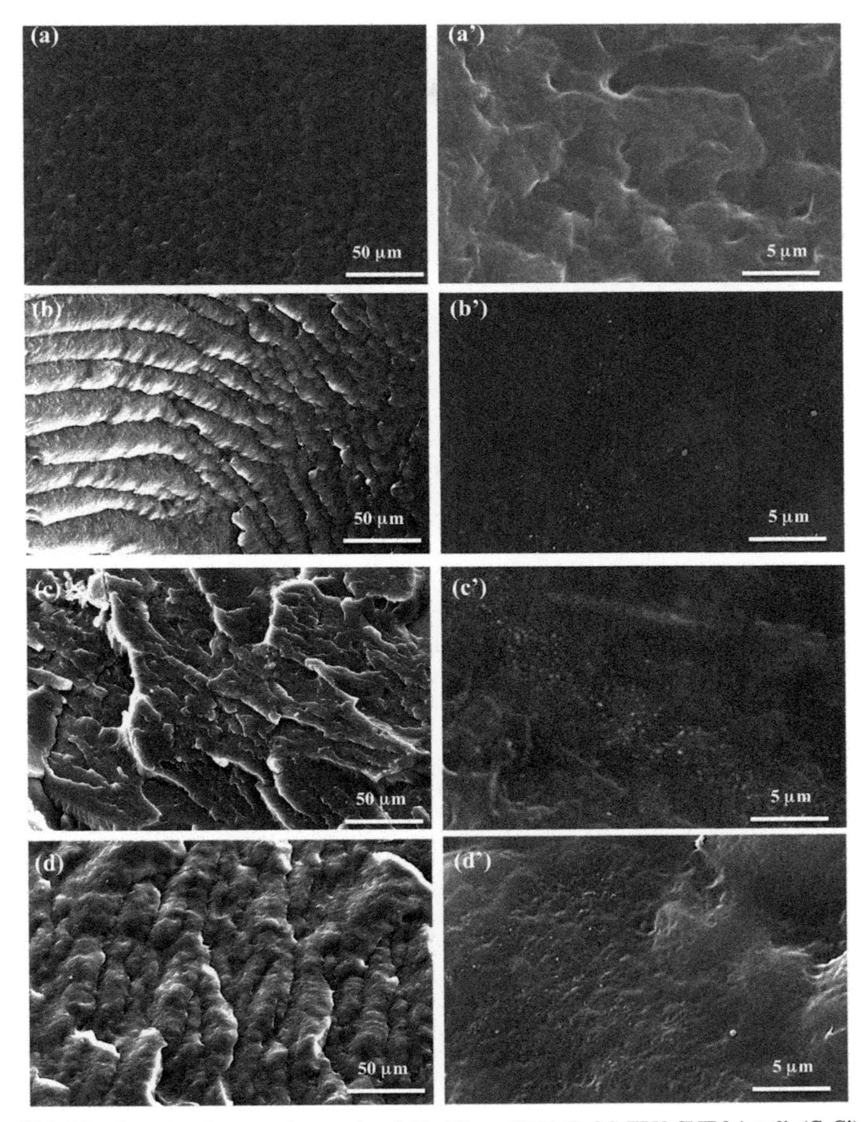

FIG. 3.2 Scanning electron micrographs of: (A, A') neat TPU, (B, B') TPU-CNT 0.1 wt%, (C, C') TPU-CNC 0.1 wt%, and (D, D') CNT 0.8 wt%-CNC 0.2 wt%. *(Reproduced with permission from Pedrazzoli D, Manas-Zloczower I. Understanding phase separation and morphology in thermoplastic polyurethanes nanocomposites. Polymer 2016;90:256-63. Copyright (2016) Elsevier.)*

bonds. This reversible behavior between the wet and dry states is associated with reversible intermolecular hydrogen bonds. As a result, CNT and CNF-C can be combined based on hydrogen bonding, and this type of crosslink leads to a three-dimensional structure inside the TPU-CNT-CNF-C [17].

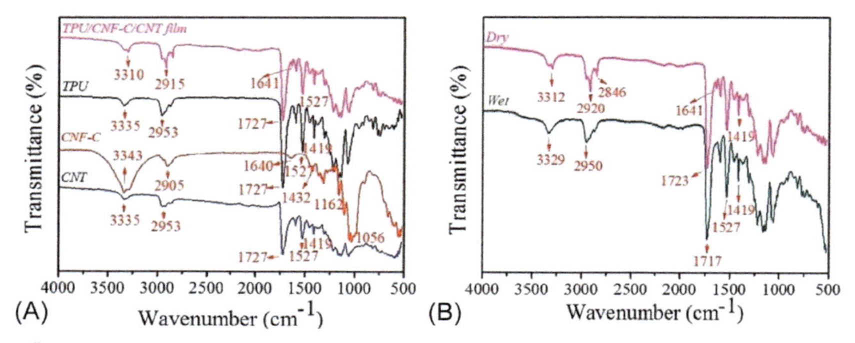

FIG. 3.3 (A) FTIR spectra of TPU-CNF-C-CNT films, TPU, CNF-C, CNT; (B) FTIR spectra of TPU-CNF-C-CNT films under original (dry) and water (wet) conditions. *(Reproduced with permission from Wu G, Gu Y, Hou X, Li R, Ke H, Xiao X. Hybrid Nanocomposites of cellulose/carbon-nanotubes/polyurethane with rapidly water sensitive shape memory effect and strain sensing performance. Polymers 2019;11(10):1-14. Copyright (2019) Multidisciplinary Digital Publishing Institute.)*

The mechanical properties of TPU-CNF-C-CNT nanocomposites were studied via tensile strain-stress curves, especially the tensile strength and elongation at break. The tensile strength was increased until an optimum CNT content was reached, and then decreased slowly. Based on the results, TPU-CNF-C-CNT (6 wt% of CNT) nanocomposites with a tensile strength of 31.8 MPa and an elongation at break of 904% was the best. For the cyclic tensile measurements, a shape fixity ratio (R_f) and shape recovery ratio (R_r) were calculate as:

$$R_f = (\mathcal{E}_u - N/\mathcal{E}_m) \times 100 \tag{3.1}$$

$$R_r = \left(\mathcal{E}_m - \mathcal{E}_p - N/\mathcal{E}_m - \mathcal{E}_p - (N-1)\right) \times 100 \tag{3.2}$$

where N is the cycle number, $\mathcal{E}_m$ is the set strain, $\mathcal{E}_u$-N is the fixed strain, and $\mathcal{E}_p$-N is the unrecovered strain after shape recovery [16, 18, 19]. The results showed that the R_f was 25.3% ($\mathcal{E}_u$-N=101%, $\mathcal{E}_m$=40%) and 49.7% ($\mathcal{E}_u$-N=19.5%, $\mathcal{E}_m$=40%) for dry and wet nanocomposites, respectively. Therefore, water is an effective plasticizer to fix the temporary shape of hybrid TPU nanocomposites. The results from the CNF-C and CNT indicates that hydrogen bonding is highly important for this nanocomposite. Furthermore, using 6 wt% CNT led to a good electrical conductivity of 0.142 S/m.

3.3 Flexible sensors

Natural rubber (NR) composites, because of their high elasticity, are interesting materials for flexible strain sensors [20]. CNT is one of the best reinforcements, especially for improving the NR electrical properties, and casting-evaporation is a common method for CNT-NR nanocomposite preparation. In this method,

CNT is mixed with the NR latex in an aqueous solution. However, because of the CNT hydrophobicity, poor dispersion is observed leading to large CNT agglomeration in the final NR nanocomposites [21]. On the other hand, nanocellulose is hydrophilic and shows excellent dispersibility in aqueous solutions [22]. But Kalashnikova et al. [23] reported that CNC has an amphiphilic character because of the high amount of hydroxyl groups on its surface and the hydrophobic character of some crystalline planes. In another study, Olivier et al. [24] confirmed that short range hydrophobic interactions are formed between the specific crystalline faces of nanocellulose and SWCNT, while long range electrostatic repulsion between the particles helped the stabilization of the SWCNT-CNC dispersion. Therefore Wang et al. [20] added CNC to the CNT suspension to increase its dispersion in NR latex. The zeta potential results confirmed the improvement in CNT suspension stability after CNC addition. SEM results also showed that the CNT-CNC hybrid system produced a continuous 3D conductive network in the NR matrix leading to significantly improved mechanical and electrical properties of CNT-CNC-NR nanocomposite, as shown in Fig. 3.4. The authors reported that this nanocomposite has a high potential for flexible strain sensor applications.

Polydimethylsiloxane (PDMS) is widely used in biomedical applications, such as microfluidic devices, contact lenses, cell bioreactors, and mammary prostheses, due to its elasticity, flexibility, optical transparency, and biocompatibility [25]. Recently, PDMS has been considered as a soft component polymer matrix for flexible devices. However, PDMS needs to be reinforced to further expand its applications. Cellulose nanofibrils (CNF) and carbon nanotubes (CNT) are known as ones of the best reinforcements to respectively increase the electrical and mechanical properties of a matrix. For example, Chen et al. [26] reinforced PDMS with CNF-CNT film as a nano-network template. The nanocomposite was prepared by immersion of CNF-CNT films in a liquid PDMS pre-polymer followed by curing with Sylgard 184 B (curing agent) at 60 °C for 2 h. The authors reported that the PDMS-CNF-CNT composite containing 30 wt% fillers showed good Young's modulus (805 MPa), tensile strength (18.3 MPa), and electrical conductivity (0.8 S/cm). The main application of these composites was for flexible sensors. In particular, two studies reported on the fabrication of PDMS-CNT-CNC imprinted electrochemical sensors using a layer-by-layer assembly for cortisol monitoring in sweat and adrenaline detection [27, 28].

Aerogels have several important properties such as high porosity, high surface area, and excellent specific mechanical properties [29, 30]. They can also have good filler distribution in the polymeric matrix [31]. Therefore, aerogels can be used in composites to have good mechanical properties. Aerogels are divided into isotropic and anisotropic categories, and most of them are used as barriers [29]. Some techniques are available to produce anisotropic aerogels, such as breath-figure self-assembly, gas inflation, direct carbonization of organic precursors, and unidirectional freeze-drying (UDF) [29, 32, 33]. But

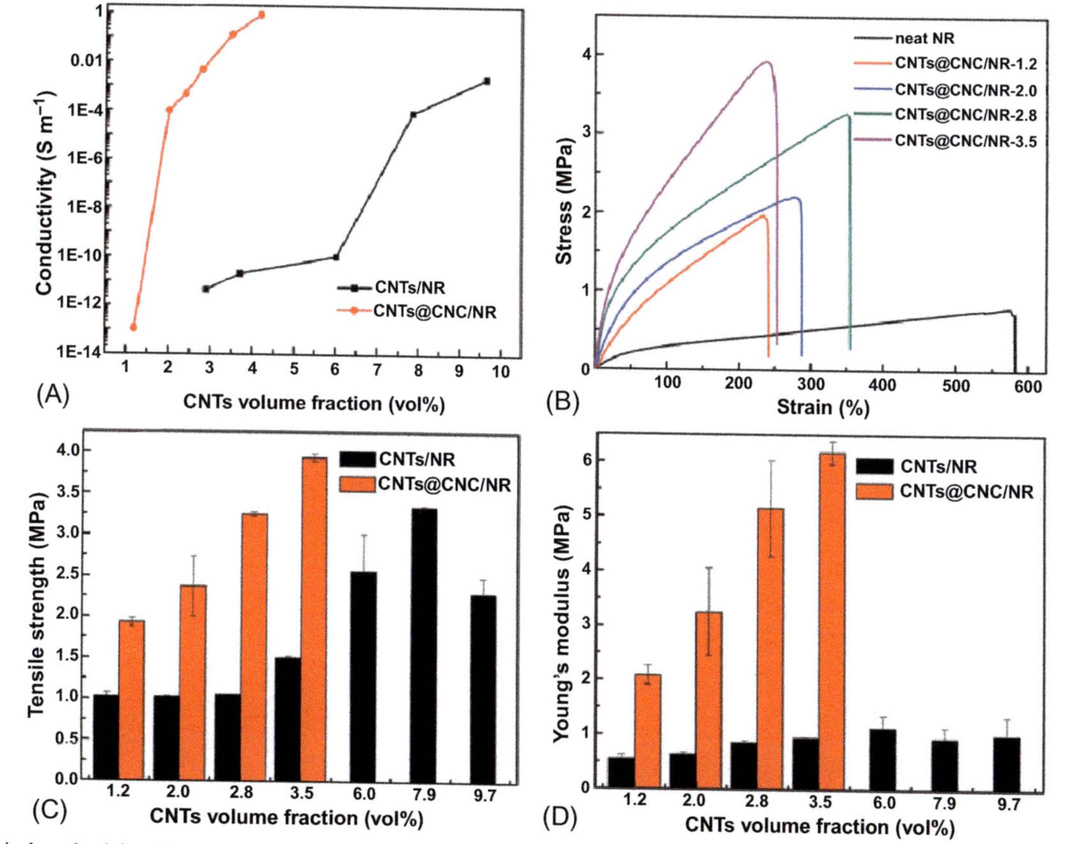

FIG. 3.4 (A) Electrical conductivity, (B) stress-strain curves, (C) tensile strength, and (D) Young's modulus of CNT-NR and CNT-CNC-NR nanocomposites as a function of CNT volume fraction. *(Reproduced with permission from Wang S, Zhang X, Wu X, Lu C. Tailoring percolating conductive networks of natural rubber composites for flexible strain sensors via a cellulose nanocrystal templated assembly. Soft Matter 2016;12(3):845-52. Copyright (2015) Royal Society of Chemistry.)*

the best method to get a defined void size is UDF. In this method, the wet gel is first frozen unidirectionally at a constant speed leading to a phase separation. Then, freeze-drying is applied to remove the dispersant. Recently, cellulose nanofiber aerogels have attracted a great deal of attention, but the hydrophilic nature of cellulose restricts the use of most polymers because the aerogel is difficult to be wet by hydrophobic polymers [29]. In fact, their insulating characteristics restrict their application in the fields of electronics [34]. In order to solve this problem, CNT addition was proposed as a second component to make the aerogel electrically conductive and improve its hydrophobicity. These new aerogels with CNT, besides micro-honeycomb cellulose (CellF), are called MCCA (micro-honeycomb cellulose-carbon nanotube aerogel). Wang et al. [29] used MCCA as a filler in PDMS and the materials showed good mechanical and electrical properties, as presented next.

Previous investigations showed that cellulose can be easily used to form a micro-honeycomb structure via UDF because of the structure-directing function of CellF [32]. Pan et al. [35] showed that CNT produced a lamella shape structure by UDF. The resulting materials showed that the aerogel pores gradually became squared with increasing CNT content.

Fig. 3.5A presents the electrical conductivity of MCCA with a honeycomb structure of cellulose and various CNT contents. An increasing trend for the electrical conductivity with CNT content can be seen as the value improved from 1×10^{-4} to 4.32 S/m between 10 wt% and 50 wt%. Fig. 3.5B also shows that the conductivity of MCCA is higher in the anisotropic state compared with the isotropic state [29]. As expected, the electrical conductivity of MCCA depends on the measuring direction with respect to the micro-channel: parallel (a-) or

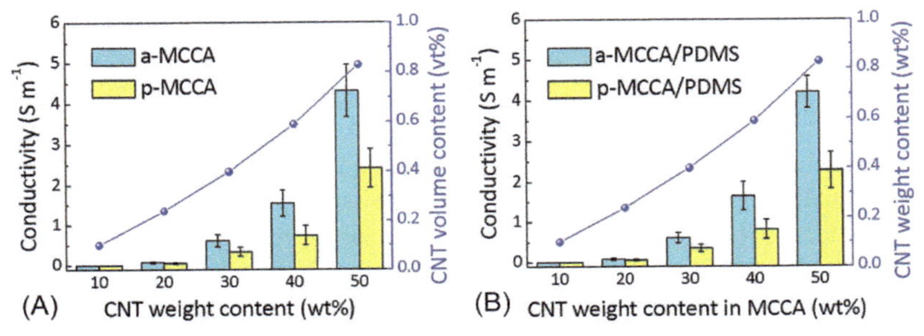

FIG. 3.5 Electrical conductivity of MCCA and MCCA-PDMS in the parallel (a-) or perpendicular (p-) direction with respect to the micro-channel direction. (A) The electrical conductivity of MCCA as a function of CNT weight content in which the *blue* line represents the CNT volume content (including the air space in MCCA), and (B) the electrical conductivity of MCCA-PDMS as a function of CNT weight content in which the *blue* line represents the CNT weight content in MCCA-PDMS. *(Reproduced with permission from Wang C, Pan ZZ, Lv W, Liu B, Wei J, Lv X, et al. A directional strain sensor based on anisotropic microhoneycomb cellulose nanofiber-carbon nanotube hybrid aerogels prepared by unidirectional freeze drying. Small 2019;15(14):1-8. Copyright (2019) John Wiley and Sons.)*

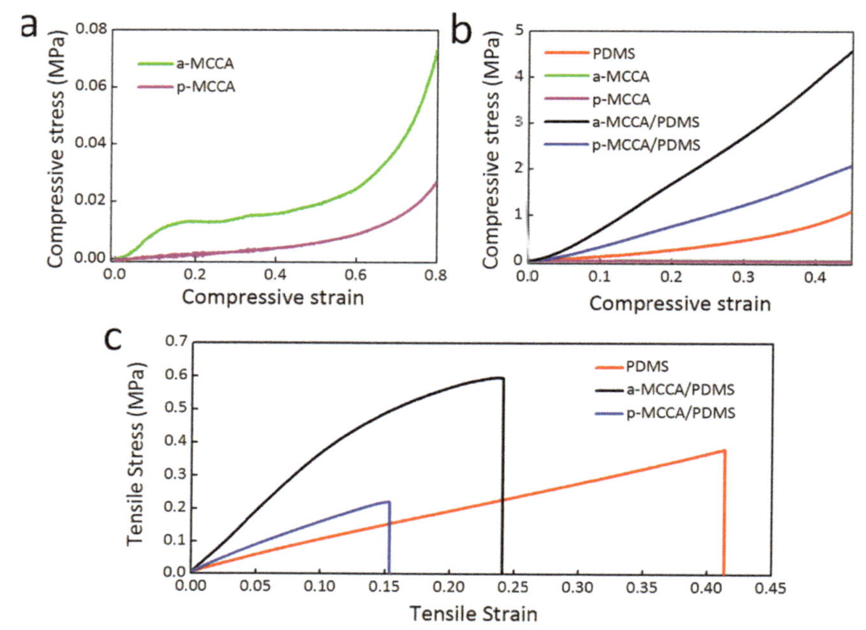

FIG. 3.6 Compressive and tensile properties of MCCA and MCCA-PDMS parallel (a-) or perpendicular (p-) to the micro-channel direction, as well as neat PDMS. The MCCA were prepared at an immersion speed of 50 cm/h with a CellF:CNT weight ratio of 1:1. (A) Compressive stress-strain curves of MCCA in both directions, (B) compressive stress-strain curves of MCCA, MCCA-PDMS and neat PDMS, and (C) tensile stress-strain curves of MCCA-PDMS and neat PDMS in both directions. *(Reproduced with permission from Wang C, Pan ZZ, Lv W, Liu B, Wei J, Lv X, et al. A directional strain sensor based on anisotropic microhoneycomb cellulose nanofiber-carbon nanotube hybrid aerogels prepared by unidirectional freeze drying. Small 2019;15(14):1-8. Copyright (2019) John Wiley and Sons.)*

perpendicular (p-). In general, the conductivity is twice as high in the a-direction compared to the p-direction. The reason is related to the shorter electron transmission path in the a-direction, leading to lower specific resistance [29].

Besides electrical properties, the mechanical properties, such as the compressive and tensile properties for both directions (a- and p-directions of MCCA), were examined. Fig. 3.6 shows that the stress-strain curves can be decomposed in three zones: a linear increase at low deformation, a plateau at intermediate deformation and a final sharp rise at high deformation. The initial slope represents the Young's modulus and MCCA-PDMS shows compressive Young's moduli of 6.8 and 3.1 MPa in the a- and p-directions, respectively, while the tensile modulus in the a- and p-directions of MCCA is 3.4 and 1.8 MPa, respectively. The continuous improvement of the MCCA-PDMS mechanical properties in the a-direction is related to the anisotropic synergic interaction between MCCA and PDMS [29].

Although CNT has excellent electrical conductivity, it has poor elasticity. Nevertheless, the honeycomb structure of cellulose can produce better elasticity. The anisotropic behavior of MCCA-PDMS under mechanical deformation with good electrical response makes it a good candidate for strain sensor applications. The external compression was applied in three directions for comparison: perpendicular, parallel and at 45 degree. The resistance of MCCA-PDMS was restricted in all directions because of the severe deformation of the conductive CNT walls. The sensitivity ratio, defined as the ratio between resistance changing rate and strain, is used to measure the response behavior of the MCCA-PDMS in different directions. These values in the a-direction, p-direction and at 45 degree are 4.81, 2.51, and 3.16, respectively. The sensitivity ratio for each direction gives more information to determine the effect of the measuring direction [29].

Chen et al. [36] worked on a pH sensor based on titanium dioxide (TiO_2)-MWCNT-cellulose (TMC) hybrid nanocomposite. CNT is known as a promising candidate for pH sensors due to its large surface area and high conductivity [37]. To start, TiO_2-MWCNT were produced by a hydrothermal process, and then blended with a cellulose solution. The adsorption sites increased due to the higher surface area of the hybrid nanocomposite leading to improved pH sensitivity. The long term stability of this type of sensor was also analyzed and the results showed high structural stability [36].

The electrical conductivity of TMC hybrid nanocomposites was measured in a pH buffer. The current-voltage (I-V) characteristics of the TMC pH sensor were assessed under fixed conditions (DC voltage of 2 V and a stabilization time of 2 min). The results indicated that the current decreased with increasing the pH level. For instance, when the sensor was dipped into a solution of pH 1 and 3, the current decreased from 4.6 to 2.1 μA.

The pH sensitivity of TMC was also reported as a function of conductivity as below [36, 38]:

$$\sigma\text{pH} = L/(WdV/I\text{pH}) \tag{3.3}$$

where σpH is the conductivity depending on the pH value, V is the applied voltage and IpH is the resulting current, while W, d and L are the width, thickness and length of the TMC hybrid nanocomposites, respectively.

The results showed that the pH sensitivity of the TMC hybrid nanocomposites can be divided in two regions based on the pH level. The TMC pH sensor has a high sensitivity for pH <6, but the sensitivity is highly reduced in an alkaline buffer solution. The durability of the TMC hybrid nanocomposites was followed by performing a measurement every 24 h for one week in different buffer solutions (pH 1, 4, and 8). The results showed that the sensor was very stable during this period under the conditions tested [36].

Meng and Manas-Zloczower [11] worked on the electrical conductivity of CNT-cellulose nanocrystals (CNC) films produced by simple vacuum filtration to investigate the synergic effect between CNT and CNC. The results showed that the hybrid films have better tensile properties and electrical conductivity compared to CNC alone. The tensile modulus and strength of CNC film with 10 wt% of CNT are 6.3 GPa and 180 MPa, respectively. On the other hand, with CNT addition, the film shows a more ductile response and three times higher tensile toughness than the CNC film (1.78 J/m^3). These CNT-CNC hybrid nanocomposites are expected to be useful for a broad range of applications, such as electromagnetic interference shielding (EMI) and electronic circuits [39].

In the work of Meng and Manas-Zloczower [11], the aspect ratio of the CNT and CNC was 158 and 60, respectively. These high values might be one of the reasons for the high mechanical properties improvement of the film. Another reason is the production of highly stable dispersion related to the negative charge of CNC in water because of some carboxylate functional groups and electrostatic repulsion effects [11, 40].

Fig. 3.7 presents the morphology of the tensile fracture surface of neat CNC film and CNC film with 1.2 wt% CNT (CNTC-1.2 wt%). It is clear that the CNC film without CNT has a smooth surface, while the CNTC-1.2 wt% film has a high number of large dents. Additionally, pulled out CNT, which is highlighted by red arrows, disturbed the structure breaking π interactions between the CNT and CNC particles instead of hydrogen bonding between CNC layers [41].

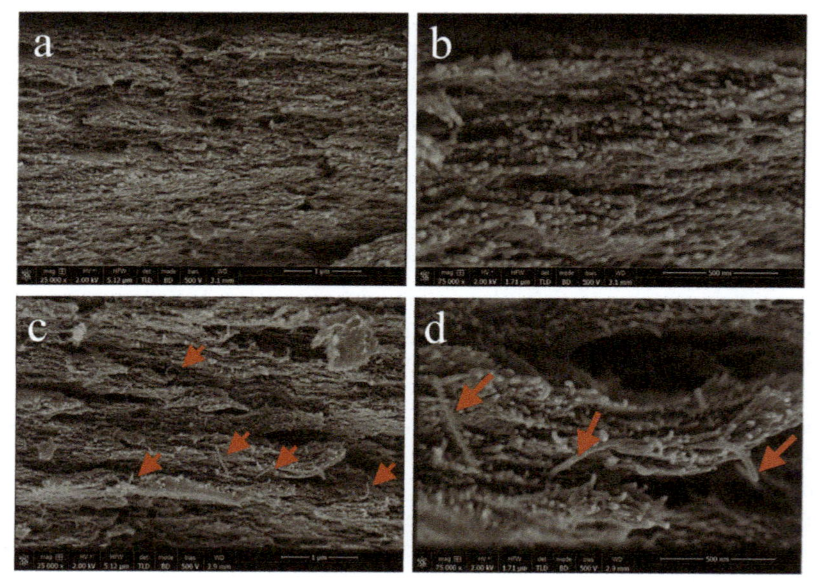

FIG. 3.7 Field emission scanning electron microscope (FESEM) micrographs of: (A–B) CNC, and (C–D) CNC films with 1.2 wt% CNT. *(Reproduced with permission from Meng Q, Manas-Zloczower I. Carbon nanotubes enhanced cellulose nanocrystals films with tailorable electrical conductivity. Compos Sci Technol 2015;120:1-8. Copyright (2015) Elsevier.)*

Fig. 3.8A presents the tensile properties of the CNT-CNC films with different CNT contents (1.2 and 2.2 wt%). Both the tensile modulus and tensile strength increased because of the reinforcement effect of CNT. Nevertheless, at higher CNT contents (5.2, 10.2, and 11.6 wt%), no significant moduli improvement was observed, mainly because of higher CNT interaction leading to particle agglomeration. The tensile properties also decreased because of possible voids between the CNT and CNC. The BET surface areas (A_{BET}) of CNT-CNC films with 20.1 wt%, and 62 wt% CNT were 0.026 m^2/g and 135.9 m^2/g respectively, indicating a significant increase in void content (higher internal surface area) leading to reduced tensile properties [11].

Fig. 3.8B illustrates the strain-stress curves of hybrid films at different CNT concentrations. The results show that CNC films are mainly brittle (linear relation with low strain at break) due to the stiffness of CNC particles and less hydrogen bonding interactions [11]. As seen in Fig. 3.9, the CNT-CNC films were characterized by a surface electrical resistivity (SER) at three different relative humidity levels (RH=22%, 59%, and 85%). The SER was 2.8×10^{11} Ω/sq at RH=22%, which dropped by a factor of five at RH=85% [11]. At the optimum CNT concentration (around 5 wt%), the CNT-CNC hybrid films have a higher probability to create a network and the conductivity is improved. In this case, the material can be used as a humidity sensor [11].

Recently, humidity measurement has attracted a great deal of attention for applications, such as weather prediction and wearable electronics [42–44]. The measurement of human body humidity with a humidity sensor is also an interesting application. However, most available humidity sensors are based on ceramics or metal oxides with rigid structures [45, 46]. But polymer-based materials are now investigated.

Zhu et al. [42] used fast vacuum filtration to produce NFC-CNT film as a flexible and highly humidity sensor to monitor human breath. They added NFC as a biodegradable material with a high humidity sensitivity because of the high amount of hydroxyl groups in its molecular chains [47]. Also, NFC can act as a surfactant to facilitate the dispersion of CNT in aqueous solutions due to its amphiphilic nature [47–50]. For example, 2,2,6,6-tetramethylpiperidinyl-1-oxyl (TEMPO)-oxidized NFC has been used to increase the dispersion of CNT in water [49]. NFC-CNT composite films with different CNT loadings (1%, 2%, 3%, 5%, and 10%, coded as CNT01, CNT02, CNT03, CNT05, and CNT10, respectively) were investigated. The results illustrate that the percolation threshold of this system is about 5 wt% CNT and the best sensitivity was obtained for CNT05. The response and recovery of CNT05 and CNT10 were characterized between 11% and 95% RH. For this range, the results showed that the maximum response of CNT05 and CNT10 is 69.9% and 66.0%, respectively, which indicates that CNT05 has a better response compared to CNT10. Additionally, CNT05 (330 s) has a shorter response time in comparison with CNT10 (401 s). All these properties make the NFC-CNT05 composite the best humidity sensor [42].

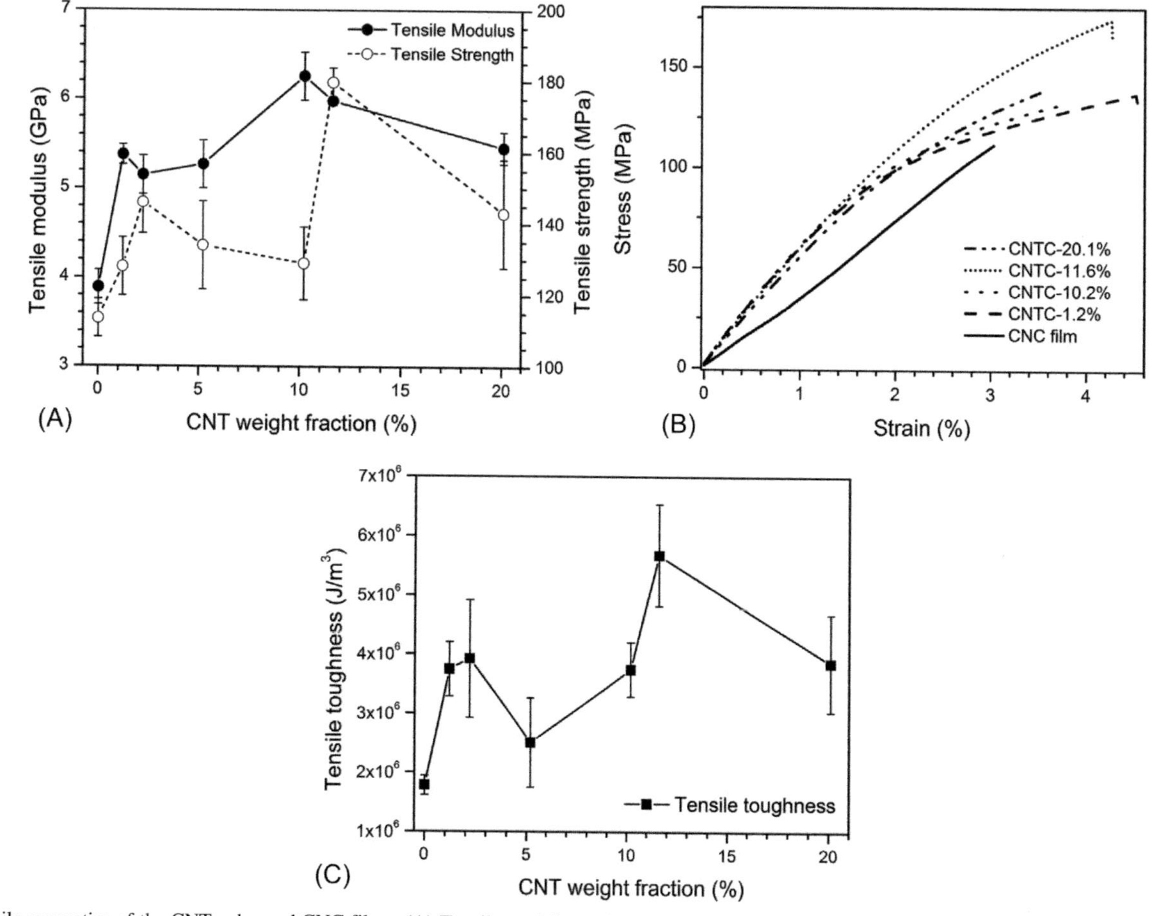

FIG. 3.8 Tensile properties of the CNT enhanced CNC films: (A) Tensile modulus and strength as a function of CNT weight fraction, (B) tensile stress-strain curves, and (C) tensile toughness as a function of CNT weight fraction. *(Reproduced with permission from Meng Q, Manas-Zloczower I. Carbon nanotubes enhanced cellulose nanocrystals films with tailorable electrical conductivity. Compos Sci Technol 2015;120:1-8. Copyright (2015) Elsevier.)*

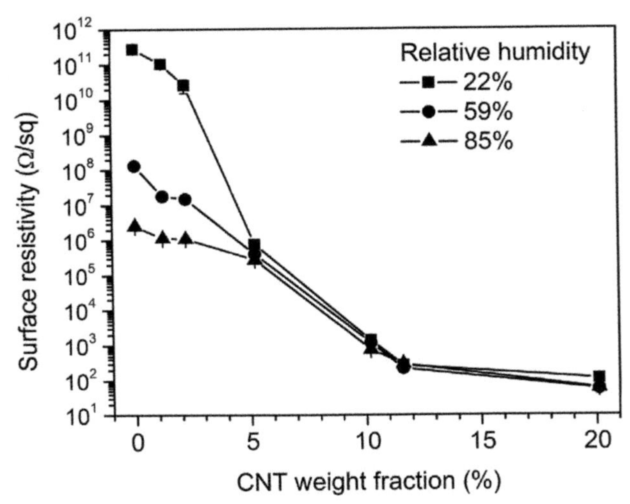

FIG. 3.9 Surface electrical resistivity of CNT enhanced CNC films as a function of CNT weight fraction at three relative humidity levels (22%, 59%, and 85%). *(Reproduced with permission from Meng Q, Manas-Zloczower I. Carbon nanotubes enhanced cellulose nanocrystals films with tailorable electrical conductivity. Compos Sci Technol 2015;120:1-8. Copyright (2015) Elsevier.)*

Safari and van de Ven [50] worked on the effect of water vapor adsorption on the electrical conductivity of CNT-nanocrystalline cellulose composites. The results showed that, contrary to the nanocrystalline cellulose (NCC) (also called cellulose nanocrystals or CNC), the electrostatically stabilized cellulose (ENCC) has an insulating behavior up to 75% RH at 1% CNT. This behavior was ascribed to the different composition of ENCC, which contains different carboxyl and hydroxyl groups bonded to one another limiting the formation of the hydrogen bond with water molecules compared with NCC [51, 52]. The moisture contents (water vapor) of CNT, NCC and ENCC at the highest relative humidity were up to 8.5%, 9.5% and 3%, respectively. These moisture content values show that, contrary to ENCC, high attraction level between the CNT powder and NCC films with water vapor coming from the differences in surface chemistry of each nanoparticle occurred [53], while the ENCC moisture adsorption remained below 1.5% when RH <80%. Argon sorption isotherm was used to evaluate the surface differences between NCC and ENCC. The results revealed that the occupied surface area by the water in NCC is 82.5 m^2/g, which is much higher than the ENCC film (14 m^2/g) [37, 50]. The electrical conductivity of the cellulose composites revealed that the electrical resistance of the composites decreased with CNT content. Based on further investigations, the direct current conductivity of NCC-CNT composites was analyzed as a function of CNT content. As seen in Fig. 3.10, the conductivity of the composites with high CNT loading (25%) decreased with

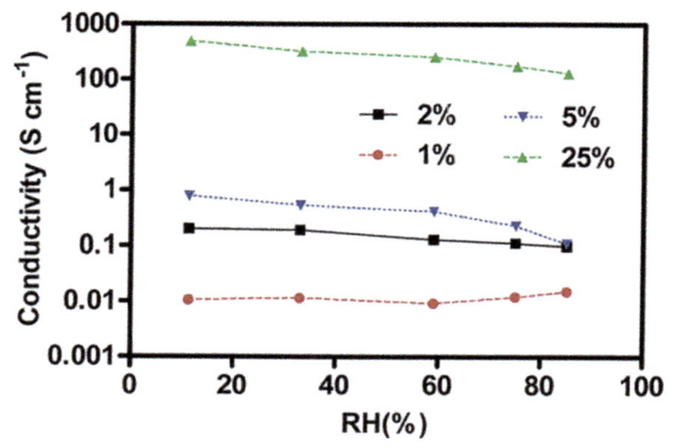

FIG. 3.10 Direct current conductivity as a function of relative humidity for CNT-NCC composites with different CNT content. *(Reproduced with permission from Safari S, van de Ven TG. Effect of water vapor adsorption on electrical properties of carbon nanotube/nanocrystalline cellulose composites. ACS Appl Mater Interfaces 2016;8(14):9483-9. Copyright (2016) American Chemical Society.)*

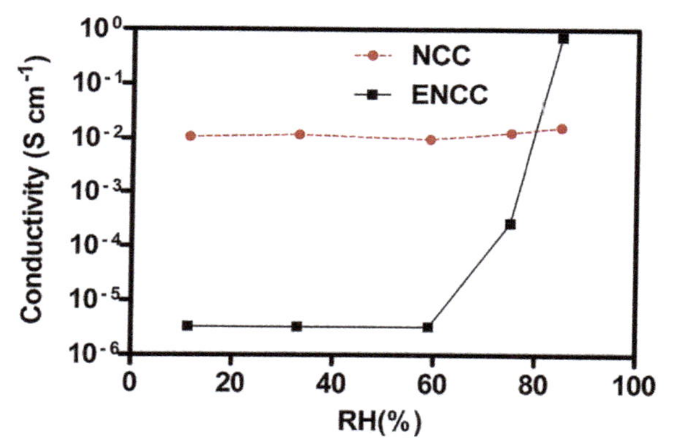

FIG. 3.11 DC electrical conductivity as a function of relative humidity (RH) for NCC and ENCC composites with 1% CNT. *(Reproduced with permission from Safari S, van de Ven TG. Effect of water vapor adsorption on electrical properties of carbon nanotube/nanocrystalline cellulose composites. ACS Appl Mater Interfaces 2016;8(14):9483-9. Copyright (2016) American Chemical Society.)*

increasing RH resulting from electrons given up by the water molecules and increased sensitivity of the composites for water sorption [54]. Between 1% and 2% of CNT, the RH sensitivity of the composite was reduced and the electrical conductivity showed less variation compared with 5% and 25% of CNT loading.

On the other hand, the electrical properties of the ENCC-CNT composites were studied with 1% CNT at various RH. As seen in Fig. 3.11, the electrical

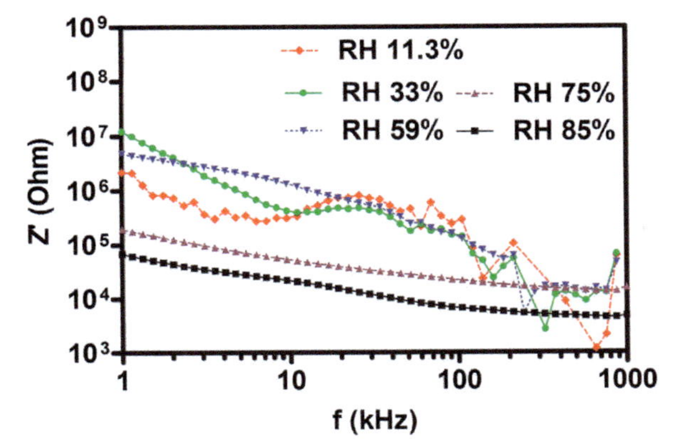

FIG. 3.12 Electrical resistance (*Z*) as a function of frequency (*f*) for ENCC composites with 1% CNT at different relative humidity (RH). *(Reproduced with permission from Safari S, van de Ven TG. Effect of water vapor adsorption on electrical properties of carbon nanotube/nanocrystalline cellulose composites. ACS Appl Mater Interfaces 2016;8(14):9483-9. Copyright (2016) American Chemical Society.)*

conductivity of CNT-NCC is steady and insensitive to moisture content, although the electrical conductivity of CNT-ENCC at 75% and 85% RH was about 2.6×10^{-4} and 0.7 S/cm, respectively. Impedance spectroscopy was used to monitor the vapor sorption on the CNT-ENCC for a frequency range of 1 kHz to 1 MHz and a voltage amplitude of 100 mV at room temperature. Fig. 3.12 reports that the resistivity of CNT-ENCC composites is reduced by about two orders of magnitude when the RH increased from 59% to 75% and 85%. As previously mentioned, this behavior is related to the breaking of hydrogen bonds in the composite and adsorption of water at higher RH. Based on these results, CNT-NCC and CNT-ENCC composites can be used in electronics and humidity sensor applications respectively, depending on their sensitivity level to the relative humidity [50].

Qi et al. [55] worked on the electrical conductivity of aerogels filled with CNT and cellulose to produce vapor detectors. The sensitivity of the sensors was examined by monitoring the electrical resistance change in contact of with the vapors of different volatile organic compounds (VOC) such as methanol, ethanol, and toluene. The results indicated that different parameters have an effect on the relative electrical resistance change: vapor composition, vapor concentration, and CNT content. Therefore, the results showed that CNT-cellulose composite aerogels have a fast response and high sensitivity in either polar or nonpolar vapors. This behavior was associated to both the nanostructured solid network (high specific surface areas between 140 and 160 m^2/g) and the nano-porous structure of the aerogel. Furthermore, the uniform CNT dispersion in the matrix played an important role [56]. The resistance change ($\Delta R = \Delta R_A + \Delta R_S$) of the CNT-cellulose composite aerogels for vapor sensing

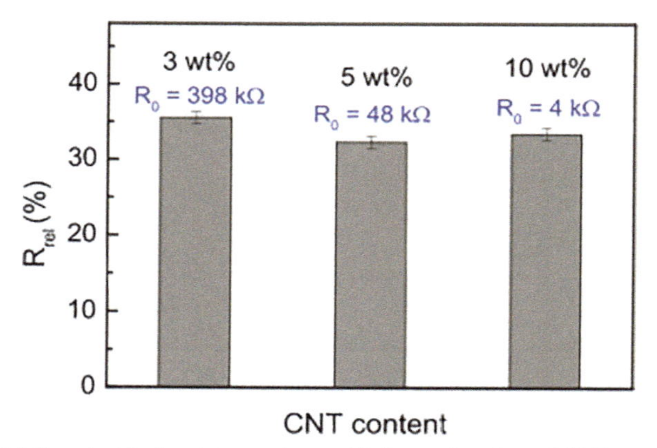

FIG. 3.13 Relative electrical resistance change (R_{rel}) as a function of CNT content for CNT-cellulose aerogels in saturated ethanol vapor at 25 °C. *(Reproduced with permission from Qi H, Liu J, Pionteck J, Pötschke P, Mäder E. Carbon nanotube-cellulose composite aerogels for vapour sensing. Sensor Actuat B Chem 2015;213:20-6. Copyright (2015) Elsevier.)*

was related to two factors: the resistance change from absorption (ΔR_A) of organic solvent between the nanotubes forming a non-conductive layer in the CNT structure [57], and the resistance change from swelling of the polymer (ΔR_S), leading to the disruption of electron transport between the CNT [58].

Fig. 3.13 shows the relative electrical resistance change (R_{rel}) to ethanol vapor of the CNT-cellulose composite aerogels with 3%, 5%, and 10% CNT. The plot presents similar resistance changes independent of the CNT loading (R_{rel} is about 35%), which is completely different from the solid CNT-polymer composites [55, 59, 60]. For example, the solid CNT-polymer composites showed 400% and 5500% R_{rel} for 10% and 2% CNT content, respectively. These significant variations are related to different sensing phenomena involved which is mainly ΔR_S for solid CNT-polymer composites [59, 61], compared to ΔR_A for CNT-cellulose composite aerogels [55].

Cellulose, as a polar polymer, has a high affinity towards polar vapor such as methanol, ethanol, acetone, and tetrahydrofuran (THF). The literature reports that cellulose has a high R_{rel} about 40%–45% to polar vapor [55]. However, it has a relatively lower response to water (9%) because of the low saturated vapor concentration [55]. Even for nonpolar vapors, such as n-hexane, the R_{rel} reported (12.5%) is higher than for coated polymers with CNT (2%) [62]. These results confirm that CNT-cellulose composite aerogels have a high potential to be used as gas (vapor) sensors, especially for VOC.

Owens [63] worked on composites based on nitrocellulose (NC) and SWCNT. The composites with 12.5% SWCNT presented an improvement in hardness from 80 to 96.8 Shore A. The results indicated that the functionalized CNT with oxygen groups are better in producing improved interactions with the matrix. Sun et al. [64] worked on composite films based on oxidized CNT (o-

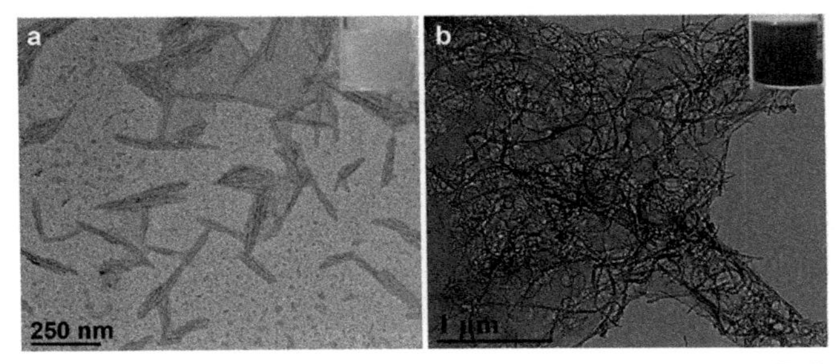

FIG. 3.14 Transmission electron microscope (TEM) images of an aqueous CNC dispersion: (A) before o-CNT addition, and (B) after o-CNT addition (0.075 wt%) (o-CNT = oxidized CNT). *(Reproduced with permission from Sun J, Zhang C, Yuan Z, Ji X, Fu Y, Li H, et al. Composite films with ordered carbon nanotubes and cellulose nanocrystals. J Phys Chem C 2017;121(16):8976-81. Copyright (2017) American Chemical Society.)*

CNT) and CNC. One interesting aspect of CNC is the ability to produce chiral nematic liquid crystals (CNLC) based on evaporation-induced self-assembly (EISA). Additionally, the CNT alignment was required to produce better opto-electronics properties, which can be achieved by using liquid crystals (LC) [65, 66]. It is generally accepted that CNT will produce the main organization of the LC phase [67]. In this study, the authors firstly oxidized CNT by a mixture of sulfuric acid and nitric acid to get hydrophilic functional groups on CNT, which was very similar as for CNC. As shown in Fig. 3.14, a homogenous dispersion was achieved after introduction of the modified CNT into a CNC aqueous solution.

Hamedi et al. [41] prepared semi-transparent conductive nano-papers from nano-fibrillated cellulose (NFC) and SWNT through a similar water-based processing route. They reported that SWCNT can be exfoliated by carboxylated NFC into individual nanotubes, which were stabilized in water. The results showed that adding 10 wt% SWCNT led to lower tensile strength, modulus, and strain to failure due to limited NFC-SWCNT interactions (poor stress transfer). However, the electrical conductivity increased at 43 wt% SWCNT to reach 174 S/cm. It was assumed that the high nonpolar interactions between both types of nanoparticles produced enough interaction to get good SWCNT dispersion. The mechanical properties of the neat and filled nano-paper with pristine SWCNT and acidified SWCNT were examined. It was shown that the tensile strength increased by 300% when pristine NFC was used compared to SWCNT-COOH. This implies that CNT-NFC composites are good candidates for electronic applications [41].

Fig. 3.15 illustrates the effect of the SWCNT volume fraction on the nanocomposite electrical conductivity. The percolation threshold (φ_c) was around 2.7% and the maximum electrical conductivity was 174 S/cm at 43 wt% SWCNT [41].

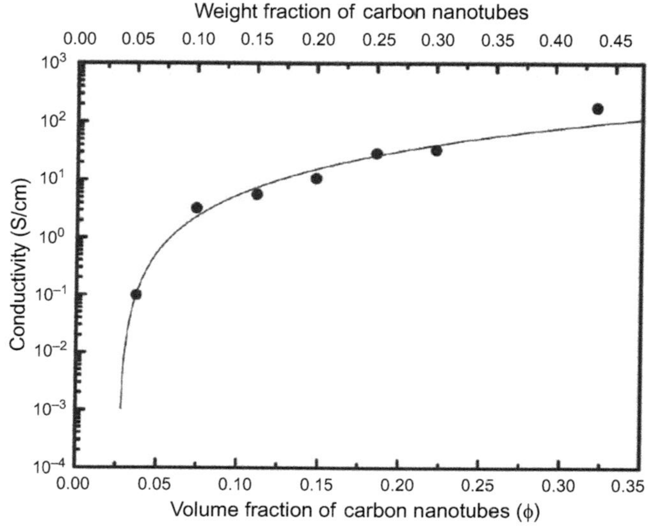

FIG. 3.15 Effect of SWCNT volume/weight fraction on the electrical conductivity of NFC. *(Reproduced with permission from Hamedi MM, Hajian A, Fall AB, Hakansson K, Salajkova M, Lundell F, et al. Highly conducting, strong nanocomposites based on nanocellulose-assisted aqueous dispersions of single-wall carbon nanotubes. ACS Nano 2014;8(3):2467-76. Copyright (2014) American Chemical Society.)*

3.4 Adsorption

Environmental issues arising from water pollution have recently become complex and threaten the ecosystems. Oil spill at the water surface is one of the most important causes of water pollution. Different methods, such as oil containment booms, oil skimmers, bioremediation, dispersants, and adsorption have been developed for oil spill cleanup [68]. Between them, adsorption is one of the most efficient methods for oil removal from the water surface [69]. Nowadays, cellulose-based adsorbents have been considered as promising candidates for oil adsorption and water treatment due to their low cost, low density, high porosity, biodegradability, and renewability [70]. However, because of the hydrophilicity of cellulose, appropriate chemical modification is required to decrease the water adsorption and increase the efficiency, selectivity, and oil separation.

Cellulose can be hydrophobically modified by chemical or physical methods. Yin et al. [71] prepared a hydrophobic cellulose-based aerogel by coating it with methyltrimethoxysilane (MTMS). The obtained aerogel showed good physical properties, such as low density (<0.0055 g/cm^3), high porosity ($>99.6\%$), as well as excellent oil adsorption capacity (58–101 g/g). Zhang et al. [72] synthesized a cellulose-based adsorbent and increased its hydrophobicity by using plasma treatment with subsequent silane modification. This aerogel produced a high oil volume adsorption capacity of 99% (cm^3/cm^3).

Zhou et al. [73] produced a superhydrophobic microfibrillated cellulose aerogel by the formation of polysiloxane on the surface of cellulose aerogels through silanization. These aerogels had an oil adsorption capacity of 159 g/g and can be reused more than 30 times. Similarly, Korhonen et al. [74] reported an adsorption capacity of 20–40 (g/g) (depending on the density of the liquid) for a cellulose aerogel modified by titanium dioxide.

Neelamegan et al. [75] used single wall carbon nanotubes (SWCNT) to increase the hydrophobicity of cellulose-based nanocomposites, thereby improving their oil adsorption capacity and recovery. SWCNT was first oxidized with a mixture of nitric acid and sulfuric acid (SWCNT-COOH) and then functionalized with thionyl chloride (SWCNT-COCl) to have better interaction with cellulose. Furthermore, the nanocomposite was immobilized with iron oxide for easier oil removal. The results showed that SWCNT addition on the cellulose significantly increased the nanocomposite hydrophobicity, as reported in Fig. 3.16. XPS and XRD analyses indicated that covalent binding was formed between cellulose and SWCNT, while SEM analysis showed that the entire cellulose surface was uniformly grafted with SWCNT leading to homogeneous hydrophobic polymeric nanocomposites having the configuration needed to be effective adsorbents. The results also showed that the chlorobenzene adsorption capacity of cellulose-based nanocomposite containing 2 wt % SWCNT was 23 g/g.

Xu et al. [76] showed that adding graphene to cellulose aerogels can decrease the volume shrinkage. Therefore because of similar characteristics between graphene and CNT, Li et al. [77] introduced CNT into cellulose

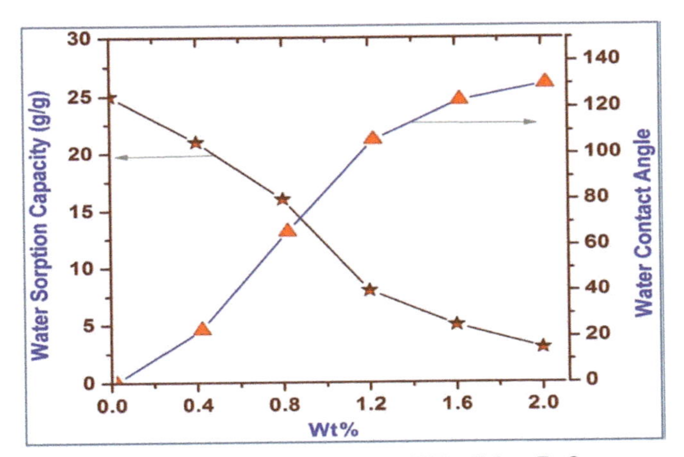

FIG. 3.16 Water adsorption and contact angle of SWCNT-cellulose-Fe_3O_4 nanocomposite as a function of SWCNT weight content. *(Reproduced with permission from Neelamegan H, Yang D-K, Lee G-J, Anandan S, Wu JJ. Synthesis of magnetite nanoparticles anchored cellulose and lignin-based carbon nanotube composites for rapid oil spill cleanup. Mater Today Commun 2020;22:1-11. Copyright (2020) Elsevier.)*

aerogels. They reported that CNT are easily agglomerated in the CNT-CNF mixed system during the freezing process. Based on the literature, a common method to improve the mixing and increase the interaction between nanocellulose and CNT is the surface modification of CNT with HNO_3-H_2SO_4 [14, 78, 79]. However, this method is not environmentally friendly and very costly. Therefore the authors mixed multiwalled carbon nanotubes (MWCNT) with poly(vinyl alcohol) (PVA) solutions before CNF addition. The idea is that the MWCNT would be well covered by PVA limiting MWCNT contacts between them. Then, Fe_3O_4 nanoparticles were added to restrict the aerogel mobility (higher mechanical strength) and improving oil removal. The results showed that CNF-PVA-MWCNT aerogel had better dimensional stability and thermal stability compared to the one without MWCNT. These aerogels also had good properties like high porosity (>90%), high hydrophobicity (water contact angle of 154 degree) and low density (0.098 g/cm^3). Furthermore, they showed an outstanding compressive strength (0.35 MPa), which was higher than several other cellulose-based aerogels. The adsorption capacities of this aerogel for different types of oils and organic solvents were in the range of 13 g/g (gasoline) to 70 g/g (pump oil). This wide range in adsorption capacity was attributed to the characteristics of the oils such as their hydrophobicity, density, and surface tension [77].

Due to the rapid development of the nuclear industry, high amounts of heavy metal ions and radionuclides have been discharged into the soil and aquatic environments [80]. These hazardous radionuclides are non-biodegradable, biologically toxic and even fatal for an ecosystem stability, aquatic organisms, and human health. Therefore it is mandatory to remove heavy metal ions and radionuclides from wastewater systems before discharging them into the environment [81]. MWCNT have already been used as potential radionuclides adsorbents because of their outstanding surface properties [82]. However, MWCNT alone do not show good separation and recovery from aqueous solutions restricting the expansion of their applications. The separation property can be improved by the incorporation of iron oxides with MWCNT (denoted as MMWCNT) [83]. Zong et al. [84] synthesized sodium carboxymethyl cellulose (CMC)-MMWCNT nanocomposites by a plasma technique for wastewater treatment. The adsorption capacity of Eu(III) on this nanocomposite was higher than that of a wide range of adsorbents, because the conjugated CMC increased the adsorption capacity of CMC-MMWCNT nanocomposite toward Eu(III) ions by providing multiple carboxyl and hydroxyl functional groups. The adsorption capacity and adsorption level of Eu(III) onto iron oxides, MMWCNT, MWCNT, and CMC-MMWCNT as a function of solid concentration are reported in Fig. 3.17.

Nano-filtration membrane is a cost-effective technology for water treatment due to several advantages like simple device operation, high efficiency, and low energy consumption [85]. However, nano-filtration membranes usually have low porosity. As a result, the pores can be easily blocked and the flux is low

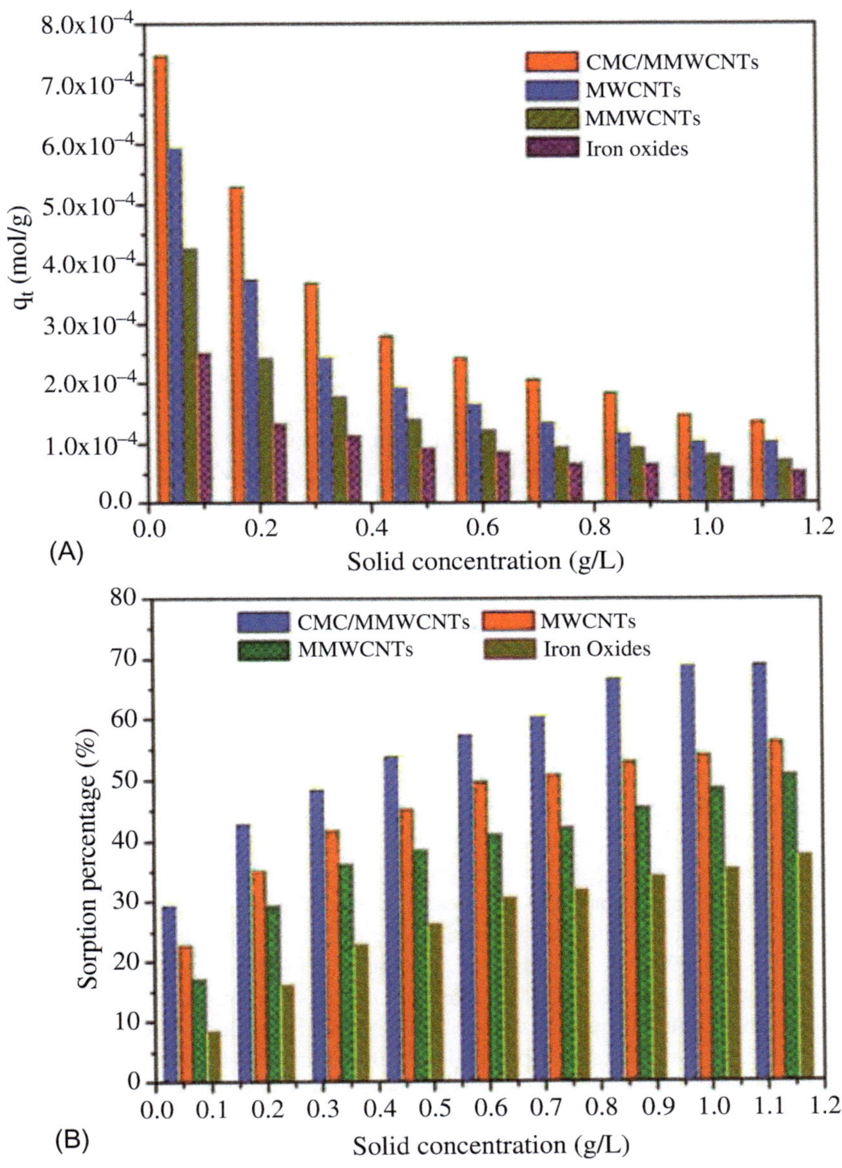

FIG. 3.17 Adsorption capacity and adsorption level of Eu(III) onto iron oxides, MWCNT, MMWCNT (MWCNT+iron oxides), and CMC-MMWCNT as a function of solid concentration. *(Reproduced with permission from Zong P, Cao D, Cheng Y, Wang S, Hayat T, Alharbi SN, et al. Enhanced performance for Eu (iii) ion remediation using magnetic multiwalled carbon nanotubes functionalized with carboxymethyl cellulose nanoparticles synthesized by plasma technology. Inorg Chem Front 2018;5(12):3184-96. Copyright (2018) Royal Society of Chemistry.)*

at low pressure [86]. Electrospinning is an effective technique to produce continuous nanofiber membranes with high porosity [87]. Therefore Zhijiang et al. [88] prepared chitosan hydrogel coated electrospun bacterial cellulose (BC) by combining electrospinning and surface coating. The CNT were then grafted onto BC molecular chains to improve the mechanical properties of the nanocomposites. The membranes were finally used for dyes removal from wastewater. Several dyes with molecular weight between 300 and 800 g/mol were prepared for this study and the results showed that the rejection rate reached up to 90% for dyes having molecular weights above 600 g/mol under 0.5 MPa.

Rare earth elements (REE) are vital components in catalysis, metallurgy, and electronics. In recent years, the increasing demand for REE led to the over-extraction of natural resources and also environmental issues. Therefore, recycling REE from consumer goods attracted high attention. Today, solid-phase extraction is the best technology for separating REE since it has high separation rates and requires less solvent compared to other methods like liquid-liquid extraction and supported liquid extraction [89, 90]. Zheng et al. [91] used imprinted cellulose nanocrystals (CNC) film as an adsorbent to selectively extract dysprosium (Dy(III)). However, it showed low adsorption capacity (17.5 mg/g). In another study [92], MWCNT and graphene oxide were added to 2,2,6,6-tetramethylpiperidine-1-oxyl (TEMPO) mediated oxidation of CNC film to provide additional binding sites with Dy(III). The oxidized CNC, CNT, and graphene all improved the stability and dispersion in water, as well as providing more carboxyl groups for coordination with the rare earth Dy(III). The results showed that the optimum pH value for Dy(III) adsorption was about 7, while the imprinted films showed the best separation at pH 4. The best Dy(III) adsorption capacities for CNC, CNC-CNT, and CNC-graphene films were 29.0, 38.7, and 41.8 mg/g, respectively.

Lanthanum (La) is another REE having special physicochemical properties. Therefore, it is widely applied in the production of catalysts, Ni-H batteries, and high-performance fluorescent devices. The increased demand for La led to a large amount of La entering the environment and resulting in severe environmental issues [93, 94]. Zhang et al. [95] used phosphorylated nanocellulose as film units and tetramethyl silicate (TEOS) as structure-oriented templates to adsorb La(III). They also added MWCNT to introduce more active sites and improve the mechanical properties of the films. Phosphorylation of CNC was performed at different conditions: (i) in aqueous media or (ii) in molten urea media. The La(III) adsorption capacity for the CNC-MWCNT film prepared in molten urea media was 118.5 mg/g, which was 83.4 mg/g higher than the one in aqueous media.

3.5 Optoelectronic applications

Another application of nanocellulose-CNT composite is for the production of high-performance transparent conductive films. These films have attracted substantial attention for a variety of products, technologies, and industries, such as organic light-emitting diodes, electromagnetic touch screens, smart windows, solar cells, solid-state lighting, transistors, etc. [96, 97]. Indium tin oxide (ITO) is the main component of conventional transparent conductive film production. However, the preparation of ITO films is complex due to the low pressure (vacuum) and high temperature used. Recently, carbon nanotubes have been considered as suitable candidates for ITO replacement due to their unique physical, mechanical, and electronic properties [98]. Kim et al. [99] reported that the current methods for CNT film production are complex. Therefore, they proposed a new and convenient method involving the adsorption of single walled carbon nanotubes (SWCNT) onto a bacterial cellulose (BC) hydrogel. The transparent conductive film containing 0.01 wt% SWCNT showed acceptable electrical surface resistance (2.8 kΩ/sq) and transparency (transmittance of 77% at 550 nm), which can be used in optoelectronic devices.

3.6 Wearable electronic devices

Zhang et al. [100] worked on the production of high-performance electromagnetic materials based on CNF modified via TEMPO-mediated oxidation and unmodified MWCNT. The CNF-MWCNT composite film was produced by a combination of vacuum filtration and hot pressing. The results showed significant tensile strength (up to 48 MPa). The high increase is related to the good dispersion and stability of the MWCNT solution when TEMPO-oxidized CNF was added. Additionally, extremely high electrical conductivity (3024 S/m) and electromagnetic interference shielding effectiveness (45.8 dB) were achieved. In another work [101], a cellulose-based filament was developed using anionic TEMPO-oxidized cellulose nanofibers (TO-CNF) and cationic chitosan (deacetylated chitin nanocrystals) via the interfacial nanoparticle complexation (INC) method. To increase the electrical conductivity, they introduced the SWCNT into a TO-CNF aqueous suspension before complexing with a chitosan aqueous suspension. The filaments showed an excellent electrical conductivity of 2056 S/m. Even if the addition of SWCNT decreased all the mechanical properties, these conductive filaments exhibited a high potential for wearable electronic devices (supercapacitors and sensors) in the biomedical field.

Yu et al. [102] used a Pickering emulsion method to develop polylactide (PLA)-cellulose nanocrystal (CNC)-CNT composites. Once again, the CNC was used as a stabilizer and dispersant for the CNT. The composites with 4.3 wt% CNT showed high mechanical properties (tensile strength of 45.5 MPa and Young's modulus of 3.15 GPa) and outstanding electromagnetic interference shielding effectiveness (41.8 dB). In another study, Wan et al. [103] developed a CNF-SWCNT filament with a robust CNT network via a SWCNT wrapping method using a combination of three-mill-roll and wet spinning techniques. The results showed a high tensile strength (421 MPa) with an excellent electrical conductivity (8643 S/m). Consequently, adequate connection between the particles was produced leading to significant electrical and mechanical properties. These filaments are considered as promising candidates for wearable bioelectronic devices, especially for therapeutics [103, 104].

3.7 Supercapacitors

Over the years, energy consumption has been continuously expanding, leading to increasing demand for efficient devices that can store and/or deliver energy. The implementation of supercapacitors represents a significant contribution to fulfill this necessity as a highly efficient storage device that can repeatedly provide high power delivery in a few seconds and act as complement for batteries in the future [105–109]. More recently, CNT has attracted high attention to be used in supercapacitors applications [105, 106, 110, 111]. On the one hand, carbon is electrically conductive and has high mechanical and thermal stability. On the other hand, it is relatively low cost and easily available with controlled porosity and high surface area [105, 112, 113]. Additionally, developing a flexible and thin supercapacitor is another critical topic in the supercapacitor production and cellulose is the best candidate to be used for the production of flexible energy storage devices [107].

Li et al. [107] worked on supercapacitors based on bacterial cellulose (BC), MWCNT, and polyaniline (PANI). In their study, vacuum filtration was used to fabricate deposited BC paper with MWCNT, and PANI was coated onto the MWCNT by using an electrochemical polymerization method. The highly conductive combination of MWCNT-PANI eliminates the necessity of other conductive additives (such as metal), leading to weight and cost reduction. Additionally, the BC paper confers high flexibility to the hybrid paper since the material was bent over 100 times and individual layer separation (delamination) was not observed. The results showed that the BC-MWCNT-PANI electrode has a very high capacitance (656 F/g at a discharge current density of 1 A/g) and very good cycling stability with less than 0.5% degradation after 1000 charge-discharge cycles (1 M H_2SO_4 electrolyte at a discharge current density of 10 A/g) at room temperature.

In another study, Lyu et al. [114] proposed CNF aerogels as the matrix with electroactive materials, such as PANI, reduced graphene oxide (RGO), and carboxylic multiwalled carbon nanotubes (CMWCNT). In their study, the PANI suspension was firstly assembled inside the CNF aerogels by vacuum filtration. Then, two types of electrodes were prepared by adding CMWCNT or RGO to the CNF-PANI aerogel via layer-by-layer assembly. The results showed that the electrode containing CMWCNT was superior to the one with RGO. For example, the specific capacitance of the CNF-PANI-CMWCNT (966 F/g) was 24% higher than that of CNF-PANI-RGO (781 F/g) electrodes in a 1 M aqueous H_2SO_4 electrolyte.

Deng et al. [108] prepared MWCNT-cellulose nanofiber by electrospinning a mixture of MWCNT and cellulose acetate solution followed by deacetylation in a 0.05 M NaOH. Then, the composite nanofibers were carbonized and activated by heat treatments to produce the activated carbon nanofiber (ACNF). The results revealed that the presence of MWNCT reduced the activation energy (from 230 to 180 kJ/mol) for the oxidative stabilization of the cellulose nanofiber. Additionally, the specific surface area of the ACNF increased from 870 to 1120 m^2/g with the addition of 6% MWCNT. The specific capacitances of the neat ACNF and ACNF-MWCNT at a current density of 10 A/g were 105 and 145 F/g, respectively. This significant improvement (38%) of the MWCNT-ACNF capacitance is related to higher surface area and electrical conductivity of the nanocomposite. It is also related to the nature of carbon nanomaterial. It is therefore possible to improve the specific capacitance by increasing the carbon nanotube content and/or adding another type of nanomaterial such as graphene [105, 106, 108].

3.8 Soy proteins reinforcement

Soy proteins, as residues of the soy oil industry, are known as a great source of natural biopolymers and ideal candidates for commercial applications such as food packaging, due to their availability, biodegradability, low cost, and non-toxicity. However, pristine soy protein-based films have some disadvantages, such as high sensitivity to moisture and poor mechanical properties, limiting their practical application [115]. This is why Jin et al. [116] reinforced soy protein nanocomposites with CNF, CNT, and graphene nanosheets as a hybrid filler system. The casting-evaporation method was used to prepare the nanocomposites. It was found that CNF addition improved the dispersion of graphene and CNT in the soy proteins, leading to a uniform dispersion of hybrid fillers in the matrix. The results also showed that the tensile strength of the nanocomposite reinforced with hybrid fillers significantly increased (79%) and the water vapor permeability decreased (32%) compared to the neat soy protein film.

3.9 Conclusion

Both nanocellulose (NC) and carbon nanotube (CNT) have their own advantages and have attracted considerable attention in several applications like sensors, electronic displays, batteries, oil recovery, biomedical field, etc. But the reinforcement of polymer matrices to produce nanocomposites is one of their most important applications. In this review, it was reported that several studies have been published showing that NC addition increased the mechanical properties, while CN addition increased both the electrical and thermal properties, as well as mechanical properties.

Over the years, NC and CNT have been used alone as a reinforcement for different polymer matrices. But more recently, due to effective synergistic effects between both nanoparticles, their combination led to the production of hybrid systems showing great potential in different polymers such as

TPU, PDMS, NR, PLA, and soy protein. In some cases, the presence of nanocellulose increased the CNT dispersion in the matrix. On the other hand, the addition of CNT increased the hydrophobicity of the hybrid fillers, therefore improving their compatibility with hydrophobic polymers. As a result, the combination of NC and CNT into the same matrix can lead to the production of high-permanence nanocomposites for different applications like flexible sensors for pH, adrenaline, cortisol, and humidity monitoring. The ratio between both fillers is the most important parameter when using a hybrid system, but the total filler content is the main option to optimize the balance between conductivity and mechanical properties.

The nanocellulose and CNT hybrid systems can also be produced into films, aerogels or filaments for applications like sensors, adsorption, wearable electronic devices, supercapacitors, and optoelectronic devices. In this case, several methods have been proposed, such as vacuum filtration, freeze-drying, layer-by-layer assembly, evaporation-induced self-assembly, interfacial nanoparticle complexation, plasma technique, and electrospinning. But more work is needed to compare these different preparation methods in terms of performance, costs, and processability.

Finally, some studies reported poor physical interaction between the unmodified NC and CNT leading to the formation of a three-dimensional conductive network. This is why chemically modifying the NC, the CNT or both would be important to increase these physical interactions. However, there is still limited information on NC and CNT modifications to form chemical bonding between both particles that would lead to stronger interaction and the development of new nano-structures for multi-functional materials.

References

[1] Badamshina E, Estrin Y, Gafurova M. Nanocomposites based on polyurethanes and carbon nanoparticles: preparation, properties and application. J Mater Chem A 2013;1(22):6509–29.

[2] Chen W, Tao X, Liu Y. Carbon nanotube-reinforced polyurethane composite fibers. Compos Sci Technol 2006;66(15):3029–34.

[3] Mi H-Y, Jing X, Salick MR, Cordie TM, Turng L-S. Carbon nanotube (CNT) and nanofibrillated cellulose (NFC) reinforcement effect on thermoplastic polyurethane (TPU) scaffolds fabricated via phase separation using dimethyl sulfoxide (DMSO) as solvent. J Mech Behav Biomed 2016;62:417–27.

[4] Saifuddin N, Raziah A, Junizah A. Carbon nanotubes: a review on structure and their interaction with proteins. J Chem 2013;1–18.

[5] Iijima S, Ichihashi T. Single-shell carbon nanotubes of 1-nm diameter. Nature 1993;363 (6430):603–5.

[6] Zhang M, Li J. Carbon nanotube in different shapes. Mater Today 2009;12(6):12–8.

[7] Motiei M, Rosenfeld Hacohen Y, Calderon-Moreno J, Gedanken A. Preparing carbon nanotubes and nested fullerenes from supercritical CO2 by a chemical reaction. J Am Chem Soc 2001;123(35):8624–5.

[8] Dresselhaus M, Dresselhaus G, Jorio A. Unusual properties and structure of carbon nanotubes. Annu Rev Mater Res 2004;34:247–78.

[9] Xia H, Song M. Preparation and characterization of polyurethane-carbon nanotube composites. Soft Matter 2005;1(5):386–94.
[10] Donaldson K, Aitken R, Tran L, Stone V, Duffin R, Forrest G, et al. Carbon nanotubes: a review of their properties in relation to pulmonary toxicology and workplace safety. Toxicol Sci 2006;92(1):5–22.
[11] Meng Q, Manas-Zloczower I. Carbon nanotubes enhanced cellulose nanocrystals films with tailorable electrical conductivity. Compos Sci Technol 2015;120:1–8.
[12] Baughman RH, Zakhidov AA, De Heer WA. Carbon nanotubes—the route toward applications. Science 2002;297(5582):787–92.
[13] Yu M-F, Files BS, Arepalli S, Ruoff RS. Tensile loading of ropes of single wall carbon nanotubes and their mechanical properties. Phys Rev Lett 2000;84(24):5552–5.
[14] Chen P, Yun YS, Bak H, Cho SY, Jin H-J. Multiwalled carbon nanotubes-embedded electrospun bacterial cellulose nanofibers. Mol Cryst Liq Cryst 2010;519(1):169–78.
[15] Pedrazzoli D, Manas-Zloczower I. Understanding phase separation and morphology in thermoplastic polyurethanes nanocomposites. Polymer 2016;90:256–63.
[16] Wu G, Gu Y, Hou X, Li R, Ke H, Xiao X. Hybrid Nanocomposites of cellulose/carbon-nanotubes/polyurethane with rapidly water sensitive shape memory effect and strain sensing performance. Polymers 2019;11(10):1–14.
[17] Xu S, Yu W, Jing M, Huang R, Zhang Q, Fu Q. Largely enhanced stretching sensitivity of polyurethane/carbon nanotube nanocomposites via incorporation of cellulose nanofiber. J Phys Chem C 2017;121(4):2108–17.
[18] Wu G, Li S, He W, Xiao X. Effect of percentage of graphene oxide on phase transition of water induced shape memory behavior of PVA-go-PEG hydrogel. Mater Res Express 2018;5(11):1–13.
[19] Zhang ZX, Qi XD, Li ST, Yang JH, Zhang N, Huang T, et al. Water-actuated shape-memory and mechanically-adaptive poly (ethylene vinyl acetate) achieved by adding hydrophilic poly (vinyl alcohol). Eur Polym J 2018;98:237–45.
[20] Wang S, Zhang X, Wu X, Lu C. Tailoring percolating conductive networks of natural rubber composites for flexible strain sensors via a cellulose nanocrystal templated assembly. Soft Matter 2016;12(3):845–52.
[21] Mohamed A, Anas AK, Bakar SA, Ardyani T, Zin WMW, Ibrahim S, et al. Enhanced dispersion of multiwall carbon nanotubes in natural rubber latex nanocomposites by surfactants bearing phenyl groups. J Colloid Interface Sci 2015;455:179–87.
[22] Kargarzadeh H, Mariano M, Huang J, Lin N, Ahmad I, Dufresne A, et al. Recent developments on nanocellulose reinforced polymer nanocomposites: a review. Polymer 2017;132:368–93.
[23] Kalashnikova I, Bizot H, Cathala B, Capron I. Modulation of cellulose nanocrystals amphiphilic properties to stabilize oil/water interface. Biomacromolecules 2012;13(1):267–75.
[24] Olivier C, Moreau C, Bertoncini P, Bizot H, Chauvet O, Cathala B. Cellulose nanocrystal-assisted dispersion of luminescent single-walled carbon nanotubes for layer-by-layer assembled hybrid thin films. Langmuir 2012;28(34):12463–71.
[25] González-Rivera J, Iglio R, Barillaro G, Duce C, Tinè MR. Structural and thermoanalytical characterization of 3D porous PDMS foam materials: the effect of impurities derived from a sugar templating process. Polymers 2018;10(6):1–13.
[26] Chen C, Bu X, Feng Q, Li D. Cellulose nanofiber/carbon nanotube conductive nano-network as a reinforcement template for polydimethylsiloxane nanocomposite. Polymers 2018;10 (9):1–10.
[27] Mugo SM, Alberkant J. Flexible molecularly imprinted electrochemical sensor for cortisol monitoring in sweat. Anal Bioanal Chem 2020;412(8):1825–33.

[28] Yu N, Mugo SM. A flexible-imprinted capacitive sensor for rapid detection of adrenaline. Talanta 2019;204:602–6.

[29] Wang C, Pan ZZ, Lv W, Liu B, Wei J, Lv X, et al. A directional strain sensor based on anisotropic microhoneycomb cellulose nanofiber-carbon nanotube hybrid aerogels prepared by unidirectional freeze drying. Small 2019;15(14):1–8.

[30] Zhai T, Zheng Q, Cai Z, Turng L-S, Xia H, Gong S. Poly (vinyl alcohol)/cellulose nanofibril hybrid aerogels with an aligned microtubular porous structure and their composites with polydimethylsiloxane. ACS Appl Mater Interfaces 2015;7(13):7436–44.

[31] Dai W, Yu J, Wang Y, Song Y, Alam FE, Nishimura K, et al. Enhanced thermal conductivity for polyimide composites with a three-dimensional silicon carbide nanowire@ graphene sheets filler. J Mater Chem A 2015;3(9):4884–91.

[32] Pan Z-Z, Nishihara H, Iwamura S, Sekiguchi T, Sato A, Isogai A, et al. Cellulose nanofiber as a distinct structure-directing agent for xylem-like microhoneycomb monoliths by unidirectional freeze-drying. ACS Nano 2016;10(12):10689–97.

[33] Chen L, Hou X, Song N, Shi L, Ding P. Cellulose/graphene bioplastic for thermal management: enhanced isotropic thermally conductive property by three-dimensional interconnected graphene aerogel. Compos Part A Appl Sci Manuf 2018;107:189–96.

[34] Pan Z-Z, Nishihara H, Lv W, Wang C, Luo Y, Dong L, et al. Microhoneycomb monoliths prepared by the unidirectional freeze-drying of cellulose nanofiber based sols: method and extensions. J Vis Exp 2018;135:1–6.

[35] Pan ZZ, Lv W, He YB, Zhao Y, Zhou G, Dong L, et al. A nacre-like carbon nanotube sheet for high performance Li-polysulfide batteries with high sulfur loading. Adv Sci 2018;5(6):1–7.

[36] Chen Y, Mun SC, Kim J. A wide range conductometric pH sensor made with titanium dioxide/multiwall carbon nanotube/cellulose hybrid nanocomposite. IEEE Sensors J 2013;13 (11):4157–62.

[37] Yang M, Kim D-H, Kim W-S, Kang TJ, Lee BY, Hong S, et al. H2 sensing characteristics of SnO2 coated single wall carbon nanotube network sensors. Nanotechnology 2010;21(21):1–7.

[38] Fulati A, Usman Ali SM, Riaz M, Amin G, Nur O, Willander M. Miniaturized pH sensors based on zinc oxide nanotubes/nanorods. Sensors 2009;9(11):8911–23.

[39] Fugestu B, Sano E, Sunada M, Sambongi Y, Shibuya T, Wang X, et al. Electrical conductivity and electromagnetic interference shielding efficiency of carbon nanotube/cellulose composite paper. Carbon 2008;46(9):1256–8.

[40] Capadona JR, Shanmuganathan K, Trittschuh S, Seidel S, Rowan SJ, Weder C. Polymer nanocomposites with nanowhiskers isolated from microcrystalline cellulose. Biomacromolecules 2009;10(4):712–6.

[41] Hamedi MM, Hajian A, Fall AB, Hakansson K, Salajkova M, Lundell F, et al. Highly conducting, strong nanocomposites based on nanocellulose-assisted aqueous dispersions of single-wall carbon nanotubes. ACS Nano 2014;8(3):2467–76.

[42] Zhu P, Liu Y, Fang Z, Kuang Y, Zhang Y, Peng C, et al. Flexible and highly sensitive humidity sensor based on cellulose nanofibers and carbon nanotube composite film. Langmuir 2019;35(14):4834–42.

[43] Xu S, Yu W, Yao X, Zhang Q, Fu Q. Nanocellulose-assisted dispersion of graphene to fabricate poly (vinyl alcohol)/graphene nanocomposite for humidity sensing. Compos Sci Technol 2016;131:67–76.

[44] Xuan W, He X, Chen J, Wang W, Wang X, Xu Y, et al. High sensitivity flexible Lamb-wave humidity sensors with a graphene oxide sensing layer. Nanoscale 2015;7(16):7430–6.

[45] Tang Q-Y, Chan Y, Zhang K. Fast response resistive humidity sensitivity of polyimide/multiwall carbon nanotube composite films. Sensor Actuat B Chem 2011;152(1):99–106.

[46] Li T, Li L, Sun H, Xu Y, Wang X, Luo H, et al. Porous ionic membrane based flexible humidity sensor and its multifunctional applications. Adv Sci 2017;4(5):1–7.
[47] Hajian A, Lindström SB, Pettersson T, Hamedi MM, Wagberg L. Understanding the dispersive action of nanocellulose for carbon nanomaterials. Nano Lett 2017;17(3):1439–47.
[48] Zeng X, Deng L, Yao Y, Sun R, Xu J, Wong C-P. Flexible dielectric papers based on biodegradable cellulose nanofibers and carbon nanotubes for dielectric energy storage. J Mater Chem C 2016;4(25):6037–44.
[49] Koga H, Saito T, Kitaoka T, Nogi M, Suganuma K, Isogai A. Transparent, conductive, and printable composites consisting of TEMPO-oxidized nanocellulose and carbon nanotube. Biomacromolecules 2013;14(4):1160–5.
[50] Safari S, van de Ven TG. Effect of water vapor adsorption on electrical properties of carbon nanotube/nanocrystalline cellulose composites. ACS Appl Mater Interfaces 2016;8(14): 9483–9.
[51] Christie J, Sylvander S, Woodhead I, Irie K. The dielectric properties of humid cellulose. J Non-Cryst Solids 2004;341(1-3):115–23.
[52] Crofton DJ, Pethrick RA. Dielectric studies of proton migration and relaxation in wet cellulose and its derivatives. Polymer 1981;22(8):1048–53.
[53] Assaf A, Haas R, Purves C. A new interpretation of the cellulose-water adsorption isotherm and data concerning the effect of swelling and drying on the colloidal surface of cellulose. J Am Chem Soc 1944;66(1):66–73.
[54] Liu L, Ye X, Wu K, Han R, Zhou Z, Cui T. Humidity sensitivity of multi-walled carbon nanotube networks deposited by dielectrophoresis. Sensors 2009;9(3):1714–21.
[55] Qi H, Liu J, Pionteck J, Pötschke P, Mäder E. Carbon nanotube-cellulose composite aerogels for vapour sensing. Sensor Actuat B Chem 2015;213:20–6.
[56] Qi H, Mäder E, Liu J. Electrically conductive aerogels composed of cellulose and carbon nanotubes. J Mater Chem A 2013;1(34):9714–20.
[57] Tournus F, Latil S, Heggie M, Charlier J-C. π-stacking interaction between carbon nanotubes and organic molecules. Phys Rev B 2005;72(7):1–5.
[58] Bouvree A, Feller J-F, Castro M, Grohens Y, Rinaudo M. Conductive Polymer nano-bioComposites (CPC): chitosan-carbon nanoparticle a good candidate to design polar vapour sensors. Sensor Actuat B Chem 2009;138(1):138–47.
[59] Qi H, Mäder E, Liu J. Unique water sensors based on carbon nanotube-cellulose composites. Sensor Actuat B Chem 2013;185:225–30.
[60] Yun S, Kim J. Multi-walled carbon nanotubes-cellulose paper for a chemical vapor sensor. Sensor Actuat B Chem 2010;150(1):308–13.
[61] Fan Q, Qin Z, Villmow T, Pionteck J, Pötschke P, Wu Y, et al. Vapor sensing properties of thermoplastic polyurethane multifilament covered with carbon nanotube networks. Sensor Actuat B Chem 2011;156(1):63–70.
[62] Parikh K, Cattanach K, Rao R, Suh D-S, Wu A, Manohar SK. Flexible vapour sensors using single walled carbon nanotubes. Sensor Actuat B Chem 2006;113(1):55–63.
[63] Owens FJ. Characterization of single walled carbon nanotube–nitrocellulose composites. J Macromol Sci Pure Appl Chem 2005;42(10):1355–60.
[64] Sun J, Zhang C, Yuan Z, Ji X, Fu Y, Li H, et al. Composite films with ordered carbon nanotubes and cellulose nanocrystals. J Phys Chem C 2017;121(16):8976–81.
[65] Mauter MS, Elimelech M, Osuji CO. Stable sequestration of single-walled carbon nanotubes in self-assembled aqueous nanopores. J Am Chem Soc 2012;134(9):3950–3.
[66] Dierking I, Scalia G, Morales P, LeClere D. Aligning and reorienting carbon nanotubes with nematic liquid crystals. Adv Mater 2004;16(11):865–9.

[67] Yadav SP, Singh S. Carbon nanotube dispersion in nematic liquid crystals: an overview. Prog Mater Sci 2016;80:38–76.

[68] Prendergast DP, Gschwend PM. Assessing the performance and cost of oil spill remediation technologies. J Clean Prod 2014;78:233–42.

[69] Sarbatly R, Krishnaiah D, Kamin Z. A review of polymer nanofibres by electrospinning and their application in oil-water separation for cleaning up marine oil spills. Mar Pollut Bull 2016;106(1-2):8–16.

[70] Meng Y, Young TM, Liu P, Contescu CI, Huang B, Wang S. Ultralight carbon aerogel from nanocellulose as a highly selective oil absorption material. Cellulose 2015;22 (1):435–47.

[71] Yin T, Zhang X, Liu X, Li B, Wang C. Cellulose-based aerogel from Eichhornia crassipes as an oil superabsorbent. RSC Adv 2016;6(101):98563–70.

[72] Zhang H, Li Y, Xu Y, Lu Z, Chen L, Huang L, et al. Versatile fabrication of a superhydrophobic and ultralight cellulose-based aerogel for oil spillage clean-up. Physiol Chem Phys 2016;18(40):28297–306.

[73] Zhou S, Liu P, Wang M, Zhao H, Yang J, Xu F. Sustainable, reusable, and superhydrophobic aerogels from microfibrillated cellulose for highly effective oil/water separation. ACS Sustain Chem Eng 2016;4(12):6409–16.

[74] Korhonen JT, Kettunen M, Ras RH, Ikkala O. Hydrophobic nanocellulose aerogels as floating, sustainable, reusable, and recyclable oil absorbents. ACS Appl Mater Interfaces 2011;3 (6):1813–6.

[75] Neelamegan H, Yang D-K, Lee G-J, Anandan S, Wu JJ. Synthesis of magnetite nanoparticles anchored cellulose and lignin-based carbon nanotube composites for rapid oil spill cleanup. Mater Today Commun 2020;22:1–11.

[76] Xu Z, Zhou H, Tan S, Jiang X, Wu W, Shi J, et al. Ultralight super-hydrophobic carbon aerogels based on cellulose nanofibers/poly (vinyl alcohol)/graphene oxide (CNFs/PVA/GO) for highly effective oil-water separation. Beilstein J Nanotechnol 2018;9(1):508–19.

[77] Li J, Zhou L, Jiang X, Tan S, Chen P, Zhou H, et al. Directional preparation of superhydrophobic magnetic CNF/PVA/MWCNT carbon aerogel. IET Nanobiotechnol 2019;13(6): 565–70.

[78] Choi H-G, Yoon SH, Son M, Celik E, Park H, Choi H. Efficacy of synthesis conditions on functionalized carbon nanotube blended cellulose acetate membrane for desalination. Desalin Water Treat 2016;57(16):7545–54.

[79] Lay M, Méndez JA, Pèlach MÀ, Bun KN, Vilaseca F. Combined effect of carbon nanotubes and polypyrrole on the electrical properties of cellulose-nanopaper. Cellulose 2016;23(6):3925–37.

[80] Gu P, Zhang S, Li X, Wang X, Wen T, Jehan R, et al. Recent advances in layered double hydroxide-based nanomaterials for the removal of radionuclides from aqueous solution. Environ Pollut 2018;240:493–505.

[81] Tan X, Fang M, Tan L, Liu H, Ye X, Hayat T, et al. Core-shell hierarchical CaNa2-Ti3O7·9H2O nanostructures for the efficient removal of radionuclides. Environ Sci Nano 2018;5(5):1140–9.

[82] Zhao G, Huang X, Tang Z, Huang Q, Niu F, Wang X. Polymer-based nanocomposites for heavy metal ions removal from aqueous solution: a review. Polym Chem 2018;9 (26):3562–82.

[83] Raja M, Suriyakumar S, Angulakshmi N, Stephan AM. High performance multi-functional trilayer membranes as permselective separators for lithium-sulfur batteries. Inorg Chem Front 2017;4(6):1013–21.

[84] Zong P, Cao D, Cheng Y, Wang S, Hayat T, Alharbi SN, et al. Enhanced performance for Eu (iii) ion remediation using magnetic multiwalled carbon nanotubes functionalized with

carboxymethyl cellulose nanoparticles synthesized by plasma technology. Inorg Chem Front 2018;5(12):3184–96.
[85] Zeng G, He Y, Zhan Y, Zhang L, Pan Y, Zhang C, et al. Novel polyvinylidene fluoride nanofiltration membrane blended with functionalized halloysite nanotubes for dye and heavy metal ions removal. J Hazard Mater 2016;317:60–72.
[86] Liang CZ, Sun SP, Li FY, Ong YK, Chung TS. Treatment of highly concentrated wastewater containing multiple synthetic dyes by a combined process of coagulation/flocculation and nanofiltration. J Membr Sci 2014;469:306–15.
[87] Ahmed FE, Lalia BS, Hashaikeh R. A review on electrospinning for membrane fabrication: challenges and applications. Desalination 2015;356:15–30.
[88] Zhijiang C, Ping X, Cong Z, Tingting Z, Jie G, Kongyin Z, et al. Preparation and characterization of a bi-layered nano-filtration membrane from a chitosan hydrogel and bacterial cellulose nanofiber for dye removal. Cellulose 2018;25(9):5123–37.
[89] Callura JC, Perkins KM, Noack CW, Washburn NR, Dzombak DA, Karamalidis AK. Selective adsorption of rare earth elements onto functionalized silica particles. Green Chem 2018;20(7):1515–26.
[90] Giret S, Hu Y, Masoumifard N, Boulanger JF. Juère E, Kleitz F, et al. Selective separation and preconcentration of scandium with mesoporous silica. ACS Appl Mater Interfaces 2018;10 (1):448–57.
[91] Zheng X, Zhang Y, Zhang F, Li Z, Yan Y. Dual-template docking oriented ionic imprinted bilayer mesoporous films with efficient recovery of neodymium and dysprosium. J Hazard Mater 2018;353:496–504.
[92] Zheng X, Zhang Y, Bian T, Zhang Y, Li Z, Pan J. Oxidized carbon materials cooperative construct ionic imprinted cellulose nanocrystals films for efficient adsorption of Dy (III). Chem Eng J 2020;381:1–10.
[93] Mattocks JA, Ho JV, Cotruvo Jr JA. A selective, protein-based fluorescent sensor with picomolar affinity for rare earth elements. J Am Chem Soc 2019;141(7):2857–61.
[94] Li F, Gong A, Qiu L, Zhang W, Li J, Liu Z. Diglycolamide-grafted Fe3O4/polydopamine nanomaterial as a novel magnetic adsorbent for preconcentration of rare earth elements in water samples prior to inductively coupled plasma optical emission spectrometry determination. Chem Eng J 2019;361:1098–109.
[95] Zhang Y, Zheng X, Bian T, Zhang Y, Mei J, Li Z. Phosphorylated-CNC/MWCNT thin films-toward efficient adsorption of rare earth La (III). Cellulose 2020;27:3379–90.
[96] Gruner G. Carbon nanotube films for transparent and plastic electronics. J Mater Chem 2006;16(35):3533–9.
[97] Zhang H, Sun X, Hubbe MA, Pal L. Highly conductive carbon nanotubes and flexible cellulose nanofibers composite membranes with semi-interpenetrating networks structure. Carbohydr Polym 2019;222:1–8.
[98] King RCY, Roussel F. Transparent carbon nanotube-based driving electrodes for liquid crystal dispersion display devices. Appl Phys A Mater Sci Process 2007;86(2):159–63.
[99] Kim Y, Kim HS, Bak H, Yun YS, Cho SY, Jin HJ. Transparent conducting films based on nanofibrous polymeric membranes and single-walled carbon nanotubes. J Appl Polym Sci 2009;114(5):2864–72.
[100] Zhang H, Sun X, Heng Z, Chen Y, Zou H, Liang M. Robust and flexible cellulose nanofiber/multiwalled carbon nanotube film for high-performance electromagnetic interference shielding. Ind Eng Chem Res 2018;57(50):17152–60.
[101] Zhang K, Ketterle L, Järvinen T, Hong S, Liimatainen H. Conductive hybrid filaments of carbon nanotubes, chitin nanocrystals and cellulose nanofibers formed by interfacial nanoparticle complexation. Mater Des 2020;191:1–9.

[102] Yu B, Zhao Z, Fu S, Meng L, Liu Y, Chen F, et al. Fabrication of PLA/CNC/CNT conductive composites for high electromagnetic interference shielding based on Pickering emulsions method. Compos Part A Appl Sci Manuf 2019;125:1–10.

[103] Wan Z, Chen C, Meng T, Mojtaba M, Teng Y, Feng Q, et al. Multifunctional wet-spun filaments through robust nanocellulose networks wrapping to single-walled carbon nanotubes. ACS Appl Mater Interfaces 2019;11(45):42808–17.

[104] Li Y, Zhu H, Wang Y, Ray U, Zhu S, Dai J, et al. Cellulose-nanofiber-enabled 3D printing of a carbon-nanotube microfiber network. Small Methods 2017;1(10):1–8.

[105] Kuzmenko V, Naboka O, Haque M, Staaf H, Göransson G, Gatenholm P, et al. Sustainable carbon nanofibers/nanotubes composites from cellulose as electrodes for supercapacitors. Energy 2015;90:1490–6.

[106] Zheng Q, Cai Z, Ma Z, Gong S. Cellulose nanofibril/reduced graphene oxide/carbon nanotube hybrid aerogels for highly flexible and all-solid-state supercapacitors. ACS Appl Mater Interfaces 2015;7(5):3263–71.

[107] Li S, Huang D, Zhang B, Xu X, Wang M, Yang G, et al. Flexible supercapacitors based on bacterial cellulose paper electrodes. Adv Energy Mater 2014;4(10):1–7.

[108] Deng L, Young RJ, Kinloch IA, Abdelkader AM, Holmes SM, De Haro-Del Rio DA, et al. Supercapacitance from cellulose and carbon nanotube nanocomposite fibers. ACS Appl Mater Interfaces 2013;5(20):9983–90.

[109] Gao K, Shao Z, Li J, Wang X, Peng X, Wang W, et al. Cellulose nanofiber-graphene all solid-state flexible supercapacitors. J Mater Chem A 2013;1(1):63–7.

[110] Bordjiba T, Mohamedi M, Dao LH. New class of carbon-nanotube aerogel electrodes for electrochemical power sources. Adv Mater 2008;20(4):815–9.

[111] Zhang LL, Zhao XS. Carbon-based materials as supercapacitor electrodes. Chem Soc Rev 2009;38(9):2520–31.

[112] Béguin F, Presser V, Balducci A, Frackowiak E. Carbons and electrolytes for advanced supercapacitors. Adv Mater 2014;26(14):2219–51.

[113] Gao Y, Pandey GP, Turner J, Westgate CR, Sammakia B. Chemical vapor-deposited carbon nanofibers on carbon fabric for supercapacitor electrode applications. Nanoscale Res Lett 2012;7(1):1–8.

[114] Lyu S, Chen Y, Zhang L, Han S, Lu Y, Chen Y, et al. Nanocellulose supported hierarchical structured polyaniline/nanocarbon nanocomposite electrode via layer-by-layer assembly for green flexible supercapacitors. RSC Adv 2019;9(31):17824–34.

[115] Martelli-Tosi M, Assis OB, Silva NC, Esposto BS, Martins MA, Tapia-Blácido DR. Chemical treatment and characterization of soybean straw and soybean protein isolate/straw composite films. Carbohydr Polym 2017;157:512–20.

[116] Jin S, Li K, Li J. Nature-inspired green procedure for improving performance of protein-based nanocomposites via introduction of nanofibrillated cellulose-stablized graphene/carbon nanotubes hybrid. Polymers 2018;10(3):1–15.

Chapter 4

Hybrid nanocomposites based on cellulose nanocrystals/nanofibrils and silver nanoparticles: Antibacterial applications

Carlo Santulli
Geology Division, Università degli Studi di Camerino, School of Science and Technology, Camerino, Italy

4.1 Introduction

4.1.1 Nanocellulose from ligno-cellulosic materials

Cellulose nanocrystals (CNCs) or nanofibrils (CNFs), amorphous nanocellulose (ANC), and cellulose nanoyarn (CNY), which are globally defined as "nanocellulose" (NC), have the potential to enable the production of highly ordered structures, fitted to different purposes and applications [1]. The characteristic of being rod-like and defect-free, therefore removed from the typical properties of ligno-cellulosic materials, offered to nanocellulose the potential for being applied in uses requiring high chemical and physical properties. These can be obtained while retaining the origin of nanocellulose from ligno-cellulosic materials, which include a countless number of possible fibrous biomass, through mechanical or hydrolytic processes. Indications on the length and diameter of the nanocellulose fibers obtained by different ligno-cellulosic materials are offered, e.g., in Ref. [2]: however, other possible sources are explored everyday and offer new indications and potential.

In practice, nanocellulose can be obtained from cellulose paste, as an industrial by-product or waste, for example from end-of-life packaging. In this case, cellulose is reasonably pure, therefore free from lignin and hemicellulose, which have been removed from the previous production process, yet it contains substantial amounts of chemicals, including especially chlorine-based products [3]. In contrast, when pre-production waste are used, such as it is the case of

Cellulose Nanocrystal/Nanoparticles Hybrid Nanocomposites. https://doi.org/10.1016/B978-0-12-822906-4.00011-6

ligno-cellulosic materials from agro-forest uses, they may be less loaded with chemicals, but on the other side cellulose is combined with other polysaccharides and with lignin and therefore the yield of nanocellulose production may be considerably lower. The use of agro-waste (e.g., raw materials of unlikely use in other productions, such as rotten fruits, shells, molasses) has proved of increased interest, also to further reduce the cost of nanocellulose obtained [4, 5]. Also for productive systems involving the extensive use of chemical treatments, it was possible to reduce the mechanical processing and obtain nanocellulose of acceptable quality. In particular, it proved feasible to avoid the pulping process of cellulose waste and obtaining proved possible from cotton linter, i.e., the fiber that cannot be used in the textile industry [6]. Also, perennial plants rich in cellulose, such as lotus, are able to be offer compacted filter cakes, which can on the other side result in the production of nanocellulose with acceptable yields [7] (Fig. 4.1).

A comprehensive review of the possible sources and of the different methods for extraction of nanocellulose has been offered in Ref. [8]. In practice,

FIG. 4.1 (A) Lotus plant, (B) Withered lotus leaf stalks decaying in wetland, (C) Dry powder of leaf lotus stalks, (D) Freeze-dried bleached lotus leaf stalk filter cake, (E) Chemically purified lotus leaf stalks under an optical microscope with ×100 magnification lens (unit bar equals to 200 μm) [7]

the possibilities are countless, since a few thousand of lignocellulosic plants offering the possibility to produce fibers are known, though those that created a productive system of some importance are only a few [9]. The raw materials span from cellulose waste, as the ones derived from paper or from cotton textiles, woody waste treated, most frequently by alkali, such as sodium hydroxide or ethanol, or untreated. A number of studies obtained nanocellulose starting from microcellulose, not always declaring the source from which the latter had been produced. This consideration, together with the availability of large amounts of cellulose-rich by-products from a number of operations linked to natural fibers or to the process of recycling of paper, makes it interesting the possibility to use these for the production of nanocellulose. This will reduce the "carbon dioxide footprint" of the operation.

Production methods combine acid hydrolysis and mechanical processing with a variety of post-treatments, including centrifugation, filtration, dialysis, etc. The chemical extraction can be performed together with mechanical treatment, in particular blending, in order to promote further self-assembling of CNCs structures. An alternative is represented by the enzymatic path, using cellulases not produced by bacteria, yet, e.g., by fungi, again with the action of a catalyst [10]. In the enzymatic process, the crucial issue appears to be obtaining a sufficient nanocellulose yield and not too sparse distribution of geometries [11]. Yarbrough et al. [12] compared the deconstructive effect of cellulose achieved using the fungal enzyme system of *Trichoderma reesei* and the complexed, multifunctional enzymes produced by a bacterial stream resident in hot springs *Caldicellulosiruptor bescii*. In particular, the difficulty appears in this case to control the process, due to the differences in the mechanisms of action of the dominant enzymes in the various systems.

The yielding in nanocellulose, as well as the morphology and chemical properties of CNCs, is also influenced by the way the extraction is performed. To this aim, a large number of possible options have been explored. In particular, the extraction of microcellulose can be carried out by the chemical path, which involves acid hydrolysis, particularly with sulfuric acid [13]. However, also formic acid [14], hydrochloric acid [15], acetic acid [16] were attempted for the purpose. This is accompanied with a necessary mechanical disintegration, which is likely though to be very energy consuming. A number of pretreatments have been proposed during the last decade to reduce energy use in passing from the microscale to the nanoscale. For example, carboxymethylation offers the possibility of application of nanocellulose as gel, which demonstrated to be suitable for 3D bioprinting with the objective of wound dressing [17]. In contrast, the catalyzed action of 2,2,6,6-tetramethylepiperidin-1-oxyl (TEMPO), also in some cases assisted by ultrasound, for cellulose oxidation offers yield increase and larger presence of carboxyl content [18]. This method allows numerous applications being experimented, especially tailoring the required porosity [19] (Fig. 4.2): this is also of interest in the case of nanocellulose.

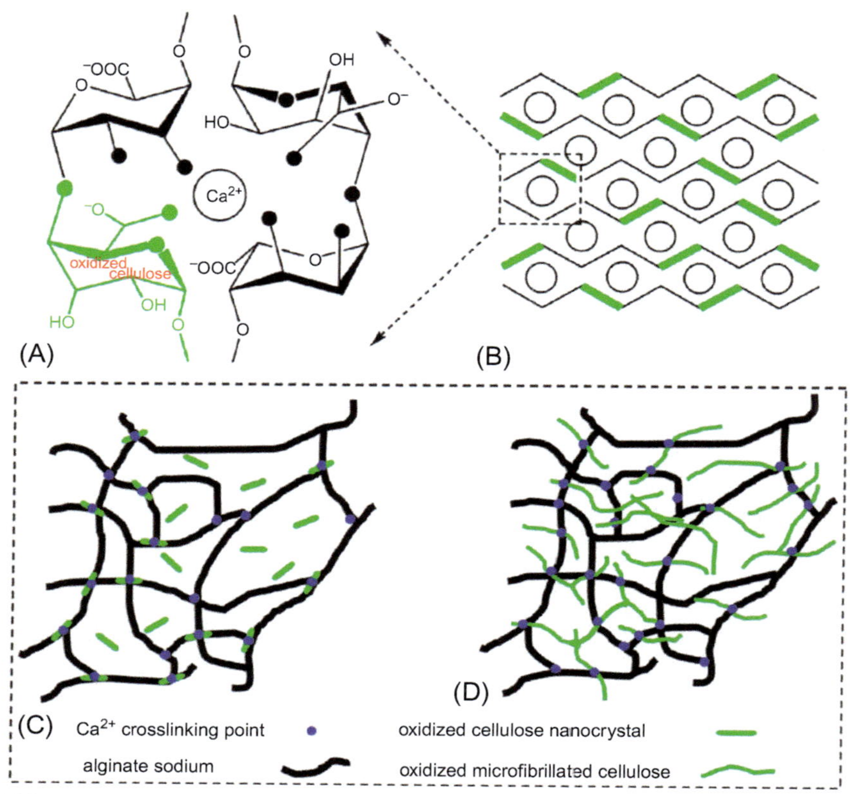

FIG. 4.2 Schematic representation of the structure of egg-box junction zones in TEMPO-mediated oxidation of cellulose/alginate/calcium system. In particular: (A) coordination of Ca^{2+} in a cavity created by a pair of guluronate sequences along alginate chains; (B) laterally associated egg-box multimer in composite sponge; (C) oxidized cellulose nanocrystals (OCN) in the cross-linking of alginate sponge; (D) construction of the semi-interpenetrating polymer network (SIPN) composed of oxidized microfibrillated cellulose (OMFC) and alginate. The *black solid circles* in panel a represent the oxygen atoms possibly involved in the coordination with Ca^{2+}. *(Reprinted with permission from Lin N, Bruzzese C, Dufresne A, ACS Appl Mater Interfaces 2012;4:4948–59.)*

4.1.2 Bacterial cellulose

Another possible origin on nanocellulose is from bacterial strains that enable the fermentation of sugars present in plant carbohydrates: here, pure fermentation can be obtained, but the costs of the process are significantly higher [20]. The principal objective of most studies over bacterial cellulose has so far, and for a few decades, elucidating which are the bacterial streams that cause the biogenesis of cellulose and which is the yield and the characteristics of the polysaccharide produced by each of them in different chemical and physical conditions. More in particular, this approach to the study of bacterial cellulose was aimed at exactly defining an optimal fermentation medium, starting from bacteria that showed ability at producing cellulose [21]. In the case of bacterial cellulose, a large amount of literature also exists over waste that could be used to be transformed into cellulose using adapted fermentation media. This spans from fibrous materials, such as wheat straw, to sugary fluids, such as fruits juices or syrups, down to wastewaters for different processes [22].

In practical terms, therefore, bacterial cellulose (BC) appears to be the only type of nanocellulose that is commercially available at affordable prices and that retains a high crystallinity and repeatability. The research has also concentrated considerably on the possible fermentation media, As the result, it has been applied with a number of possible reinforcements, including hydroxyapatite, graphite nanoplatelets, or in matrices, such as polyurethane [23]. In particular, *Acetobacter xylinum* produces cellulose I (ribbon-like polymer) and cellulose II (thermodynamically stable polymer) [24]. The synthesis process involves the secretion of polysaccharide protofibrils through bacteria cell wall, which are then aggregated forming nanofibrils cellulose ribbons [25]. Bacterial cellulose has a web-shaped structure and appears to be highly porous, which enables a facile interaction with other elements, such as water and degrading agents, and generally a high specific liquid carrying possibility, in a number of different situations [26]. This implies the diffuse presence on the surface of hydroxyl groups, which explains hydrophilicity, biodegradability, and chemical-modifying capacity [3]. BC is also for example produced as a nanomaterial by *Gluconoacetobacter* species cultivated in a medium with carbon and nitrogen sources, using agricultural and food industry waste [27]. Obtaining antibacterial properties is of paramount importance for nano-cellulose: with this purpose, some inherent characteristics of nanocellulose, such as the large surface area coupled with a low coefficient of thermal expansion, make nanocellulose very suitable for biomedical applications. In particular, nanocellulose is effective in active blending, such as it is the case in fillers for injectable hydrogels or drug carriers, or the applications of substrates with other properties, such as nanosilver for anti-bacterial properties [28].

The possible uses of nanocellulose range in very different fields: a possibility is also employing it as the reinforcement of polymer composites [29]. This is in principle limited to polar biopolymers. However, chemical treatment of nanocellulose with treatments typical for ligno-cellulosic fibers, such as acetylation, silanization, or more specific, such as TEMPO mediated oxidation, or physical treatments, e.g., surface fibrillation, electric discharge (corona, cold plasma), ultrasonic, irradiation, and the application of electrical currents [30].

The application of nanocellulose can also include the development of printed electronics, through the combination of printing processes and ink chemistry for the manufacture of electronic components, blending conductive particles with nanocellulose or using it as a substrate [31]. An important field is the nutrition sector: for example in food science, nanocellulose finds potential use as emulsifier and functional food ingredient [32]. In the case of food packaging, nanocellulose is particularly suitable for the production of bio-based films, in which case bacterial cellulose can be used as such or in a disintegrated form [33]. The case of food packaging is particularly suggestive, because most properties needed in this field are of interest also for many other applications, such as barrier properties offering controlled passage of oxygen and water, surface smoothness and optical properties, other than obviously mechanical performance [34].

Applications of nanocellulose are very widespread, spanning also to uses defined as generally "biomedical". This involves a number of large fields. In general terms, nanocellulose does appear to present a very low ecotoxicity, as demonstrated by a number of studies, summarized, e.g., in the review by Seabra et al. [35], which indicates the potential of nanocellulose, in particular as a safe drug carrier support. However, as far as an active anti-bacterial action is concerned, some concerns were raised, in particular in terms of the use of bacterial cellulose in wound dressing to ease and accelerate the healing process [36]. To obtain a sufficient wound dressing capability and bearing in mind the difficulties connected with the excessive cost of bacterial cellulose, also the development of hybrids has been investigated, which would include cotton gauze together with bacterial cellulose [37]. These hybrids could be immersed in a nanosilver solution, thus enabling a proper distribution of nanosilver without agglomerations at surface and inside nanobiocomposite which caused improved antimicrobial efficiency.

The two actions of wound healing and drug-loading can also be combined, a process in which the high purity, good shape retention, and considerable water binding capacity on nanocellulose will be of significant assistance [38]. Another, even more ambitious, possibility is the fabrication of biocompatible devices, starting for example from the coupling of nanocellulose with hydrogels, such as polyvinylalcohol (PVA), which offers multifunctionality together with adaptation to body implant [39]. Most recent developments do also involve, in terms of scaffolding, also the development of production via 3D printing and the fabrication of magnetically responsive materials [40].

The relevant crystallinity influences the relation of the polysaccharides with water and fluids in general and modifies also the mechanical and thermal properties of the obtained films [41]. Thoroughly controlling the extraction process, it is possible to obtain highly crystalline nanofibers, with a density of up to 1.6 g/cm^3 and a stiffness reaching up to 220 GPa [42]. In general terms, nanocellulose can be obtained from a large number of natural fibers and agro-forest waste, though the original raw material has been wood pulp. Of course, the use of different lignocellulosic biomass will result in different yields for synthesis, which is inherently connected with the initial amount of cellulose included in each source, and on its degree of crystallinity. Bacterial cellulose is a possible competitor for nanocellulose, with the advantage that the latter can be obtained in dimensions thinner than the former, yet it may present also anti-bacterial properties in wound dressing [43]. It needs also to be noticed that the action of bacteria can interfere with some applications of the resulting cellulose, in particular in sectors such as production of biomedical devices and food packaging industry.

Anti-bacterial action is therefore connected with the relation of the different types of nanocellulose with water, in terms of cycles of water absorption/desorption and moisture regain over time, which in turn in related with the structural geometry that has been obtained during the fabrication of the material. The following part of the chapter concentrates on anti-bacterial properties, in particular reasoning on the involvement of nanosilver as a coating for nanocellulose.

4.2 Antibacterial properties of nanosilver

Silver has been known from a very long time for its antibiotic properties, when in nanometric dimensions, due to the action of its ions against a very broad range of bacteria, yeasts, fungi and viruses, performed even when the metal is at very low concentration [44]. The number of supports whose application was investigated to exploit these characteristics of nanosilver was considerable, spanning, e.g., from bone cement [45] to chitin [46]. In particular, among the supports employed to host a nanosilver particles coating, an obvious starting point for biomedical applications has been the combination with poly(methylmetacrylate) (PMMA) for its properties of compatibility with bone structures minimizing adverse reactions with it [47]. The development of nanosilver to create some coating films had also some wider objectives for their use as catalysts in the partial oxidation of some organic compounds and additionally for their electronic and optical properties [48].

Some techniques that proved suitable for the synthesis of nanosilver particles are based on metal evaporation in an inert gas atmosphere, with subsequent condensation resulting in the development of nanoparticles [49]. Other routes are possible though, such as for example the biosynthesis of silver nanoparticles using natural resources, such as algae and sea weeds. These contain various phytochemicals, in particular hydroxyl, carboxyl and amino functional groups, which are suitable for metal reduction and work also as capping agents to provide a robust coating on the nanoparticles [50]. On the other side, the biosynthesis has also created some concern for the possible toxicity of silver, in the case some amount of it is released from the nanometric film [51]. It is well known on the other side that nanosilver particles may present toxicity over the aquatic organisms, such as for example inducing oxidative stress in some protozoa, such as *Tetrahymena thermophila* [52].

It is therefore of paramount importance that the nanosilver impregnation is stable under different conditions. In wound healing, it is essential that silver is preserved inside a solid support so to be applied on the affected area. In this case, it needs to be ascertained whether the possible release of silver does not bring any part of the wound dressing device to release silver to an extent that in the tissue the maximum allowable amount is exceeded. Conversely, the antibacterial action is retained as the function of pH, concentration of nanosilver and presence of organic matter [53]. It has also been proposed that it is possible to act to improve the anti-bacterial properties simply by modifying the geometries of the nanosilver particles, a procedure which can be of interest in presence of moist environments and aggressive conditions [54].

4.3 Application of nanosilver on nanocellulose

The application of nanosilver on nanocellulose films by impregnation improves their antibacterial properties [55]. The possible applications of the combination nanocellulose-nanosilver are spanning from paper, barrier or biodegradable films, especially for packaging [56], to medical devices, such as wound dressings [57]: a typical distribution of diameters is reported in Fig. 4.3 [58]. These devices exhibited strong activity against *Escherichia coli and Staphylococcus aureus,* and bioactive implants [46]. Different methods have been employed for the impregnation of nanosilver on bacterial cellulose for the purpose to improve antimicrobial characteristics, essential requirement for wound dressing. In particular, sodium borohydride ($NaBH_4$) method reduces silver nitrate to neutral silver, while Tollens' reaction, provided aldehydes are available in a solution, will reduce silver (I) ions to silver (0) [43]. Some questions have to be evaluated, in particular the lowest concentration able to produce an anti-bacterial effect, the size distribution of nanosilver particles and the storage characteristics, hence the preservation time in moist environment [59].

The use of cellulose or lignin pulp allows forming and strongly binding nanosilver and nanogold particles: the experience started from wool fibers for textile coloring and then provided effective antimicrobial and catalytic properties also for cellulose or lignin pulp [60]. One of the first studies that enabled forming silver nanoparticles on cellulose fiber surface was carried out on cellulose acetate electrospun fibers with silver starting from silver nitrate and acting by UV rays irradiation [61].

This has been also confirmed in the insertion in biopolymers, such as poly (vinylalcohol) (PVA), in comparative study using nanocellulose obtained from different sources, in particular commercial microcellulose, *Phormium tenax* and flax fibers scraps [62]. Another possibility is the formation of antimicrobial regenerated cellulose fibers coated with nanosilver using ionic liquid solvents such as 1-butyl-3-methylimidazolium chloride, which showed high bioactivity, in particularly against *Escherichia coli,* with very limited leaching problems [63].

In this task, nanosilver is in competition with other additives with similar characteristics, such as drug carrier capabilities. These are in particular montmorillonite (MMT) [64], zinc oxide [65] and other chemicals having specific medical properties, such as benzalkonium chloride [66]. It is also suggested that nanocellulose would serve as a stabilizing agent for nanosilver particles [67].

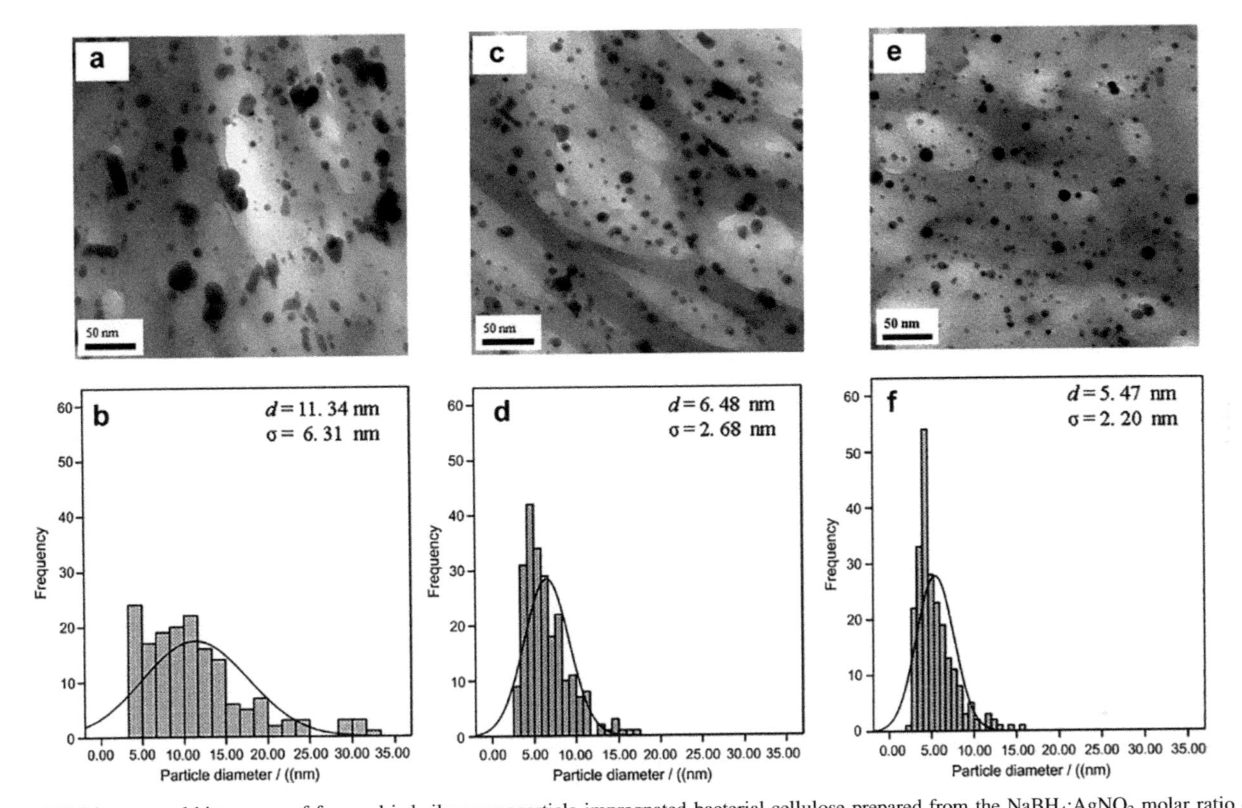

FIG. 4.3 TEM images and histograms of freeze-dried silver nanoparticle-impregnated bacterial cellulose prepared from the $NaBH_4$:$AgNO_3$ molar ratio of 1:1 (A, B), 10:1 (C, D) and 100:1 (E, F) [58].

Tests are also required to assess the bioactive efficacy of the cellulose/nanosilver fibers using the standard method ASTM E 2149-10. In some cases, hybrid structures between nanocellulose and biopolymers, such as poly(lactic acid) (PLA) are also used for the application of nanosilver [68].

Another possibility is to combine bacterial cellulose with silver nanoparticles to obtain antimicrobial properties through the inherent porosity of the material. Silver nanoparticles (AgNPs) are increasingly immobilized on the 3D structure according to the amounts of silver nitrate ($AgNO_3$) introduced [69]. Also the production of aerogels with di-aldehyde cellulose (DAC) and silver nanoparticles (SNPs) has been recently attempted, leading to a high specific area and a more controlled dimension of pores [70].

4.4 Novel preparation methods for improved biocompatibility

To be able to assess the practical application of nanocellulose-nanosilver hybrid biomaterials, it is necessary to perform studies on their biocompatibility. The starting point is that bacterial cellulose structures were recognized to have characteristics similar to implants for replacement of natural tissues as regards moldability, biophysical and chemical possibilities, and of course biocompatibility, on which studies have been limited so far, especially as concerns nanosilver-nanocellulose hybrids [71]. The definition of "biocompatibility" includes a few requirements: that the host response to the new material is appropriate in every specific situation of use, that there are no toxic effects or tissue injury [72]. To be suitable for application, biocompatibility is reached through a mixture of different characteristics, in particular low friction coefficient, suitable surface topography, appropriate chemical composition and hydrophilicity [73]. Biocompatibility tests can be performed after the minimal amount of nanosilver is introduced in the material with no cytotoxicity effect. In particular, this proved effective also in the case of sanitary napkins produced introducing an amount of up to 0.4 g of nanosilver in pulp cellulose, which passed the in-vivo vaginal irritation and the intracutaneous vaginal tests in rabbit [74]. The results are depicted in Fig. 4.4.

In practice, biocompatibility studies of nanosilver-nanocellulose hybrids are still in their early development, because they need to be performed starting from a stable structure and showing relatively low variation of properties between different regions of the sample, therefore limited nanoparticles agglomeration. A number of limitations can be observed in this regard: in particular, the amount of silver nanoparticles that can use nanocellulose as their template is limited due to the very large dimensions of the latter. This poses the issue of immobilization of nanosilver in the structure, which can be possibly addressed by hydrothermal development of the material in autoclave [75]. A possible solution is enhancing the colloidal stability of the solution, as achieved by Shi et al. [76], by modifying nanocellulose, e.g., with dopamine, or else immobilizing silver nanoparticles by

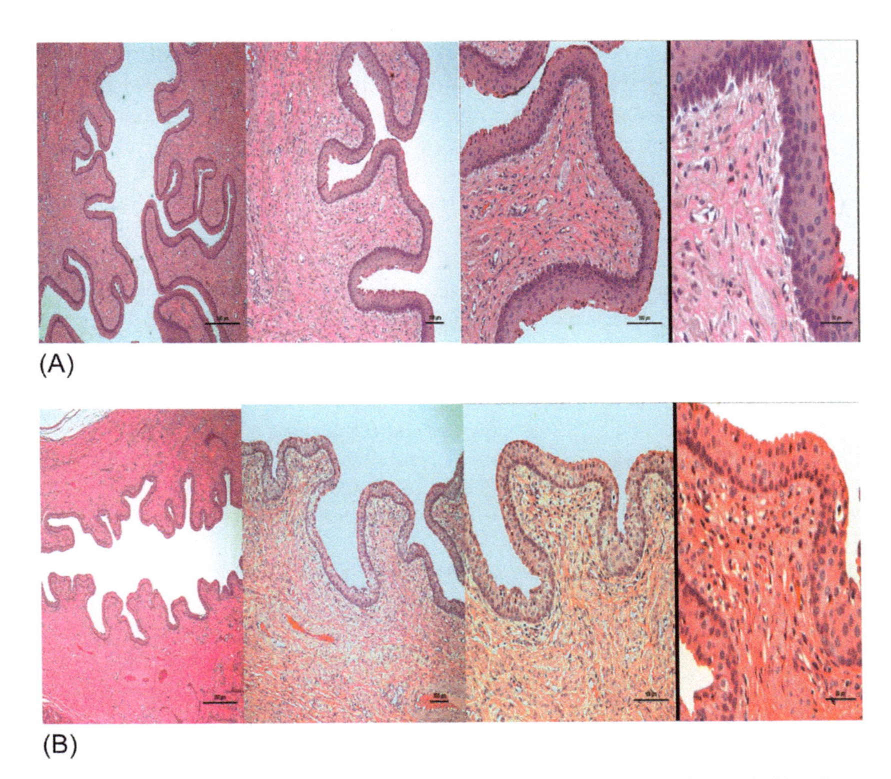

FIG. 4.4 Hematoxylin and eosin stained sections of rabbit vaginal tissue of control (A) and treated with sample extract (B). No visible signs of vaginal inflammation (epithelial ulceration, leukocyte infiltration, vascular congestion or edema) in the control or silver nanocellulose pulp exposed vaginal tissue sections. Photographs are representative of control and test animal. *(Reprinted with permission from Kavitha Sankar PC, Ramakrishnan R, Rosemary MJ, Biological evaluation of nanosilver incorporated cellulose pulp for hygiene products, Mater Sci Eng C 2016;61:631-7.)*

providing nanocellulose with star-shaped modifiers, such as fourth generation poly(amido-amine) dendrimers [77].

4.5 Conclusions

The development of nanocellulose-nanosilver composites appears promising, combining the unique aspects of crystallinity, high performance and tailored water absorption, of the former, as a template for the demonstrated antibacterial properties of the latter. Also in terms of costs, the composite obtained could be of interest, especially whenever nanocellulose is obtained from byproducts of large productive systems, starting basically from agrowaste. What appears still under definition is though the optimization of the amount of nanosilver particles to be introduced, with possible geometrical control or modification and size distribution. This would be required to obtain a sound and reliable interface with nanocellulose, by exploring procedures of chemical or physical methods for compatibilisation of the two materials.

References

[1] Kargarzadeh H, Ioelovich M, Ahmad I, Thomas S, Dufresne A. Methods for extraction of nanocellulose from various sources. Chapter 1, In: Kargarzadeh H, Ahmad I, Thomas S, Dufresne A, editors. Handbook of nanocellulose and cellulose nanocomposites, vol. 1. Wiley Verlag GmbH & Co. KGaA; 2017. p. 1–49.

[2] Spence K, Habibi Y, Dufresne A. Nanocellulose-based composites. Chapter 7, In: Kalia S, et al., editors. Cellulose fibers: bio- and nano-polymer composites. Berlin-Heidelberg: Springer-Verlag; 2011. p. 179–213. https://doi.org/10.1007/978-3-642-17370-7_7.

[3] Klemm D, Kramer F, Moritz S, Lindström T, Ankerfors M, Gray D, et al. Nanocelluloses: a new family of nature-based materials. Angew Chem 2011;50:5438–66.

[4] Habibi Y, Lucia LA, Rojas OJ. Cellulose nanocrystals: chemistry, self-assembly, and applications. Chem Rev 2010;110:3479–500.

[5] Jozala AF, de Lencastre-Novaes LC, Lopes AM, de Carvalho S-EV, Mazzola PG, Pessoa-Jr A, et al. Bacterial nanocellulose production and application: a 10-year overview. Appl Microbiol Biotechnol 2016;100:2063–72.

[6] Saraiva Morais JP, de Freitas RM, Souza Filho MM, Dias Nascimento L, Magalhães do Nascimento D, Ribeiro Cassales A. Extraction and characterization of nanocellulose structures from raw cotton linter. Carbohydr Polym 2013;91:229–35.

[7] Chen Y, Wu Q, Huang B, Huang M, Ai X. Isolation and characteristics of cellulose and nanocellulose from lotus-leaf stalks agro-wastes. Bioresources 2015;10:684–96.

[8] Trache D, Hussin MH, Haafiz MKM, Thakur VK. Recent progress in cellulose nanocrystals: sources and production. Nanoscale 2017;9:1763–86.

[9] Chirayil CJ, Mathew L, Thomas S. Review of recent research in nanocellulose preparation from different lignocellulosic fibers. Rev Adv Mater Sci 2014;37:20–8.

[10] Serizawa T, Kato M, Okura H, Sawada T, Wada M. Hydrolytic activities of artificial nanocellulose synthesized via phosphorylase-catalyzed enzymatic reactions. Polym J 2016;48:539–44.

[11] Ifuku S, Tsuji M, Morimoto M, Saimoto H, Yano H. Synthesis of silver nanoparticles templated by TEMPO-mediated oxidized bacterial cellulose nanofibers. Biomacromolecules 2009;10:2714–7.

[12] Yarbrough JM, Zhang R, Mittal A, Wall TV, Bomble YJ, Decker SR, et al. Multifunctional cellulolytic enzymes outperform processive fungal cellulases for coproduction of nanocellulose and biofuels. ACS Nano 2017;11:3101–9.

[13] Jiang F, Hsieh Y-L. Chemically and mechanically isolated nanocellulose and their self-assembled structures. Carbohydr Polym 2013;95:32–40.

[14] Liu C, Li B, Du H, Lv D, Zhang Y, Yu G, et al. Properties of nanocellulose isolated from corncob residue using sulfuric acid, formic acid, oxidative and mechanical methods. Carbohydr Polym 2016;151:716–24.

[15] Araki J, Wada M, Kuga S, Okano T. Flow properties of microcrystalline cellulose suspension prepared by acid treatment of native cellulose. Colloid Surf A 1998;142:75–82.

[16] Chung NH, Van Binh N, Dien LQ. Preparation of nanocellulose acetate from bleached hardwood pulp and its application for seawater desalination. Vietnam J Chem 2020;58:281–6.
[17] Rees A, Powell LC, Chinga-Carrasco G, Gethin DT, Syverud K, Hill KE, et al. 3D bioprinting of carboxymethylated-periodate oxidized nanocellulose constructs for wound dressing applications. Biomed Res Int 2015;, 925757. 7 pages https://doi.org/10.1155/2015/925757.
[18] Mishra SP, Thirree J, Manent A-S, Chabot B, Daneault C. Ultrasound-catalyzed tempo-mediated oxidation of native cellulose for the production of nanocellulose: effect of process variables. Bioresources 2011;6:121–43.
[19] Lin N, Bruzzese C, Dufresne A. TEMPO-oxidized nanocellulose participating as crosslinking aid for alginate-based sponges. ACS Appl Mater Interfaces 2012;4:4948–59.
[20] Klemm D, Cranston ED, Fischer D, Gama M, Kedzior SA, Kralisch D, et al. Nanocellulose as a natural source for groundbreaking applications in materials science: today's state. Mater Today 2018;21:720–48.
[21] Rajwade JM, Paknikar KM, Kumbhar JV. Applications of bacterial cellulose and its composites in biomedicine. Appl Microbiol Biotechnol 2015;99:2491–511.
[22] Tsouko E, Kourmentza C, Ladakis D, Kopsahelis N, Mandala I, Papanikolaou S, et al. Bacterial cellulose production from industrial waste and by-product streams. Int J Mol Sci 2015;16:14832–49.
[23] Esa F, Masrinda Tasirin S, Abd Rahman N. Overview of bacterial cellulose production and application. Agric Agric Sci Procedia 2014;2:113–9.
[24] Chawla PR, Bajaj IB, Survase SA, Singhal RS. Microbial cellulose: fermentative production and applications. Food Technol Biotechnol 2009;47:107–24.
[25] Ross P, Mayer R, Benziman M. Cellulose biosynthesis and function in bacteria. Microbiol Rev 1991;55:35–58.
[26] Dahman Y. Nanostructured biomaterials and biocomposites from bacterial cellulose nanofibers. J Nanosci Nanotechnol 2009;9:5105–22.
[27] Castro C, Zuluaga R, Putaux J-L, Caro G, Mondragon I, Gañán P. Structural characterization of bacterial cellulose produced by Gluconacetobacter swingsii sp. from Colombian agroindustrial wastes. Carbohydr Polym 2011;84:96–102.
[28] Li J, Cha R, Mou K, Zhao X, Long K, Luo H, et al. Nanocellulose-based antibacterial materials. Adv Healthc Mater 2018;7:1800334.
[29] Aitomäki Y, Oksman K. Reinforcing efficiency of nanocellulose in polymers. React Funct Polym 2014;85:151–6.
[30] Islam MT, Alam MM, Zoccola M. Review on modification of nanocellulose for application in composites. Int J Innov Res Sci Eng Technol 2013;2:5444–51.
[31] Hoeng F, Denneulin A, Bras J. Use of nanocellulose in printed electronics: a review. Nanoscale 2016;8:13131–4.
[32] Gómez C, Serpa HA, Velásquez-Cock J, Gañán P, Castro C, Vélez L, Zuluaga R. Vegetable nanocellulose in food science: a review. Food Hydrocoll 2016;57:178–86.
[33] Azeredo HMC, Rosa MF, Mattoso LHC. Nanocellulose in bio-based food packaging applications. Ind Crop Prod 2017;97:664–71.
[34] Li F, Mascheroni E, Piergiovanni L. The potential of nanocellulose in the packaging field: a review. Packag Technol Sci 2015;28:475–508.
[35] Seabra AB, Bernardes JS, Fávaro WJ, Paula AJ, Durána N. Cellulose nanocrystals as carriers in medicine and their toxicities: a review. Carbohydr Polym 2018;181:514–27.
[36] Fu L, Zhou P, Zhang S, Yang G. Evaluation of bacterial nanocellulose-based uniform wound dressing for large area skin transplantation. Mater Sci Eng C 2013;33:2995–3000.

[37] Meftahi A, Khajavi R, Rashidi A, Sattari M, Yazdanshenas ME, Torabi M. The effects of cotton gauze coating with microbial cellulose. Cellulose 2010;17:199–204.
[38] Artem Ataide J, Mendes de Carvalho N, de Araújo Rebelo M, Vinícius Chaud M, Grotto D, Gerenutti M, et al. Bacterial nanocellulose loaded with bromelain: assessment of antimicrobial, antioxidant and physical-chemical properties. Sci Rep 2017;7, 18031.
[39] Ding Q, Xu X, Yue Y, Mei C, Huang C, Jiang S, Wu Q, Han J. Nanocellulose-mediated electroconductive self-healing hydrogels with high strength, plasticity, viscoelasticity, stretchability, and biocompatibility toward multifunctional applications. Appl Mater Interfaces 2018;10:27987–8002.
[40] Dumanli AG. Nanocellulose and its composites for biomedical applications. Curr Med Chem 2017;24:512–28.
[41] Islam MT, Alam MM, Patrucco A, Montarsolo A, Zoccola M. Preparation of nanocellulose: a review. AATCC J Res 2014;5:17–23.
[42] Phanthong P, Reubroycharoen P, Hao X, Xu G, Abudula A, Guana G. Nanocellulose: extraction and application. Carbon Resourc Convers 2018;1:32–43.
[43] Mohite BV, Patil SV. In situ development of nanosilver impregnated bacterial cellulose for sustainable released antimicrobial wound dressing. J Appl Biomater Funct Mater 2016;14 (1):e53–8.
[44] Fortunati E, Rinaldi S, Peltzer M, Bloise N, Visai L, Armentano I, Jiménez A, Latterini L, Kenny JM. Nano-biocomposite films with modified cellulose nanocrystals and synthesized silver nanoparticles. Carbohydr Polym 2014;101:1122–33.
[45] Wekwejt M, Michno A, Truchan K, Pałubicka A, Świeczko-Żurek B, Osyczka AM, Zieliński A. Antibacterial activity and cytocompatibility of bone cement enriched with antibiotic, nanosilver, and nanocopper for bone regeneration. Nanomaterials 2019;9:1114.
[46] Madhumathi K, Sudheesh Kumar PT, Abhilash S, Sreeja V, Tamura H, Manzoor K, Nair SV, Jayakumar R. Development of novel chitin/nanosilver composite scaffolds for wound dressing applications. J Mater Sci Mater Med 2010;21:807–13.
[47] Petrochenko E, Zheng J, Casey BJ, Bayati MR, Narayan RJ, Goering PL. Nanosilver-PMMA composite coating optimized to provide robust antibacterial efficacy while minimizing human bone marrow stromal cell toxicity. Toxicol in Vitro 2017;44:248–55.
[48] Kibis LS, Stadnichenko AI, Pajetnov EM, Koscheeva SV, Zaykovskii VI, Boronin AI. The investigation of oxidized silver nanoparticles prepared by thermal evaporation and radio-frequency sputtering of metallic silver under oxygen. Appl Surf Sci 2010;57:404–13.
[49] Baker C, Pradhan A, Pakstis L, Pochan DJ, Shah SI. Synthesis and antibacterial properties of silver nanoparticles. J Nanosci Nanotechnol 2005;5:244–9.
[50] El-Sheekh MM, El-Kassas HY. Algal production of nano-silver and gold: their antimicrobial and cytotoxic activities: a review. J Genetic Eng Biotechnol 2016;14:299–310.
[51] McShan D, Ray PC, Yu H. Molecular toxicity mechanism of nano silver. J Food Drug Anal 2014;22:116–27.
[52] Juganson K, Mortimer M, Ivask A, Pucciarelli S, Miceli C, Orupõld K, Kahrua A. Mechanisms of toxic action of silver nanoparticles in the protozoan Tetrahymena thermophila: from gene expression to phenotypic events. Environ Pollut 2017;225:481–9.
[53] Fabrega J, Fawcett SR, Renshaw JC, Lead JR. Silver nanoparticle impact on bacterial growth: effect of pH, concentration, and organic matter. Environ Sci Technol 2009;43:7285–90.
[54] Sadeghi B, Garmaroudi FS, Hashemi M, Nezhad HR, Nasrollahi A, Sima A, Sahar A. Comparison of the anti-bacterial activity on the nanosilver shapes: nanoparticles, nanorods and nanoplates. Adv Powder Technol 2012;23:22–6.

[55] Wu J, Zheng Y, Song W, Luan J, Wen X, Wu Z, Chen X, Wang Q, Guo S. In situ synthesis of silver-nanoparticles/bacterial cellulose composites for slow-released antimicrobial wound dressing. Carbohydr Polym 2014;102:762–71.

[56] Amini E, Azadfallah M, Layeghi M, Talaei-Hassanloui R. Silver-nanoparticle-impregnated cellulose nanofiber coating for packaging paper. Cellulose 2016;23:557–70.

[57] Ye D, Zhong Z, Xu H, Chang C, Yang Z, Wang Y, Ye Q, Zhang L. Construction of cellulose/nanosilver sponge materials and their antibacterial activities for infected wounds healing. Cellulose 2016;23:749–63.

[58] Maneerung T, Tokura S, Rujiravanit R. Impregnation of silver nanoparticles into bacterial cellulose for antimicrobial wound dressing. Carbohydr Polym 2008;72:43–51.

[59] Pal S, Nisi R, Stoppa M, Licciulli A. Silver-functionalized bacterial cellulose as antibacterial membrane for wound-healing applications. ACS Omega 2017;2:3632–9.

[60] Johnston JH, Nilsson T. Nanogold and nanosilver composites with lignin-containing cellulose fibres. J Mater Sci 2012;47:1103–12.

[61] Son WK, Youk JH, Ho Park W. Antimicrobial cellulose acetate nanofibers containing silver nanoparticles. Carbohydr Polym 2006;65:430–4.

[62] Fortunati E, Luzi F, Puglia D, Terenzi A, Vercellino M, Visai L, Santulli C, Torre L, Kenny JM. Ternary PVA nanocomposites containing cellulose nanocrystals from different sources and silver particles: part II. Carbohydr Polym 2013;97:837–48.

[63] Chen JY, Sun L, Jiang W, Lynch VM. Antimicrobial regenerated cellulose/nano-silver fiber without leaching. J Bioact Compat Polym 2015;30:17–33.

[64] Ul-Islam M, Khan T, Khattak WA, Park JK. Bacterial cellulose-MMTs nanoreinforced composite films: novel wound dressing material with antibacterial properties. Cellulose 2013;20:589–96.

[65] Soltani RDC, Mashayekhi M, Naderi M, Boczkaj G, Jorfi S, Safari M. Sonocatalytic degradation of tetracycline antibiotic using zinc oxide nanostructures loaded on nano-cellulose from waste straw as nanosonocatalyst. Ultrason Sonochem 2019;55:117–24.

[66] Wei B, Yang G, Hong F. Preparation and evaluation of a kind of bacterial cellulose dry films with antibacterial properties. Carbohydr Polym 2011;84:533–8.

[67] Ogundare SA, van Zyl WE. Nanocrystalline cellulose as reducing- and stabilizing agent in the synthesis of silver nanoparticles: application as a surface-enhanced Raman scattering (SERS) substrate. Surf Interfaces 2018;13:1–10.

[68] Gan I, Chow WS. Antimicrobial poly(lactic acid)/cellulose bionanocomposite for food packaging application: a review. Food Packag Shelf Life 2018;17:150–61.

[69] Berndt S, Wesarg F, Wiegand C, Kralisch D, Müller FA. Antimicrobial porous hybrids consisting of bacterial nanocellulose and silver nanoparticles. Cellulose 2013;20:771–83.

[70] Zhang L, Li Q, Duan L, Liu R. Preparations and characterisation of a nanosilver-loaded aerogel based on nanocellulose. Micro Nanolett 2020;15:409–14.

[71] Ludwicka K, Jedrzejczak-Krzepkowska M, Kubiak K, Kolodziejczyk M, Pankiewicz T, Bielecki S. Chapter 9. Medical and cosmetic applications of bacterial nanocellulose. In: Bacterial nanocellulose. From biotechnology to bio-economy. Elsevier; 2016. p. 145–65.

[72] Williams DF. The Williams dictionary of biomaterials. Liverpool University Press; 1999.

[73] Khajavi R, Meftahi A, Alibakhshi S, Samih L. Investigation of microbial cellulose/cotton/silver nanobiocomposite as a modern wound dressing. Adv Mater Res 2013;829:616–21.

[74] Kavitha Sankar PC, Ramakrishnan R, Rosemary MJ. Biological evaluation of nanosilver incorporated cellulose pulp for hygiene products. Mater Sci Eng C 2016;61:631–7.

[75] Zhang X, Sun H, Tan S, Gao J, Fu Y, Liu Z. Hydrothermal synthesis of Ag nanoparticles on the nanocellulose and their antibacterial study. Inorg Chem Commun 2019;100:44–50.

[76] Shi Z, Tang J, Chen L, Yan C, Tanvir S, Anderson WA, Berry RM, Tam KC. Enhanced colloidal stability and antibacterial performance of silver nanoparticles/cellulose nanocrystal hybrids. J Mater Chem B 2015;3:603–11.

[77] Ramaraju B, Imae T, Destaye AG. Ag nanoparticle-immobilized cellulose nanofibril films for environmental conservation. Appl Catal A 2015;492:184–9.

Chapter 5

Hybrid materials from cellulose nanocrystals for wastewater treatment

Hanane Chakhtouna[a,b], Hanane Benzeid[b], Nadia Zari[a], Abou el Kacem Qaiss[a], and Rachid Bouhfid[a]
[a]*Moroccan Foundation for Advanced Science, Innovation and Research (MAScIR), Composites and Nanocomposites Center, Rabat Design Center, Rabat, Morocco,* [b]*Laboratoire de Chimie Analytique et de Bromatologie, Faculté de Médecine et de Pharmacie, Université Mohamed V de Rabat, Rabat, Morocco*

5.1 Introduction

In recent years, extensive efforts have been taken for the utilization of abundant, biodegradable, and green materials for the uptake of different pollutants from sewage [1–6]. Cellulose, as the most abundant biopolymer available on the earth, has shown higher efficacy toward a wide range of pollutants [5, 7–9], due to its large availability in nature, nontoxicity, unique interconnected network structure, and robust physico-chemical properties [10, 11]. However, it was observed that converting cellulose to nanoscale structures improves its adsorption behavior, especially toward heavy metals via its functional surface groups, mainly OH groups [12, 13]. According to the literature, nanocellulose can be divided in three type of nanomaterials: cellulose nanocrystals (CNC), cellulose nanofibrils (CNF), and bacterial nanocellulose (BNC), differing from each other in their properties (dimension and crystallinity), as well as in their production methods [14–16]. Among these three types, cellulose nanocrystals (CNC) have received high interest in the wastewater treatment field as a powerful adsorbent for several pollutants such as: organic dyes, pharmaceutical compounds, and heavy metals [17–19]. This higher adsorption capacity is attributed to their unique structure and exceptional physico-chemical properties including large surface area, rich hydroxyl groups, high crystallinity index, low density, and excellent mechanical properties [20, 21]. But to expand their applicability and avoid the problem of particles aggregation, extensive works have been undertaken in coupling cellulose nanocrystals with other materials

Cellulose Nanocrystal/Nanoparticles Hybrid Nanocomposites. https://doi.org/10.1016/B978-0-12-822906-4.00001-3

such as: biopolymers, polymers, metal or metal oxide, magnetic materials, and many others [22–25]. This chapter aims to provide an overview on recent developments of different hybrid materials based on cellulose nanocrystals, with special attention to their synthesis methods, functional properties, and application in the wastewater field as powerful materials for the removal of a wide range of pollutants.

5.2 Cellulose nanocrystals generalities: From synthesis to application as a potential adsorbent in wastewater treatment field

5.2.1 Synthesis, structure, and morphology

Cellulose nanocrystals (CNC), also known as cellulose nanowhiskers (CNW) or nanocrystalline cellulose (NCC), can be easily extracted from a variety of renewable sources via acid hydrolysis under controlled conditions of acid concentration, temperature, and reaction time [26, 27]. The acid hydrolysis preparation of CNC can be performed via different mineral acids such as: sulfuric acid (H_2SO_4), hydrochloridric acid (HCl), phosphoric acid (H_3PO_4), and hydrobromic acid (HBr) [28]. However, it has been found that the acid hydrolysis of cellulose with sulfuric acid makes it possible to prepare a more stable CNC than those obtained using other acids, due to the formation of negatively charged sulfate groups on the CNC surface preventing their agglomeration [29]. Generally, the hydrolysis process with sulfuric acid requires a concentration of around 60% to 65%, a reaction temperature of 40 °C to 50 °C, and a reaction time of 30 to 60 min. Under such conditions, hydrogen ions from the acid penetrate the amorphous regions of the cellulose chains resulting in hydrolytic cleavage of 1,4-β-glycosidic bonds, releasing individual crystallites, while the crystalline region of cellulose is recovered (Fig. 5.1) [28]. After complete reaction, the CNC is collected by centrifugation followed by dialyzing against distilled water until a neutral pH is reached.

Cellulose nanocrystals are generally rigid and short nanocrystals with a rod-shaped structure [30]. The size, as well as the morphology of those rod-shaped, are found to vary widely depending on the cellulosic source and the operating conditions under which the hydrolysis is performed. Typical dimensions of 3–35 nm in dimeter and 100 to 250 nm in length are reported [31, 32].

Kargarzadeh et al. studied the effect of acid-hydrolysis conditions on the morphology, crystallinity, and thermal of cellulose nanocrystals extracted from kenaf bast fibers via sulfuric acid hydrolysis (65 wt% at 45 °C) for reaction time between 20 and 120 min [33]. Such treatments showed their effectiveness in eliminating all the CNC amorphous zones, which was confirmed by Fourier transform Infrared spectrum. Moreover, it was observed from TEM images that all the CNC produced under different hydrolysis times exhibited a needle shape with a diameter or around 11–13 nm and a length of 124–166 nm (Fig. 5.2).

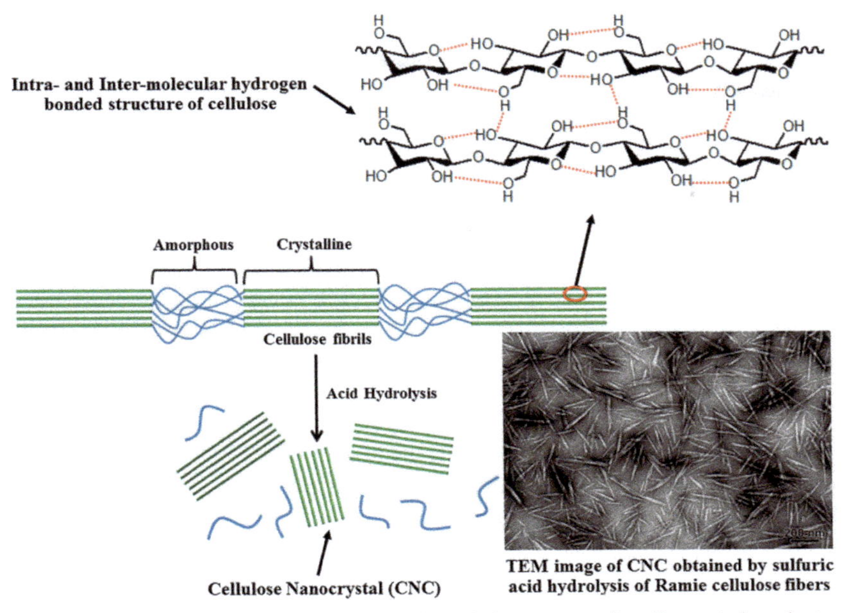

FIG. 5.1 CNC preparation by acid hydrolysis and the corresponding Transmission electron microscopy image [14].

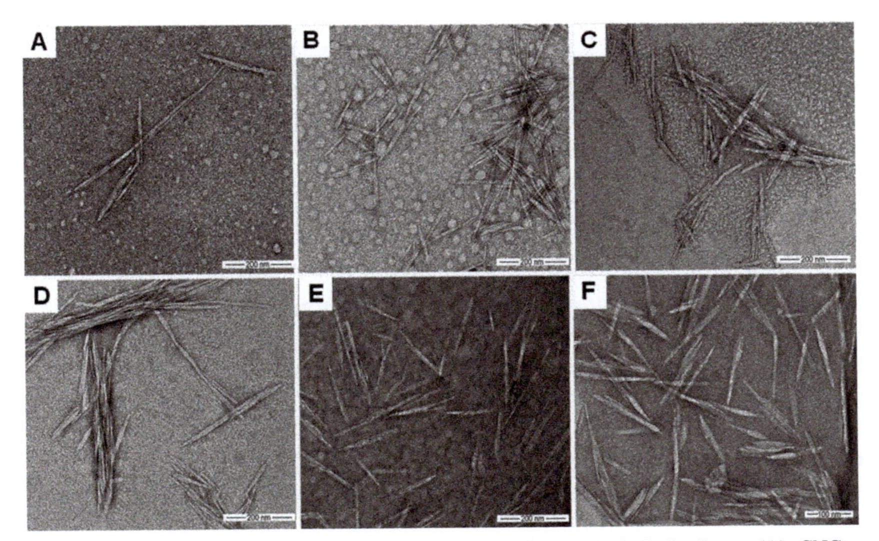

FIG. 5.2 TEM micrographs of CNC produced for different hydrolysis times: (A) CNC_{20}, (B) CNC_{30}, (C) CNC_{40}, (D) CNC_{60}, (E) CNC_{90}, and (F) CNC_{120} [33].

Both diameter and length decreased with increasing hydrolysis time from 20 to 120 min, which was expected since more contact is necessary between the CNC amorphous regions and the acid ions. Moreover, 40 min can be considered as the optimal hydrolysis time to obtain CNC with a good yield value (41%), high aspect ratio (13.2), and high crystallinity index (81.8%). The authors also assessed the stability of different CNC prepared in aqueous media. It was found that the CNC produced at 20 min exhibited the lowest zeta potential value (−8.77 mV) compared to 120 min, which presented the highest value (−95.3 mV. This difference was attributed to the presence of a larger amount of negatively charged sulfate groups on the CNC_{120} surface increasing the number of surface charges on the particles. However, it was observed that the presence of such groups on the CNC surface presented a negative effect on the CNC thermal stability. The thermal stability of cellulose nanocrystals prepared by sulfuric acid hydrolysis was lower than that of the original cellulose. Similar results were found for CNC isolated from hardwood pulp under different acid-hydrolysis conditions [34]. The authors have attributed this decrease to the higher number of sulfate groups acting as flame-retardants.

5.2.2 Cellulose nanocrystals as a potential adsorbent in wastewater treatment

Owing to their unique and potentially useful properties, cellulose nanocrystals (CNC) have aroused great interest as a potential adsorbent to remove contaminants, such as organic dyes, oils, pharmaceuticals, and heavy metals in wastewater [35–38]. Yang et al. [39] have successfully prepared cellulose nanocrystals from carex meyeriana kunth for the adsorption of methylene blue (MB) dye from aqueous solutions. The obtained CNC had a short-rod shape with an average diameter and length of 33 ± 1 nm and 175 ± 3 nm, respectively. Thanks to its high crystallinity and smaller size, the resulting CNC presented high adsorption capacity 217.4 mg/g toward MB molecules under optimum conditions of 10 mL of H_2O_2 assay, 30 mg of CNC content, 200 mg/L as initial MB concentration and 180 min as contact time. The adsorption kinetics and isotherms data were well fitted to the pseudo-second-order and Langmuir models respectively, suggesting that the MB adsorption occurred as a monolayer on the external surface of CNC via chemisorption. Similar results were found by Abiaziem et al. [40] for the removal of lead ions (Pb^{2+}) using cellulose nanocrystals from cassava peel (CPCNC) produced via acid-hydrolysis using 64% concentrated sulphuric acid. The characterization results confirmed the complete dissolution of the amorphous regions and the increase of both crystallinity degree and specific surface area providing more adsorption active sites. The CPCNC exhibited a maximum adsorption capacity of 6.4 mg/g, which was dependent on different environmental parameters such as: pH, temperature, initial concentration of lead ions, and contact time. The isotherm adsorption data well fitted to both the Langmuir and Freundlich isotherm models. The first model suggests that adsorption was due to the formation of Pb^{2+} in contact with the CPCNC surface, while the Freundlich isotherm indicates a good surface heterogeneity of the active sites.

Cellulose nanocrystals can be chemically modified to enhance their adsorption capacity and expand their application to large scale [41–43]. In this context, Ranjbar and his co-authors [44] have successfully modified pristine CNC with various amounts of a positively-charged surfactant cetyltrimethylammonium bromide (CTAB) for the adsorption of Congo red (CR) as an anionic dye. The advantages of using this surfactant was attributed to its availability, biodegradability, and low cost. The characterization results confirmed the successful modification of the CNC surface and showed that the surfactant molecules have been absorbed on the CNC surface via electrostatic interactions between the positively charged headgroup of cetyltrimethylammonium bromide and the negatively charged sulfate ester groups on the CNC surface. Furthermore, it was found that the surfactant amount strongly affected the CNC adsorption properties. It was shown that increasing the CTAB concentration up to 0.25 w/w increased the adsorption capacity of CR on CNC up to 220 mg/g. This behavior can be attributed to the additional active sites formed after CTAB addition improving CR adsorption. Adsorption isotherm studies revealed that the CR adsorption onto modified CNC is exothermic, spontaneous, and well described by the Langmuir model, while the kinetics data were well fitted by both a pseudo-second-order and intra-particle. In fact, various adsorption mechanisms can be involved during the CR adsorption as reported in Fig. 5.3. FTIR spectra of CNC before and after adsorption tests showed that the adsorption process involved both chemisorption and physisorption through several mechanisms including electrostatic attraction, hydrogen bonding, and hydrophobic attractions, while the prevalence of each mechanism depends on the amount of surfactant added.

5.3 Hybrid materials from cellulose nanocrystals for wastewater treatment

Although cellulose nanocrystals have proved their effectiveness for the elimination of a range of organic and inorganic pollutants from sewage, they usually suffers from their difficult separation from aqueous solutions and their agglomeration limiting their applications [45, 46]. Therefore, CNC with other materials, such as carbonaceous materials, metal or metal oxides, and many others, appears as an efficient strategy to overcome these drawbacks and combine their advantages to obtain hybrid composites with higher performances. The following section aims to present the recent developments in hybrid materials based cellulose nanocrystals, from design to processing and applications for wastewater purification.

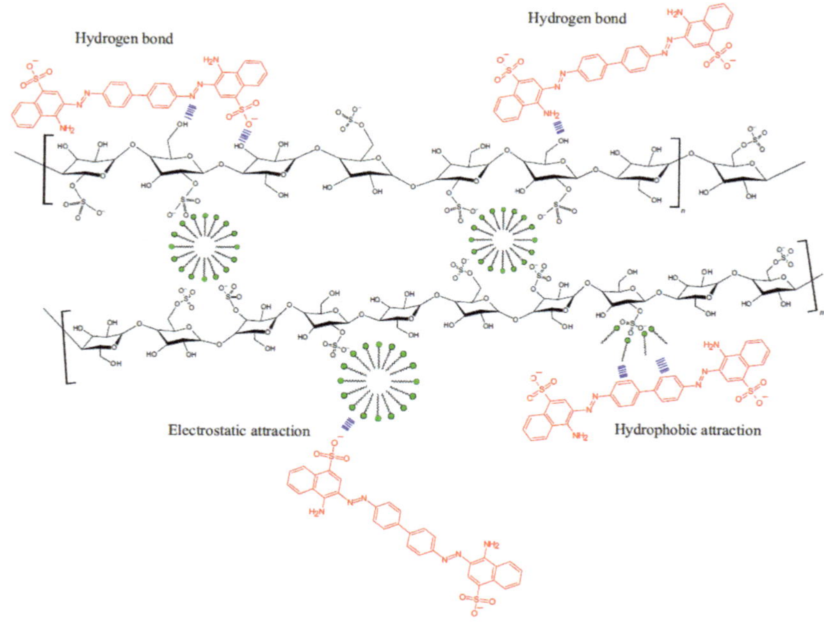

FIG. 5.3 Schematic representation of different adsorption mechanisms involved during the adsorption of CR on CNC [44].

5.3.1 CNC/polymer hybrid materials

Since the first utilization of cellulose nanocrystals as reinforcing fillers in poly (styrene-co-butyl acrylate) by Favier et al. [47] in 1995, there has been an increased interest in the preparation of cellulose nanocrystals based polymer hybrid materials in several areas, especially in the treatment of polluted water [14]. Although both natural and synthetic polymers are used for the preparation of CNC/polymer hybrid materials, the main challenges of such hybrid materials lie in attaining homogeneous dispersion and achieving strong interactions taking into account that the polymers are nonpolar and hydrophobic unlike CNC which are highly polar and hydrophilic [14]. Environmentally friendly nanocomposites based on cellulose nanocrystals and polydopamine (CNC@PDA), with uniform distribution and higher surface area (107.2 m^2/g), have been prepared via self-polymerization of dopamine on the CNC surface and used as a highly efficiency adsorbent for the removal of methylene blue dye from aqueous solutions [23]. TEM images showed that the resulting composite exhibited a core-shell structure with a very thin PDA coating of 4 nm on the CNC external surface. The presence of such layer is more beneficial for the adsorption of cationic dye due to the presence of various functional groups serving as active sites for the MB molecules adsorption. Likewise, both FTIR analysis and X-Ray

Photo-electron Spectroscopy (XPS) were used to confirm the successful PDA modification of CNC. As shown in Fig. 5.4, the XPS spectrum of CNC@PDA clearly shows the C1s, O1s, and N1s peaks, while that of CNC shows only C1s and O1s peaks, indicating that the PDA has been successfully coated onto CNC. The high resolution N1s reveals the presence of tertiary (—NH_2), secondary (—NH—) and primary (—N=) amines related to the oxidative polymerization of dopamine.

The CNC@PDA nanocomposites exhibited an exceptionally high adsorption capacity of up to 2066 mg/g for a very high initial concentration of MB (500 mg/L) and within a very short time (5 min). This higher adsorption capacity of MB onto CNC@PDA nanocomposites can be ascribed to its higher surface area and the presence of abundant functional groups, such as catechol, quinone, and amine groups, as well as aromatic moieties easily interacting with the aromatic rings of the MB molecules. Equilibrium data were well described by the Langmuir isotherm model, suggesting a monolayer adsorption of MB onto the CNC@PDA nanocomposite, while the kinetics data followed a pseudo-second-order model indicating that the adsorption of the MB on CNC@PDA nanocomposites can be attributed to chemical adsorption. Moreover, the CNC@PDA nanocomposites were effectively regenerated and maintained high adsorption capacity (900 mg/g) after four adsorption-desorption

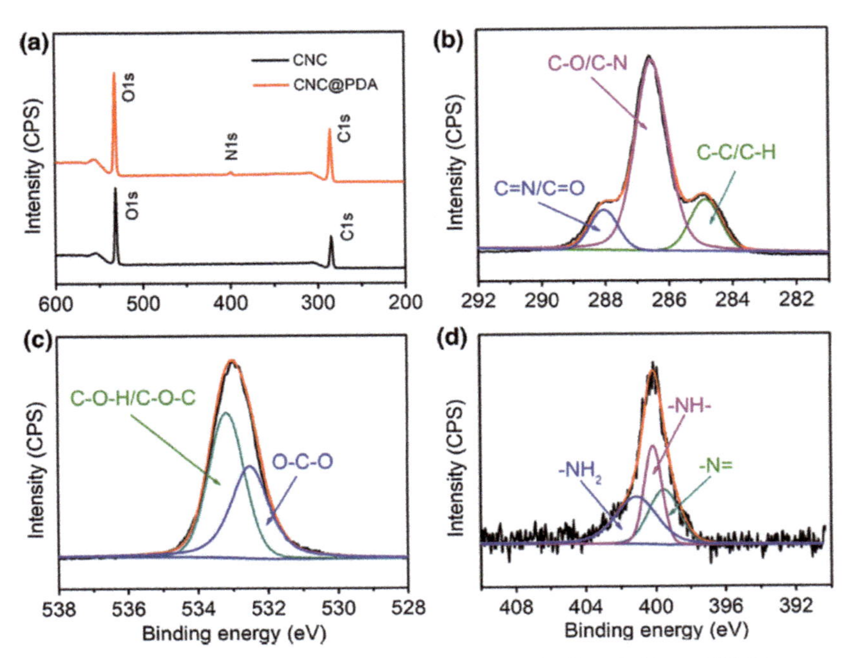

FIG. 5.4 (A) XPS spectra of CNC and CNC@PDA with the (B) C1s, (C) O1s, and (D) N1s spectra of CNC@PDA nanocomposites [23].

cycles. XPS spectra of CNC@PDA before and after MB adsorption was carried out to understand the mechanism involved during the MB adsorption (Fig. 5.5). The shift of different peaks of C=N/C=O, C—O/C—N, C—C/C—H, —NH_2, —NH— and —N= after MB adsorption suggests that the MB adsorption on CNC@PDA nanocomposites occurred via two majors mechanisms, namely electrostatic and strong π-π interactions between the aromatic MB molecule and the rich aromatic groups of CNC@PDA. The deprotonate phenolic groups on the CNC@PDA surface can provide negatively charged adsorption sites for the cationic MB dye through electrostatic interaction, while the aromatic MB molecule is assumed to have π-π stacking interaction with the CNC@PDA with rich aromatic moieties.

Chen et al. [48] have successfully synthesized cellulose nanocrystal-*g*-poly (acrylic acid-co-acrylamide) aerogels (CNC-*g*-P(AA/AM), with different amount of CNC for the removal of Pb(II) via free radical graft polymerization followed by freeze-drying. The successful grafting of P(AA/AM) chains onto the CNC surface was confirmed via FTIR analysis which displayed additional infrared vibrations compared to both neat CNC and P(AA/AM). Moreover, TEM images and SEM micrographs (Fig. 5.6) revealed the good CNC dispersion with a needle-shaped structure into the matrix. The CNC-*g*-P(AA/AM)

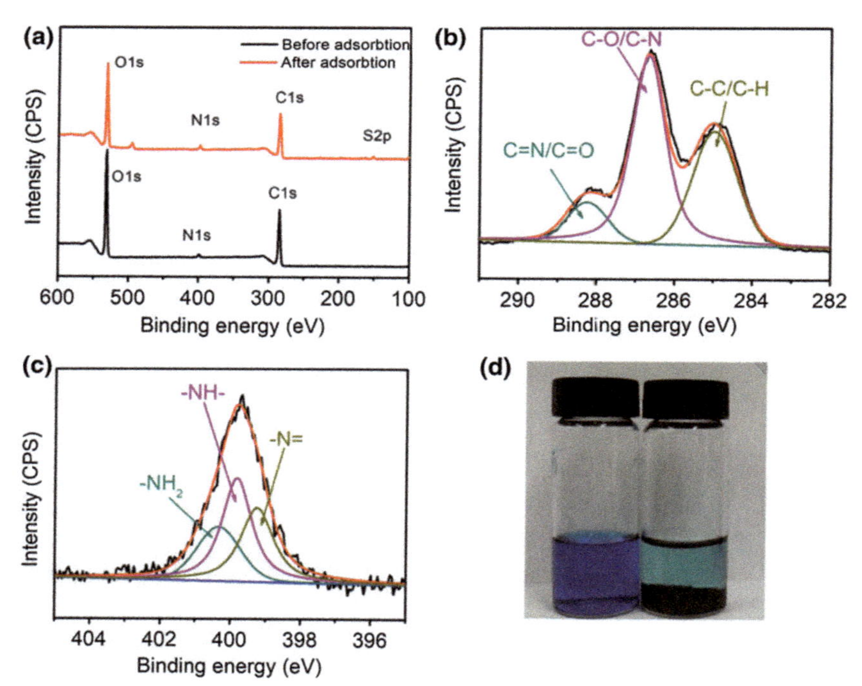

FIG. 5.5 XPS spectra (A, B, C) and (D) MB solution before and after the addition of the CNC@PDA adsorbent [23].

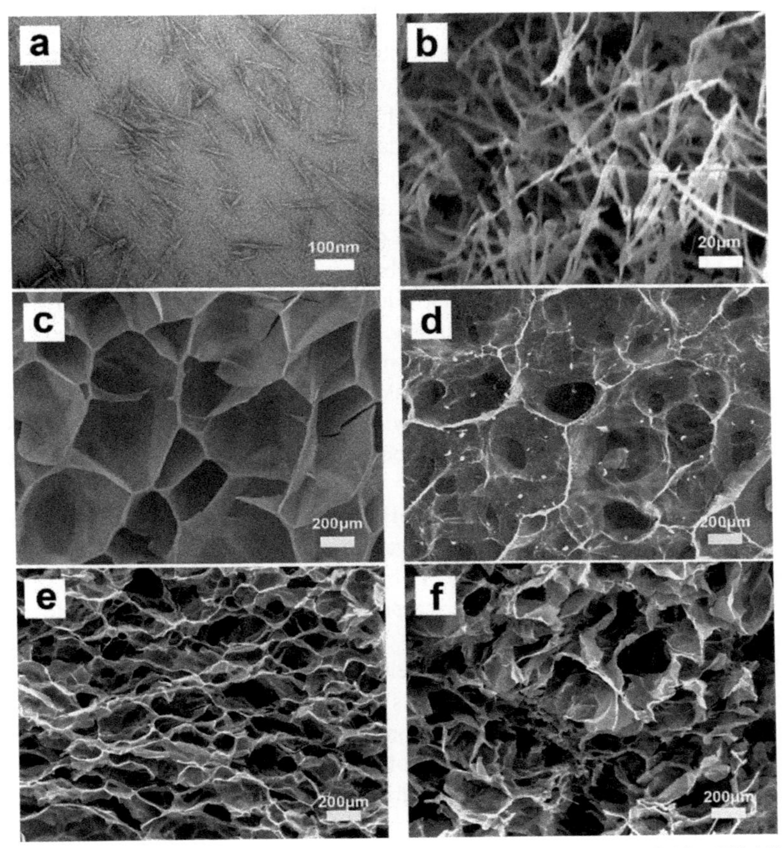

FIG. 5.6 (A) TEM image and (B) and SEM images of: (C) CNC, P(AA/AM), (D) CNC-*g*-P(AA/AM)-0.5, (E) CNC-*g*-P(AA/AM)-1, and (F) CNC-*g*-P(AA/AM)-2 [48].

aerogels had an open and macroporous honeycomb structure, high porosity, huge specific surface area, and abundant hydroxyl/sulfonic acid groups, leading to excellent adsorption capacities toward Pb(II) compared to the neat CNC. The CNC-*g*-P(AA/AM) composite with 1 wt% CNC presented the highest adsorption capacity (366 mg/g) compared with other adsorbent aerogels, and was easily regenerated by a simple immersion in HCl solution giving a 81.3% removal efficiency which was maintained after five consecutive adsorption-desorption cycles. Moreover, the adsorption of Pb(II) on CNC-*g*-P(AA/AM)-1 aerogel suggested a monolayer adsorption and chemisorption binding mechanism.

Zhao et al. [49] developed a series of polyethylenimine (PEI) cross-linked cellulose nanocrystal (PEI-CNC) hybrid materials, with different PEI content for rare earth elements (REE) uptake. The composites were prepared via an environmentally friendly approach based on the amidation of carboxylic groups of TEMPO-CNC with the amino groups of PEI acting as cross-linkers and chelating agents. The hybrid materials displayed different internal nano and microstructures morphologies. CNC, with an average size ranging from 10 to 100 nm, was well dispersed in the PEI network. FTIR analysis (Fig. 5.7A) shows that the PEI-CNC exhibited more intense bands at 1642, 1561, and 1452 cm^{-1}, compared to the neat CNC, confirming the successful formation of peptide bonds and the introduction of amino groups during PEI cross-linking. These results were also confirmed by XPS analysis (Fig. 5.7B and C). Compared to CNC and TEMPO-CNC, the PEI-CNC hybrid material showed the presence of N_{1S} peaks proving the successful introduction of nitrogen into the hybrid composite. The three main forms (primary, secondary, and tertiary amines) were present which played an interesting role in the adsorption behaviors of the resulting composite: fast and high uptake capacity toward three rare earth elements (La(III), Eu(III), and Er(III)) with maximum adsorption capacities of 0.611, 0.670, and 0.719 mmol/L respectively, indicating a higher adsorption affinity for Er(III).

The effect of adsorbent dosage, pH, and temperature, on the adsorption capacity of the composite was also investigated. It was observed that the adsorption efficiency of the PEI-CNC hybrid material increased with increasing adsorbent dose from 0.1 to 1.0 g/L, temperature from 20°C to 40°C, while maintaining the pH in the range of 4 to 6.5. Adsorption isotherm studies revealed that the adsorption of the three rare earth elements onto PEI-CNC was well described by the Sips model [50], which is a combination of both Langmuir and Freundlich models, suggesting both mono and multilayers adsorption. The kinetics data were also well fitted by both the pseudo-second-order and intraparticle models, suggesting that both the chemisorption and diffusion affected the REE adsorption on the PEI-CNC hybrid material surface. The comparison of FTIR and XPS spectra of PEI-CNC before and after adsorption led to conclude that amino groups (N–H and $-NH_2$), with a lone pair of electrons, are the active sites responsible for the higher REE adsorption (Fig. 5.8).

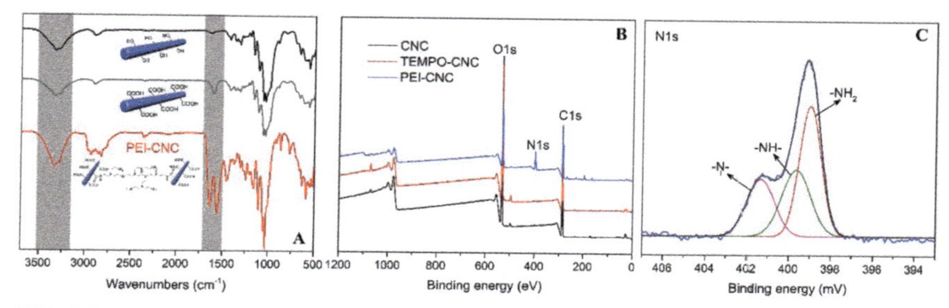

FIG. 5.7 (A) FTIR and (B,C) XPS spectra of CNC, TEMPO-CNC, and PEI-CNC3 [49].

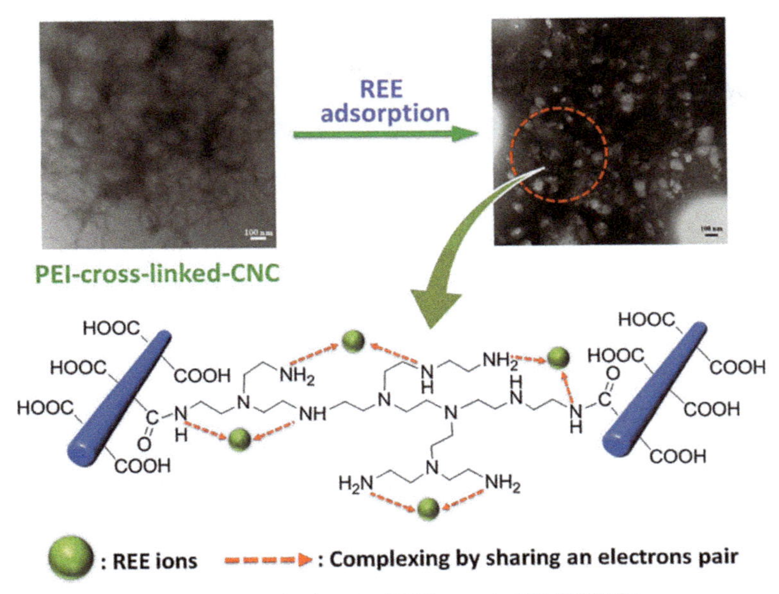

FIG. 5.8 Possible adsorption mechanism for REE ions onto PEI-CNC [49].

Recently, PVA/SA/CNC@PEI core-shell microspheres, composed of polyvinyl alcohol (PVA), sodium alginate (SA), and cellulose nanocrystal (CNC), have been synthesized and modified with polyethyleneimine (PEI) to introduce a high number of active amino sites onto the surface for diclofenac sodium (DS) for adsorption [22]. The PVA/SA/CNC@PEI was formed via PEI grafting onto PVA/SA/CNC microspheres with optimal CNC, SA, and PVA contents of 0.5, 1, and 2 g, respectively. The characterization results confirmed the successful modification of the PVA/SA/CNC microspheres with PEI, thanks to the strong chemical interactions formed between the functional groups of SA and PEI. Furthermore, the prepared microspheres exhibited maximum adsorption capacity of 418 mg/g, which is higher than that of SA, SA/CNC, and SA/CNC/PVA, confirming the role of amino groups on improving the composite adsorption behaviors, which provided more active sites on the composite surface to catch more DS molecules. The effect of different environmental parameters was studied and the results revealed that most of the DS was successfully removed at 303 K in solutions with a pH of 4.5, after 50 min using 20 mg of PVA/SA/CNC@PEI. The adsorption capacity of the PVA/SA/CNC@PEI composite showed a slight loss (11%) after five adsorption-desorption cycles, but the DS removal rate remained above 80%. The PVA/SA/CNC@PEI adsorption process of DS was well fitted to the Langmuir adsorption isotherm suggesting that the DS adsorption was homogenous, forming a monolayer of DS on the microspheres surface, and also followed to the pseudo-second-order model,

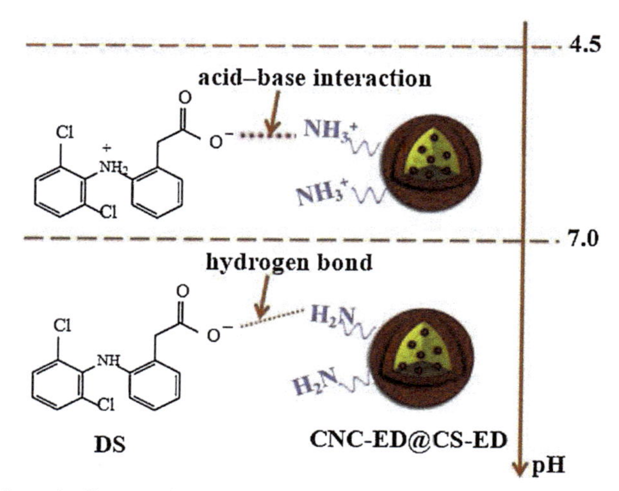

FIG. 5.9 Schematic diagram of the DS adsorption mechanism [51].

involving chemical adsorption, either through chemical bonding or via ion exchange. Similar results were obtained by Hu and his collaborators using CNC-ED@CS-ED nanohybrid composite composed of cellulose nanocrystals (CNC) and chitosan (CS) modified by ethylenediamine (ED) in both layers for the removal of the same compound (DS) and using a synthesis method similar to that used previously by Fan and his collaborators [51]. The CNC-ED@CS-ED nanohybrid composite exhibited a maximum adsorption capacity of 444 mg/g, which is slightly higher than that found using PVA/SA/CNC@PEI microspheres under the same conditions of pH, contact time, and microbeads dosage. The DS adsorption on CNC-ED@CS-ED also followed the pseudo-second-order kinetic and Langmuir models, suggesting that the DS adsorption can be achieved through either acid-base or hydrogen bond interactions, according to the pH of the solution as shown in Fig. 5.9.

5.3.2 CNC/metal or metal oxide hybrid materials

Owing to their high surface area, thermal stability, functionalized surface, and environmental benefits, CNC have attracted a great deal of attention as support for metal or metal oxide nanoparticles to hinder their aggregation [52, 53]. Various approaches have been proposed for the immobilization of metallic nanoparticles onto cellulose nanocrystals including precipitation, sol-gel, hydro/solvothermal methods, in-situ reduction, and many others [53]. An et al. [54] developed cellulose nanocrystal/hexadecyltrimethylammonium bromide/silver nanoparticle (CNC/CTAB/Ag) nanohybrid catalysts with small size and narrow particle distribution though a facile method for the reduction of 4-nitrophenol. The CNC/CTAB/Ag nanohybrid composite was produced by reduction of Ag^+

ions onto Ag nanoparticles in the presence of sodium borohydride solution, CNC, and CTAB surfactant with an alkyl chain and quaternary ammonium groups playing an important role in stabilizing and dispersing silver nanoparticles. UV-vis spectra showed a localized surface plasmon (LSP) resonance peak at 410 nm attributed to the silver nanoparticles, confirming their presence in the final product, with a highly homogeneous dispersion due to the presence of CTAB limiting their aggregation while improving their stability via electrostatic interactions between the hydrophilic heads of CTAB and the hydroxyl groups in silver. The results of the photocatalytic reaction experiments showed that the CNC/CTAB/Ag nanohybrid composite displayed higher photodegradation efficiency towards 4-nitrophenol compared to silver nanoparticles and CNC/Ag. The photo-reduction was completed within 20 min at room temperature. This higher catalytic performance was attributed to silver ions since the reduction of 4-nitrophenol did not occur in the presence of both CNC and CTAB. Recently, a novel Fe-Cu alloy coated cellulose nanocrystals (Fe-Cu@CNC) with strong antibacterial ability and efficient Pb^{2+} removal was synthesized via a simple oxidation/reduction reaction, as shown in Fig. 5.10 [25].

CNC, with rod-like shape and a diameter of around 20 nm, was randomly covered by both quasi-spherical zero-valent Fe^0 and Cu^0 sheets. XRD analysis showed that the crystallinity index of CNC decreased with metal nanoparticles addition due to the breakdown of hydrogen bonds between CNC macromolecules. The obtained nanocomposite also displayed lower thermal stability compared to the neat CNC and Fe-CNC, which suggested that Cu^0 had a negative effect on the composite thermal stability. Moreover, the Fe-Cu@CNC composite showed fast Pb^{2+} ions uptake, reaching a 71% removal within 5 min and 94%

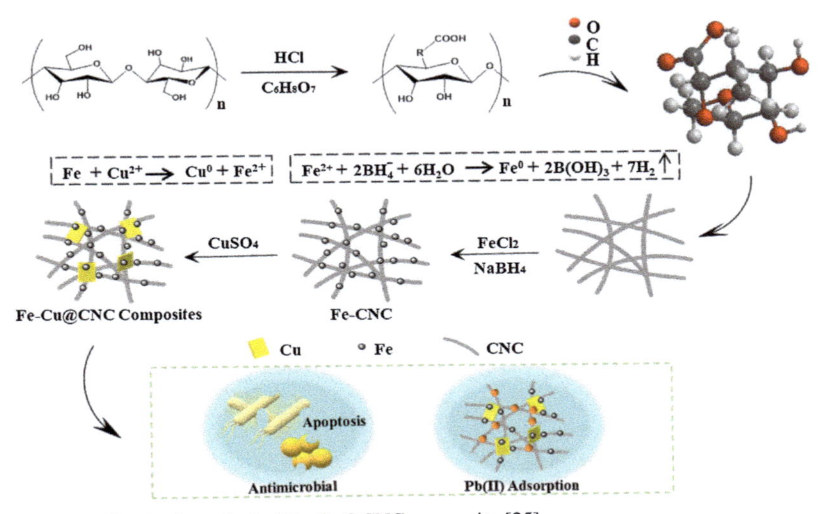

FIG. 5.10 Synthesis method of Fe-Cu@CNC composite [25].

after 1 h. The adsorption isotherms and kinetics showed that Pb^{2+} ions adsorption on the Fe-Cu@CNC composite strongly corresponded to the Langmuir and pseudo-second-order models, suggesting that the Pb^{2+} ions adsorption on the Fe-Cu@CNC was homogenous, monolayer, and strictly belong to a chemical adsorption process. The authors also investigated the antibacterial activity of the composite against both *Escherichia coli* and *Staphylococcus aureus* bacteria. It was observed that the Fe-Cu@CNC composite showed stronger antibacterial activity against both bacteria, with antibacterial ratios of 99.9% and 99.5% respectively, unlike the CNC alone that did not provide any effect. These results confirmed that the bactericidal activity was only due to the presence of both Fe^0 and Cu^0 in the Fe-Cu@CNC composite having a higher affinity toward gram-negative bacteria than gram-positive ones. This observation is in good agreement with other results in the literature attributing this difference to the membrane cell characterizing each species [55]. The peptidoglycan layer of gram-negative bacteria is thinner (2–3 nm) than that of gram-positive bacteria (30 nm) and is therefore easier to destroy by metallic ions released. Different results were observed Azizi et al. [56] using a series of ZnO-Ag/CNC hybrid materials with different silver loading content from 1 to 10 wt% against *Salmonella choleraesuis* and *Staphylococcus aureus* bacteria. The composite was successfully prepared via a precipitation method involving the absorption of Zn^{2+} cations onto the negatively charged hydroxyl functional groups through electrostatic interactions followed by the reduction of Ag^+ ions to Ag nanoparticles in alkaline suspension. The resulting composites exhibited small spherical particles with a narrow size distribution ranging from 34 to 8.8 nm with increasing Ag amount (Fig. 5.11), leading to high thermal stability, high specific surface

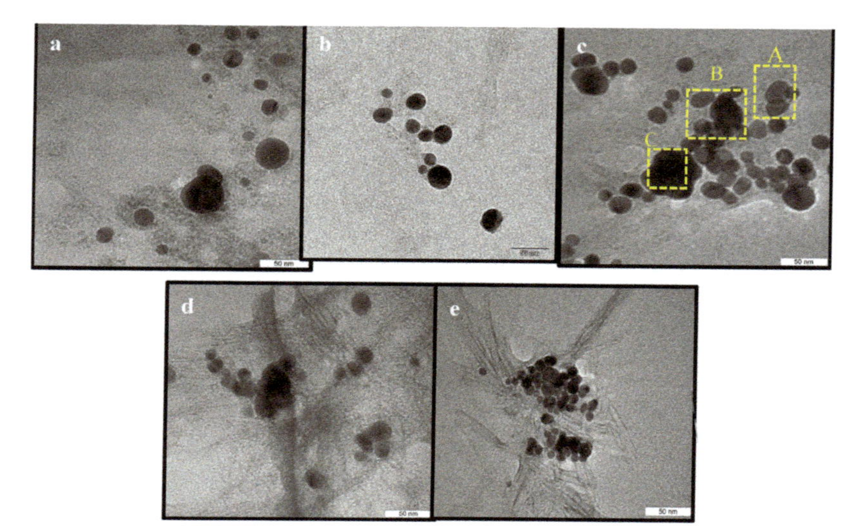

FIG. 5.11 TEM images of ZnO-Ag/CNC with different silver concentrations [56].

area, and excellent bactericidal activity against both bacterial species *Salmonella choleraesuis* and *Staphylococcus aureus* compared to CNC and ZnO-Ag. In addition, it was observed that the antibacterial ability of the ZnO-Ag/CNC increased with increasing Ag content. This was attributed to the small crystal size, large surface area of the heterostructure nanoparticles, the silver ions released, and the reactive oxygen species (ROS) generated by ZnO. Moreover, it was observed that the antibacterial ability of the ZnO-Ag/CNC samples was stronger against the Gram-positive *Staphylococcus aureus* compared to the Gram-negative *Salmonella choleraesuis*. The authors explained these results by the fact that the cell walls of Gram-negative bacteria contain an external lipopolysaccharide (LPS) membrane shielding the peptidoglycan layer. Furthermore, it helps bacteria to survive in surroundings where exterior materials can damage them.

Elfeky and his co-authors [13] prepared CNC/ZnO/CuO hybrid nanocomposite through in-situ solution casting technique, where the cellulose nanocrystal was extracted by acid-hydrolysis, while the ZnO/CuO nanostructures were synthesized via a sol-chemical method using $Zn(CH_3COO)_2{\cdot}2H_2O$ and $Cu(CH_3COO)_2{\cdot}2H_2O$ as a precursor for Zn and Cu, respectively. The resulting composite exhibited an average size ranging from 33.8 to 81.6 nm with various shapes, highly crystalline structure, and high dye degradation ability of 99.7% for Rose Bengal (RB) dye which was degraded with only 40 min of irradiation. Likewise, the resulting composite presented greater antibacterial activities against both Gram-positive (*Bacillus subtilis* and *Staphylococcus aureus*) and Gram-negative (*Escherichia coli* and *Pseudomonas aeruginosa*) bacteria, as a result of cell membrane damage. It was observed that the ZnO/CuO nanoparticles are the main responsible elements in the antibacterial activity of the composite, and unlike the results presented previously, it was observed that the inhibition efficiency of the composite had no relation with the type and nature of the bacteria. The CNC/ZnO/CuO nanocomposite showed high antibacterial activity toward *Bacillus subtilis*, which is Gram-positive followed by *Escherichia coli* (Gram-negative), *Staphylococcus aureus* (Gram-positive), and finally *Pseudomonas aeruginosa* (Gram-negative), but was rather attributed to the bacteria resistance and ZnO/CuO nanoparticles antibacterial selectivity. The authors also tested the larvicidal activity of CNC/ZnO/CuO hybrid nanocomposite against the third instar larvae of *Anopheles stephensi*. It was found that the lowest larval mortality was achieved by CNC (51.3%), followed by ZnO/CuO (70.3%), and finally the CNC/ZnO/CuO hybrid nanocomposite had the highest larval mortality (99.7%).

It was also found that CNC not only enhanced the adsorption capacity of semiconductor photocatalysts, but also extended their light response into the visible region due to the electron-rich hydroxyl and sulfuric groups on the CNC surface. The presence of such groups induced the formation of ligand-to-metal charge transfer (LMCT) complex improving its photo-response into the visible light region. In this context, Li and his coauthors successfully prepared TiO_2/CNC nanocomposites with anatase TiO_2 crystals and large specific surface area via a facile in situ hydrolysis method for the reduction of hexavalent chromium ions Cr(VI) under visible light [57]. SEM micrographs and EDX mapping (Fig. 5.12) revealed the distribution of a large amount of uniform TiO_2 nanocrystals with a diameter of less than 10 nm on the CNC surface, exhibiting a necklace-like structure with no aggregation. The effect of both reaction time and temperature on the resulting morphology of TiO_2/CNC

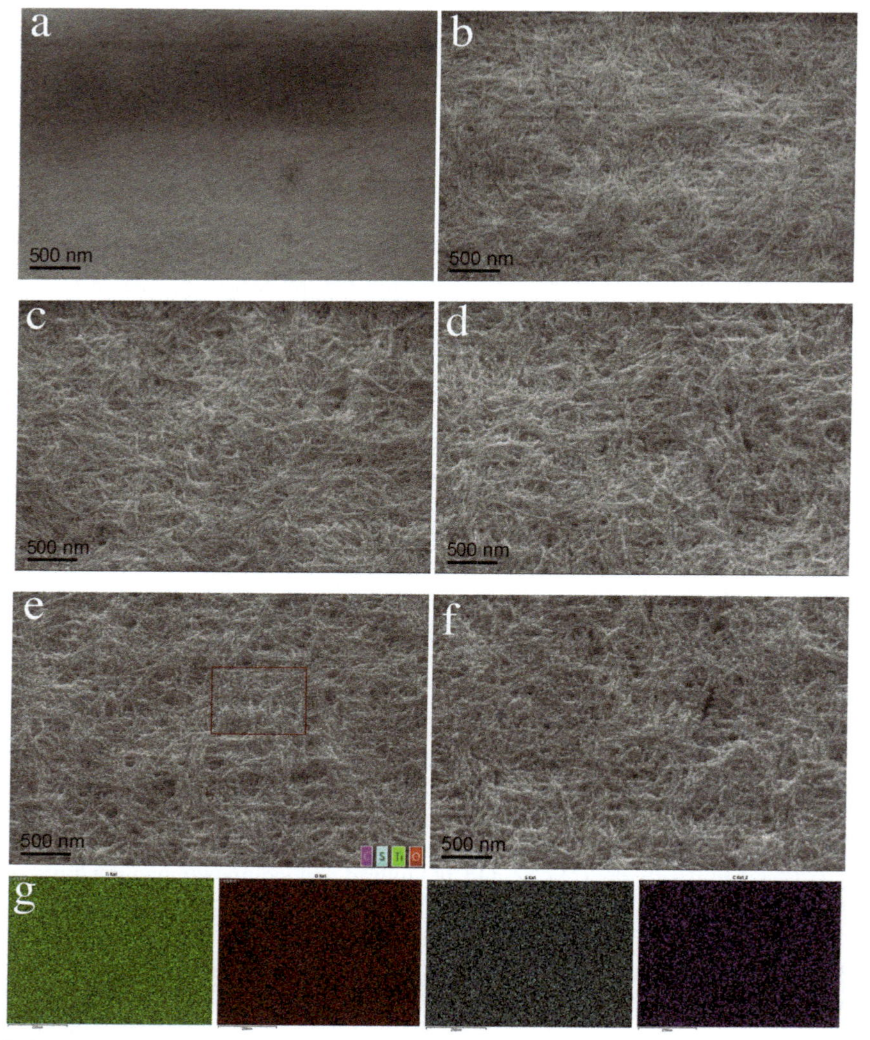

FIG. 5.12 (A) SEM images of pure CNC and (B) TiO_2/CNC composites obtained at 20 min, (C) 40 min, (D) 1 h, (E) 2 h, and (F) 3 h [57].

nanocomposites was investigated. It was observed that the optimum synthesis conditions for producing TiO_2/CNC nanocomposite with uniform TiO_2 nanocrystals were a reaction temperature of 70 °C, a reaction time of 2 h, and a CNC concentration of 0.1 wt%.

Moreover, TiO_2/CNC nanocomposite showed superior photocatalytic activity for Cr(VI) reduction under visible light as the Cr(VI) ions were completely removed after only 80 min. This efficient removal can be attributed to both the mesoporous structure of the composite and to its high specific surface area (112.7 m^2/g), which are extremely beneficial for photocatalysis applications. In addition, the photocatalyst maintained more than 90% Cr(VI) removal efficiency without significant decrease of photocatalytic activity after 5 cycles, suggesting that the adsorption sites can be regenerated by simple light irradiation.

5.3.3 CNC/magnetic hybrid materials

In recent years, several studies reported on the effectiveness of magnetic materials for the removal of different pollutants from sewage, due to their available active sites, specific surface area, pore volume, and pore size distribution [58]. CNC functionalizing with magnetic nanoparticles is a powerful strategy for easy recovery after treatment with the help of an external magnetic field. Different synthesis methods can be used for the preparation of CNC/magnetic hybrid materials. But the main ones are: co-precipitation and hydrothermal methods [59, 60]. In this context, $MnFe_2O_4$/CNC nanocomposites with magnetically recoverable photocatalytic performances have been prepared for the degradation of methylene blue dye in H_2O_2 aqueous solution, through a hydrothermal method using manganese carbonate and iron chloride hexahydrate as precursors (Fig. 5.13) [24]. The resulting nanoparticles exhibited lower band gap, higher specific surface, and smaller particle sizes than those of neat $MnFe_2O_4$, confirming the efficient role of CNC in suppressing $MnFe_2O_4$ nanoparticles aggregation leading to interesting benefits for its photocatalytic activity. A $MnFe_2O_4$/CNC composite containing 20 wt% of CNC produced a 99% degradation degree under natural sunlight, which represents more than 60% increase over the neat $MnFe_2O_4$ nanoparticles. This enhancement of MB degradation may be due to the higher specific surface area and $MnFe_2O_4$ smaller particle sizes providing more accessible active sites and promoting MB sorption, in addition to its lower ban gap shifting its response to visible light. In particularly, $MnFe_2O_4$/CNC composites, featuring excellent magnetic properties, exhibited excellent recycling catalytic performances.

Similarly, Dang et al. [61] prepared Ag@Fe_3O_4@cellulose nanocrystals nanocomposites via a facile and green microwave-assisted hydrothermal method for MB uptake. XRD analysis confirmed the successful incorporation of Ag and CNC in the Fe_3O_4 without disruption of its crystal structure, while SEM micrographs showed that the introduction of Ag and CNC preserved

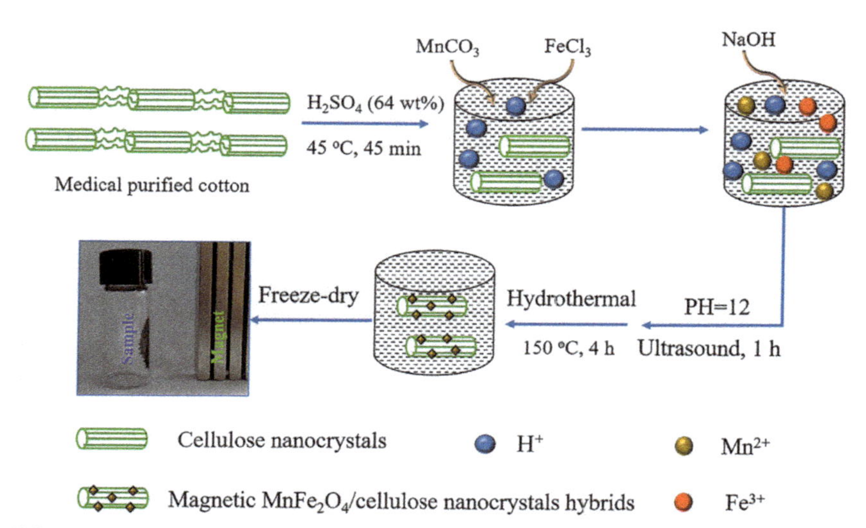

FIG. 5.13 Schematic illustration for the preparation of $MnFe_2O_4$/CNC nanocomposites [24].

the globular structure of Fe_3O_4 crystals, but increased the nanocomposite surface regurosity. Moreover, it was observed that longer reaction time were favored for the synthesis of more Ag and larger Ag@Fe_3O_4 particles (Fig. 5.14), due to the presence of CNC acting as both reductant and stabilizer. The magnetic hysteresis loops of the synthesized nanocomposites showed that the Ag@Fe_3O_4@CNC nanocomposites display both superparamagnetic and ferromagnetic behaviors, which can be ascribed to the presence of Fe_3O_4 nanoparticles which can be easily recovered upon its contact with a magnet. Furthermore, the as-prepared nanocomposite showed good adsorption performance for methylene blue dye and demonstrated high antimicrobial activities against *Escherichia coli* and *Staphylococcus aureus* bacteria compared to the neat CNC, due to the presence of Ag@Fe_3O_4 nanoparticles on the CNC external surface, confirming the role of Ag and Fe_3O_4 nanoparticles in both photocatalytic and antibacterial effects.

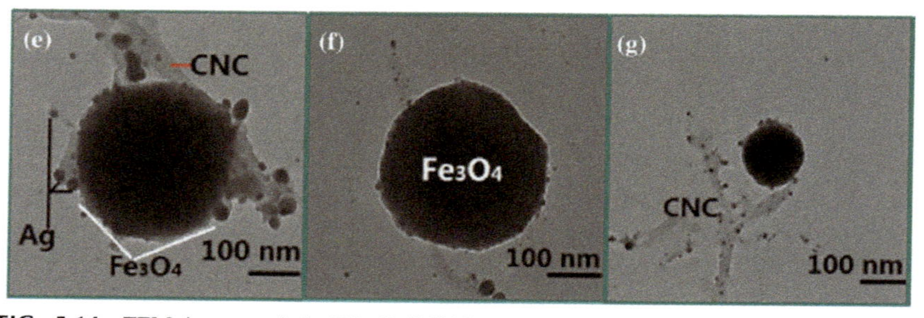

FIG. 5.14 TEM images of Ag@Fe_3O_4@CNC nanocomposites synthesized by microwave-hydrothermal method for 30 min [61].

Wang and his collaborators [62] fabricated a magnetic cellulose nanocrystal hybrid material (MCNC@Zn-BTC) composed of CNC, Fe_3O_4, and metal-organic framework (MOF) based on Zn(II) and benzene-1,3,5-tricarboxylic acid, via a simple mechanical agitation method, for Pb(II) ions uptake. The characterization results confirmed the successful preparation and distribution of Zn-BTC, with an average length of 800 nm, on the MCNC surface. The resulting composite exhibited fast magnetic separability as it was separated within only 30 s by an external magnetic field after a vigorous shake, due to the encapsulation of Fe_3O_4 nanoparticles. Moreover, it presented high adsorption capacity (558 mg/g) for Pb(II) ions, which can be ascribed to its high specific surface area and abundant hydroxyl and carboxyl groups on its surface having free pairs of electrons reacting with Pb(II) via coordinative bonds. Investigating the impact of different environmental parameters on Pb(II) adsorption showed that optimal Pb(II) ions adsorption can be achieved when the Pb(II) concentration, MCNC@Zn-BTC dose, pH, temperature, and adsorption time were 200 mg/L, 20 mg, 5.45, 298.2 K and 30 min, respectively. The removal efficiency remained above 80% after five recycled times.

5.3.4 CNC/carbonaceous hybrid materials

Carbonaceous materials are also widely used as supports for cellulose nanocrystals due to their abundance, chemical inertness, high surface area, adsorption capacity and excellent electronic, as well as high thermal and mechanical properties [63]. In this context, a novel three-dimensional graphene oxide/cellulose nanocrystals (GO/CNC) hybrid composite, with 30% of GO, was successfully synthesized and used as an adsorbent for the removal of ionic liquid [BMIM][Cl] from an aqueous solution [64]. SEM micrographs showed that the cellulose nanocrystals were well embedded within the graphene oxide nanosheets, indicating the successful synthesis of the composite. XRD analysis revealed some changes in the CNC XRD spectrum after GO introduction. The diffraction peak at 22.6 degree gradually became weak and the peak at 34.2 degree disappeared, indicating that the CNC crystalline structure was affected by GO introduction. On the other hand, it was observed that GO introduction can positively affect the adsorption behavior of the resulting composite. The addition of 30% GO resulted in an increase of [BMIM][Cl] removal rate from 2.7% to 55%. This large difference can be attributed to the homogenous distribution of rod-like CNC on GO nanosheets, which was confirmed by AFM analysis (Fig. 5.15), to the highly microporous structure of the composite or to the rich functional groups abundant on its surface. The equilibrium isotherm data fitted well with the Langmuir isotherms with a maximum sorption capacity of 0.455 mmol/g at 50 °C, while the adsorption kinetics were better described by the pseudo-second-order and the Elovich models [65], suggesting a monolayer adsorption of $[BMIM^+][Cl^-]$ on the three-dimensional structure via strong chemical interactions between the $[BMIM^+][Cl^-]$ cation and the oxygen-containing groups of the adsorbent.

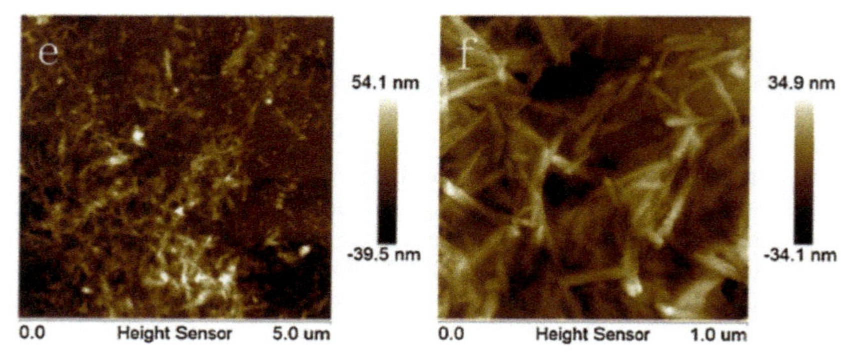

FIG. 5.15 AFM images GO/CNC composite [64].

Wu et al. [66] recently prepared a polyvinyl alcohol/cellulose nanocrystal/graphene composite aerogel (PCGA) for the methylene blue dye uptake from aqueous solutions. It was observed that the addition of PVA did not destroy the skeleton of CGA as both composites had the typical interconnected three-dimensional network structures showing porosity, disorder, and uneven network surface. However, the pore size distribution of both aerogels was different. The average pore size of CGA was more than 5 μm with mostly large holes, while the pore size of PCGA was within 1 to 2 μm with mostly mesopores and micropores (Fig. 5.16). The addition of PVA increased the solution viscosity, compacted the aerogel surface, and therefore decreased the pore size structure of PCGA. FTIR spectrum showed strong interaction between the PVA layer, the graphene sheet, and CNC, resulting in a more stable and uniform structure of PCGA. Moreover, it was observed that the incorporation of PVA improved the composite aerogel structure, increased the crystallinity, and decreased both thermal stability and specific surface area of the cellulose nanocrystal/graphene composite (CGA). Under the conditions of 120 min adsorption time, an initial MB solution pH value of 11, an initial MB concentration of 100 mg/L, and a PCGA addition of 20 mg, the removal rate of MB by PCGA reached up to the maximum value of 98%. The adsorption of MB by PCGA followed well the Langmuir and pseudo-second-order models, indicating that the adsorption of MB by PCGA was a monolayer via chemical adsorption.

Recently, CNC was modified by phosphorylation using two different solvents (water and molten urea) and coupled with multi-walled carbon nanotubes (MWCNT) for the adsorption of rare earth La(III) [67]. The phosphate groups had high affinity towards La(III) easily forming $LaPO_4$ by coordinate covalent bonds, while multi-wall carbon nanotube was added to improve both thermal and adsorption properties. The presence of carboxyl groups were also involved in the coordination of La(III). The resulting *p*-CNC(water)/MWCNT and *p*-CNC(urea)/MWCNT composites displayed a spiral twisted layered structure. SEM micrographs confirmed the successful preparation of the composite films

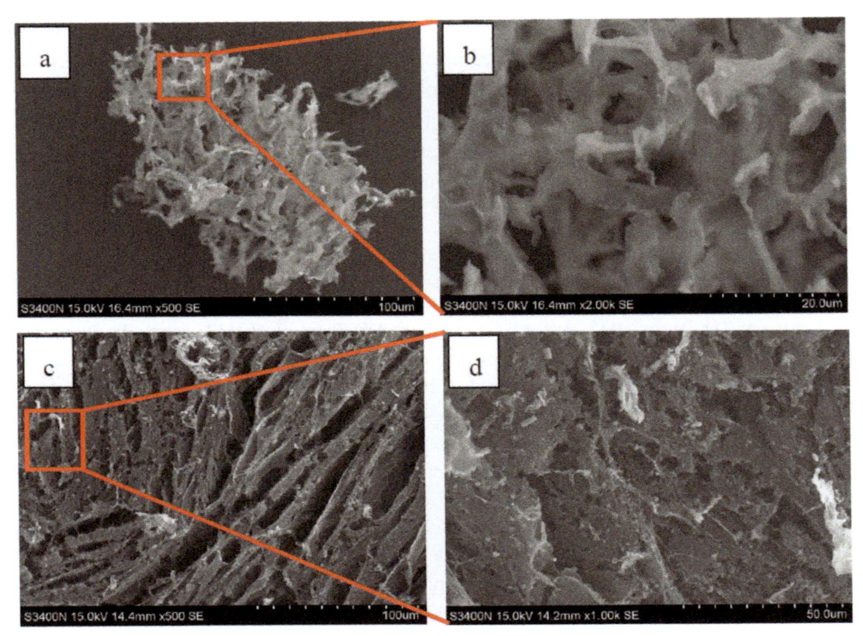

FIG. 5.16 SEM images of: (A,B) CGA and (C,D) PCGA [66].

and showed that the addition of carbon nanotubes did not affect the phosphorylated-CNC structure. The maximum adsorption capacity of *p*-CNC (urea)/MWCNT was 118 mg/g, which is six times higher than that of p-CNC (water)/MWCNT composite (32 mg/g). This enhancement was attributed to the large number of phosphate and carboxyl groups contributing to the coordination of La(III).

5.4 Conclusion

No one can deny the great potential of cellulose nanocrystals as effective adsorbents for the uptake of a wide range of organic and inorganic contaminants present in wastewater. CNC with its unique structure, high surface area, and a wide variety of functional groups can interact easily with pollutants, leading to their withdrawal. However, like all materials at the nanoscale, CNC suffers from agglomeration tendency and recovery difficulty that limit their applications and decrease their efficiency. Therefore, coupling CNC with other materials known for their interesting properties such as carbonaceous materials, bio/polymers, metals, or metal oxides has proven to be an effective way to overcome these problems and to widen their applicability. The interesting data presented in this work clearly demonstrate that all CNC based hybrid materials exhibited high performance and improving properties compared to the starting materials, making them more effective in removing pollutants and bacteria from wastewater. Therefore, although its forcefulness as a new environmentally-friendly adsorbent for a wide range of pollutants has been revealed in the laboratory scale more investigations are requisite for the practical transition from laboratory scale to industrial or real environmental applications, without forgetting the optimization of the preparation method while reducing time and energy-consumption.

References

[1] Šoštarić TD, Petrović MS, Pastor FT, Lončarević DR, Petrović JT, Milojković JV, Stojanović MD. Study of heavy metals biosorption on native and alkali-treated apricot shells and its application in wastewater treatment. J Mol Liq 2018;259:340–9.

[2] Abouzeid RE, Khiari R, El-Wakil N, Dufresne A. Current state and new trends in the use of cellulose nanomaterials for wastewater treatment. Biomacromolecules 2019;20:573–97.

[3] Kulkarni P, Dixit M. Sources of cellulose and their applications—a review. Int J Drug Formul Res 2015;2:19–38.

[4] Peng B, Yao Z, Wang X, Crombeen M, Gsweene D, Tam KC. Cellulose-based materials in wastewater treatment of petroleum industry. Green Energy Environ 2020;5:37–49.

[5] Syazwani N, Rahman A, Firdaus M, Baharin Y. Utilisation of natural cellulose fibres in wastewater treatment. Cellulose 2018;25:4887–903.

[6] Crini G, Lichtfouse E, Wilson LD, Morin-Crini N. Conventional and non-conventional adsorbents for wastewater treatment. Environ Chem Lett 2019;17:195–213.

[7] Varghese AG, Paul SA, Latha MS. Remediation of heavy metals and dyes from wastewater using cellulose-based adsorbents. Environ Chem Lett 2019;17:867–77.

[8] Olivera S, Muralidhara HB, Venkatesh K, Guna VK, Gopalakrishna K. Potential applications of cellulose and chitosan nanoparticles/composites in wastewater treatment: a review. Carbohydr Polym 2016;153:600–18.

[9] Jamshaid A, Hamid A, Muhammad N, Naseer A, Ghauri M, Iqbal J, Rafiq S, Shah NS. Cellulose-based materials for the removal of heavy metals from wastewater—an overview. ChemBioEng Rev 2017;4:240–56.

[10] Gupta VK, Carrott PJM, Singh R, Chaudhary M, Kushwaha S. Cellulose: a review as natural, modified and activated carbon adsorbent. Bioresour Technol 2016;216:1066–76.

[11] El Achaby M, Ruesgas-Ramon M, Fayoud N-EH, Figueroa-Espinoza MC, Maria Cruz Draoui K, Ben Youcef H. Bio-sourced porous cellulose microfibrils from coffee pulp for wastewater treatment. Cellulose 2019;26:3873–89.

[12] Zheng M, Wang P-L, Zhao S-W, Guo Y-R, Li L, Yuan F-L, Pan Q-J. Cellulose nanofiber induced self-assembly of zinc oxide nanoparticles: theoretical and experimental study on interfacial interaction. Carbohydr Polym 2018;195:525–33.

[13] Elfeky AS, Salem SS, Elzaref AS, Medhat E, Eladawy HA, Saeed AM, Awad MA, Abou-zeid RE, Fouda A. Multifunctional cellulose nanocrystal/metal oxide hybrid, photo-degradation, antibacterial and larvicidal activities. Carbohydr Polym 2019;230:115711.

[14] Chakrabarty A, Teramoto Y. Recent advances in nanocellulose composites with polymers: a guide for choosing partners and how to incorporate them. Polymers (Basel) 2018;10:517.

[15] Wang L, Okada K, Sodenaga M, Hikima Y, Ohshima M, Sekiguchi T, Yano H. Effect of surface modification on the dispersion, rheological behavior, crystallization kinetics, and foaming ability of polypropylene/cellulose nanofiber nanocomposites. Compos Sci Technol 2018;168:412–9.

[16] Curvello R, Raghuwanshi VS, Garnier G. Engineering nanocellulose hydrogels for biomedical applications. Adv Colloid Interf Sci 2019;267:47–61.
[17] Huang R, Liu Z, Sun B, Fatehi P. Preparation of dialdehyde cellulose nanocrystal as an adsorbent for creatinine. Can J Chem Eng 2016;94:1435–41.
[18] Mahfoudhi N, Boufi S. Nanocellulose as a novel nanostructured adsorbent for environmental remediation: a review. Cellulose 2017;24:1171–97.
[19] Villares A, Moreau C, Dammak A, Capron I, Cathala B. Kinetic aspects of the adsorption of xyloglucan onto cellulose nanocrystals. Soft Matter 2015;11:6472–81.
[20] Heinze T. Cellulose: structure and properties. Adv Polym Sci 2015;271:1–52.
[21] Tshikovhi A, Mishra SB, Mishra AK. Nanocellulose-based composites for the removal of contaminants from wastewater. Int J Biol Macromol 2020;152:616–32.
[22] Fan L, Lu Y, Yang L, Huang F, Ouyang X. Fabrication of polyethylenimine-functionalized sodium alginate/cellulose nanocrystal/polyvinyl alcohol core-shell microspheres ((PVA/SA/CNC)@PEI) for diclofenac sodium adsorption. J Colloid Interface Sci 2019;554:48–58.
[23] Wang G, Zhang J, Lin S, Xiao H, Yang Q, Chen S, Yan B, Gu Y. Environmentally friendly nanocomposites based on cellulose nanocrystals and polydopamine for rapid removal of organic dyes in aqueous solution. Cellulose 2019;27:2085–97.
[24] Zhan Y, Meng Y, Li W, Chen Z, Yan N, Li Y. Magnetic recoverable $MnFe_2O_4$/cellulose nanocrystal composites as an efficient catalyst for decomposition of methylene blue. Ind Crop Prod 2018;122:422–9.
[25] Chen L, Yu H, Deutschman C, Yang T, Tam KC. Novel design of Fe-Cu alloy coated cellulose nanocrystals with strong antibacterial ability and efficient Pb^{2+} removal. Carbohydr Polym 2020;234:115889.
[26] Lizundia E, Puglia D, Nguyen TD, Armentano I. Cellulose nanocrystal based multifunctional nanohybrids. Prog Mater Sci 2020;112:100668.
[27] Wang X, Zhang Y, Jiang H, Song Y, Zhou Z, Zhao H. Fabrication and characterization of nano-cellulose aerogels via supercritical CO_2 drying technology. Mater Lett 2016;183:179–82.
[28] Xie H, Du H, Yang X, Si C. Recent strategies in preparation of cellulose nanocrystals and cellulose nanofibrils derived from raw cellulose materials. Int J Polym Sci 2018;2018:1–25.
[29] Hamad WY. Cellulose nanocrystals: properties, production and applications. Chichester, UK: Wiley; 2017.
[30] Trache D, Hussin MH, Haafiz MKM, Thakur VK. Recent progress in cellulose nanocrystals: sources and production. Nanoscale 2017;9:1763–86.
[31] Nagalakshmaiah M, Rajinipriya M, Afrin S. Cellulose nanocrystals-based nanocomposites. In: Bio-based polymers and nanocomposites. Springer; 2019. p. 49–65.
[32] Carpenter AW, Lannoy D, Wiesner MR. Cellulose nanomaterials in water treatment technologies. Environ Sci Technol 2015;49:5277–87.
[33] Kargarzadeh H, Ahmad I, Abdullah I, Dufresne A, Zainudin SY, Sheltami RM. Effects of hydrolysis conditions on the morphology, crystallinity, and thermal stability of cellulose nanocrystals extracted from kenaf bast fibers. Cellulose 2012;19:855–66.
[34] Lin K, Enomae T, Chang F. Cellulose nanocrystal isolation from hardwood pulp using various hydrolysis conditions. Molecules 2019;24:1–15.
[35] Yu X, Tong S, Ge M, Wu L, Zuo J, Cao C, Song W. Adsorption of heavy metal ions from aqueous solution by carboxylated cellulose nanocrystals. J Environ Sci 2013;25:933–43.
[36] Rathod M, Haldar S, Basha S. Nanocrystalline cellulose for removal of tetracycline hydrochloride from water via biosorption: equilibrium, kinetic and thermodynamic studies. Ecol Eng 2015;84:240–9.

[37] Salama A, Shukry N, El-Sakhawy M. Carboxymethyl cellulose-G-poly(2-(dimethylamino) ethyl methacrylate) hydrogel as adsorbent for dye removal. Int J Biol Macromol 2015;73:72–5.
[38] Sehaqui H, Mautner A, de Larraya U, Pfenninger N, Tingaut P, Zimmermann T. Cationic cellulose nanofibers from waste pulp residues and their nitrate, fluoride, sulphate and phosphate adsorption properties. Carbohydr Polym 2016;135:334–40.
[39] Yang X, Liu H, Han F, Jiang S, Liu L, Xia Z. Fabrication of cellulose nanocrystal from Carex meyeriana Kunth and its application in the adsorption of methylene blue. Carbohydr Polym 2017;175:464–72.
[40] Abiaziem CV, Williams AB, Ibijoke A. Adsorption of lead ion from aqueous solution unto cellulose nanocrystal from cassava peel. J Phys Conf Ser 2019. https://doi.org/10.1088/1742-6596/1299/1/012122.
[41] Emam HE, Shaheen TI. Investigation into the role of surface modification of cellulose nanocrystals with succinic anhydride in dye removal. J Polym Environ 2019;27:2419–27.
[42] Rafieian F, Jonoobi M, Yu Q. A novel nanocomposite membrane containing modified cellulose nanocrystals for copper ion removal and dye adsorption from water. Cellulose 2019;26:3359–73.
[43] Tang J, Sisler J, Grishkewich N, Tam KC. Functionalization of cellulose nanocrystals for advanced applications. J Colloid Interface Sci 2017;494:397–409.
[44] Ranjbar D, Raeiszadeh M, Lewis L, Maclachlan MJ, Hatzikiriakos SG. Adsorptive removal of Congo red by surfactant modified cellulose nanocrystals: a kinetic, equilibrium, and mechanistic investigation. Cellulose 2020;27:3211–32.
[45] Batmaz R, Mohammed N, Zaman M, Minhas G, Merry R, Tam KC. Cellulose nanocrystals as promising adsorbents for the removal of cationic dyes. Cellulose 2014;21:1655–65.
[46] Sobolčiak P, Tanvir A, Popelka A, Moffat J, Mahmoud KA, Krupa I. The preparation, properties and applications of electrospun co-polyamide 6,12 membranes modified by cellulose nanocrystals. Mater Des 2017;132:314–23.
[47] Favier V, Chanzy H, Cavaille JY. Polymer nanocomposites reinforced by cellulose whiskers. Macromolecules 1995;28:6365–7.
[48] Removal PII, Chen Y, Li Q, Li Y, Zhang Q, Huang J, Wu Q. Fabrication of cellulose nanocrystal-g-poly (acrylic acid-co-acrylamide) aerogels for efficient Pb(II) removal. Polymers (Basel) 2020;12:333.
[49] Zhao F, Yin D, Tam KC, Sillanpää M. Polyethylenimine-cross-linked cellulose nanocrystals for highly efficient recovery of rare earth elements from water and a mechanism study. Green Chem 2017;19:4816–28.
[50] Sips R. On the structure of a catalyst surface. J Chem Phys 1948;16:490–5.
[51] Hu D, Jiang R, Wang N, Xu H, Wang Y. Adsorption of diclofenac sodium on bilayer amino-functionalized cellulose nanocrystals/chitosan composite. J Hazard Mater 2019;369:483–93.
[52] Eisa WH, Abdelgawad AM, Rojas OJ. Solid-state synthesis of metal nanoparticles supported on cellulose nanocrystals and their catalytic activity. ACS Sustain Chem Eng 2018;6:3974–83.
[53] Kaushik M, Moores A. Review: nanocelluloses as versatile supports for metal nanoparticles and their applications in catalysis. Green Chem 2016;18:622–37.
[54] An X, Long Y, Ni Y. Cellulose nanocrystal/hexadecyltrimethylammonium bromide/silver nanoparticle composite as a catalyst for reduction of 4-Nitrophenol. Carbohydr Polym 2017;156:253–8.
[55] Pinto RJB, Daina S, Sadocco P, Neto CP, Trindade T. Antibacterial activity of nanocomposites of copper and cellulose. Biomed Res Int 2013;2013:1–6.

[56] Azizi S, Ahmad MB, Hussein MZ, Ibrahim NA. Synthesis, antibacterial and thermal studies of cellulose nanocrystal stabilized ZnO-Ag heterostructure nanoparticles. Molecules 2013; 18:6269–80.

[57] Li Y, Zhang J, Zhan C, Kong F, Li W, Yang C, et al. Facile synthesis of TiO_2/CNC nanocomposites for enhanced Cr(VI) photoreduction: synergistic roles of cellulose nanocrystals. Carbohydr Polym 2020;233:115838.

[58] Hassan M, Naidu R, Du J, Liu Y, Qi F. Critical review of magnetic biosorbents: their preparation, application, and regeneration for wastewater treatment. Sci Total Environ 2019;702:134893.

[59] Majidi S, Sehrig FZ, Farkhani SM, Goloujeh MS, Akbarzadeh A. Current methods for synthesis of magnetic nanoparticles. Artif Cells Nanomed Biotechnol 2016;44:722–34.

[60] Ali N, Zaman H, Bilal M, Shah A, Nazir MS, HMN I. Environmental perspectives of interfacially active and magnetically recoverable composite materials—a review. Sci Total Environ 2019;670:523–38.

[61] Dong Y, Liu S, Liu Y, Meng L, Ma M. Ag@Fe_3O_4@cellulose nanocrystals nanocomposites: microwave-assisted hydrothermal synthesis, antimicrobial properties, and good adsorption of dye solution. J Mater Sci 2017;52:8219–30.

[62] Wang N, Ouyang X, Yang L, Omer AM. Fabrication of a magnetic cellulose nanocrystal/metal-organic framework composite for removal of Pb(II) from water. ACS Sustain Chem Eng 2017;5:10447–58.

[63] Srikanth B, Goutham R, Badri Narayan R, Ramprasath A, Gopinath KP, Sankaranarayanan AR. Recent advancements in supporting materials for immobilised photocatalytic applications in waste water treatment. J Environ Manag 2017;200:60–78.

[64] Zhou H, Gao B, Zhou Y, Qiao H, Gao W, Qu H, Liu S, Zhang Q, Liu X. Facile preparation of 3D GO/CNCs composite with adsorption performance towards [BMIM][Cl] from aqueous solution. J Hazard Mater 2017;337:27–33.

[65] Elovich SY, Larinov OG. Theory of adsorption from solutions of non electrolytes on solid (I) equation adsorption from solutions and the analysis of its simplest form, (II) verification of the equation of adsorption isotherm from solutions. Izv Akad Nauk SSSR, Otd Khim 1962;2:209–16.

[66] Wu Y, Wu X, Yang F, Xu L, Sun M. Study on the preparation and adsorption property of polyvinyl alcohol/cellulose nanocrystal/graphene composite aerogels (PCGAs). J Renew Mater 2019;7:1181–95.

[67] Zhang Y, Zheng X. Phosphorylated-CNC/MWCNT thin films-toward efficient adsorption of rare earth La (III). Cellulose 2020;27:3379–90.

Chapter 6

Hybrid nanocomposites based on cellulose nanocrystals/ nanofibrils and titanium oxide: Wastewater treatment

Ateeq Rahman[a], V.S.R. Rajasekhar Pullabhotla[b], Likius Daniel[a], and Veikko Uahengo[a]
[a]*Faculty of Science, Department of Chemistry and Biochemistry, University of Namibia, Windhoek, Namibia,* [b]*Department of Chemistry, University of Zululand, Richards Bay, South Africa*

6.1 Introduction

Water scarcity is one of the major concerns to be faced by the majority of countries around the world. The increasing pollution level and the amount of new pollutants emerging in different water sources, from seawater, municipal, and industrial wastewater, are promoting to explore the efficient and specific water reuse, recycling, and treatment methods to assure that the appropriate water quality is required for human consumption. Hence, the pollutants present in wastewater are classified into three broad groups: microorganisms (bacteria and fungi), viruses, and toxic organic pollutants (pharmaceuticals, tanning, dyes, pesticides, petroleum products, personal care products, endocrine disruptors, and detergents). For example cadmium, lead, cobalt, copper, mercury, chromium, nickel, selenium, and zinc fertilizers, nitrates, chlorides, sulfates phosphates, and radionuclides are carcinogenic to human beings if consumed in high quantities leading to many diseases (Hlongwane et al. [1]; Peters [2]). The conventional cellulose-based adsorbents are challenging to recover from treated wastewater. Hence the reclamation step generally requires filtration or high-speed centrifugation. In comparison to conventional method, magnetic cellulose nanocomposites can easily overcome the recovery issue by application of an external magnetic field. As a result, magnetic nanomaterials have drawn increasing attention. When nanocellulose is incorporated with other magnetic nanomaterials, it produces exceptional composite adsorbent with magnetic properties (Wang et al. [3]). For example, a core-shell cellulose

Cellulose Nanocrystal/Nanoparticles Hybrid Nanocomposites. https://doi.org/10.1016/B978-0-12-822906-4.00004-9

magnetite Fe_3O_4 polymeric ionic liquid magnetic biosorbent has been developed and applied for congo red adsorption. A very novel property of cellulose is that it can also act as a stabilizer for Fe_3O_4 nanoparticles to prevent crystal growth and aggregation of particles (Shak et al. [4]). However, with the higher solubility and bioavailability rate of heavy metals in aquatic environments, they can be absorbed by living organisms. The removal of pollutants from water resources is normally carried out by adsorption, flocculation, or catalytic approaches (Macedo et al. [5]). The traditional well known activated carbons *Acacia Pods*, (Rahman et al. [6]) charcoal and activated carbon (Hager [7]; Sun [8]), carbon nanotubes (Fei et al. [9]), activated sludge (Lian et al. [10]), bamboo charcoal, and graphene oxide acacia pods (Rahman et al. [6]) are traditionally used as effective adsorbents for the removal of pollutant's due to their physicochemical characteristics (porosity and surface chemistry) (Macedo et al. [5]). Nonetheless, adsorption only transfers the pollutants from effluents to the adsorbents, which must be later regenerated using up time and resources. From these perspectives, advanced oxidation processes (AOP) have proven to be efficient for water purification because of total mineralization of pollutants (to CO_2 and H_2O) or leading to the formation of less toxic compounds followed by generating oxidizing radicals such as hydroxyls ($HO^{\bullet}$). This radical has a high oxidation potential (2.8 eV) and is able to oxidize recalcitrant organic compounds (Legrini et al. [11]). Numerous technologies are possible for water purification via the generation of hydroxyl radicals (Poyatos et al. [12]). Hence, it is essential to treat metal contaminated wastewater before it is discharged into the environment. Numerous treatment technologies are available to eliminate the pollutants' concentrations in wastewater including ammonia, hydrochloric acid, alum, ferric salts, corrosion control chemicals, chemical oxidation chlorine, chlorine dioxide, potassium permanganate, ozone or/and hydrogen peroxide, semiconductors (titanium dioxide or zinc oxide) or catalysts (transition metal ions), which can lead to the recontamination of water, as well as irradiation (ultraviolet and/or visible, sunlight, or ultrasounds) (Hokkanen et al. [13]). Pathogens are major issues for human health in developing and underdeveloped countries, where most diseases are spread from water acting as a carrier for microbes or microorganisms which results into fever, cholera, typhoid, diarrhea, sanitation, water polluted by microorganisms such as Algae, *bacterium Escherichia coli, Cryptosporidium* and *Giardia lamblia*, pathogenic bacteria, pathogenic viruses, and pathogenic protozoa. About 4 million people die every year due to contact with microorganisms, mostly due to contaminated water from microorganisms (Fisher et al. [14]). Mercury and methyl mercury, which is consumed by fish, is a toxic chemical as man consumes fish in many parts of the world and likely to get infected by mercury leading to dreadful diseases (Henriques et al. [15]). Radioactive elements, such as cesium, plutonium, and uranium (Minnesota Department of Health [16]) is also of concern. These radionuclides enter from the soil due to the disposal and storage of radioactive wastes, or from mining practices of phosphorus or uranium. Some people drink water contaminated with alpha emitters, such as uranium or radium, in excess over the past years increasing the risk of getting cancer (Minnesota Department of Health [16]). There is a clear need for the development of cost-effective stable materials and challenges facing humanity so that high quality water is delivered to the people facing challenges in under developed and developing countries on poor unsafe water, poor sanitation, and insufficient hygiene. However, these methods differ in their efficiency and economic parameters. The main advantages and disadvantages of the various physicochemical methods for water treatment are presented in Table 6.1 (Minnesota Department of Health [16]). This chapter highlight's the key constituents on the recent developments of nanocellulose hybrid materials and process for the treatment of waste water that are afflicted by contaminants, such as toxic metals, organic and inorganic compounds, drugs, bacteria, and viruses. After the introduction on water sources and pollutants, the chapter provides a brief summary on the current water purification processes, which will benefit the reader's understanding and evaluating nanomaterials compared against current technologies. Application possibilities are not limited. However, this chapter will mainly focus on nanocellulose hybrids and nanocellulose-TiO_2 as examples of applications in waste water purification (Yu et al. [17]). Cellulose and nanocellulose (NC) are for a few years of scientific interest in waste water treatment. Furthermore, cellulose-based materials have been around for over 16 years (Ortelli et al. [18]; Marinho et al. [19]). A total of 475 patents were granted depending on the type of cellulose: nanocellulose (112 patents), cellulose nanocrystals (CNC) (38 patents), microfibrillated cellulose (125 patents), bacterial nanocellulose

TABLE 6.1 Advantages and disadvantages of the various physicochemical methods of water treatment (Mannesota Department of Health [16]).

Treatment method	Advantages	Disadvantages
Chemical Precipitation	Low capital cost simple operation	Sludge generation, extra operational cost for sludge disposal
Membrane filtration	Small space requirement, low pressure, high separation selectivity	Small space requirement, low pressure, high separation selectivity
Electrodialysis	High separation selectivity	High operational cost due to membrane fouling and energy consumption
Adsorption	Low cost, easy operating conditions, high metal binding capacities	Low selectivity, production of waste products

(5 patents), cellulose nanowhiskers (5 patents), and cellulose nanofibers (CNF) (190 patents) (Wieh [20]). Hence, one fifth of all the NC patents filed are related to environmental applications including membrane technology, with a total of 97 patents linking the above keywords describing NC with at least one of the following terms: membrane, adsorbent, environment, and filter (Wieh [20]). These celluloses are obtained in different forms: crystallites, nanocrystals, and whiskers which can be extracted from wood tissues 40–45%, linen fibers 60–85%, cotton 95–98%, non-wood fibers, industrial algae, tunicates, and agro-biomass, among other sources (Liu et al. [21]). Hence, wood and cotton are the main raw materials for the commercial production of cellulose (Table 6.2). Cellulose, in its natural state, serves as a structural material within the complex architecture of plant cell walls with variation in its content.

Large amount of cellulose is present in banana stalks enabling its use as an adsorbent for heavy metals after chemical initiation grafting. The adsorbent (polyacrylamide-grafted bearing sulfur COOH) has a carboxylate functional group at its chain end, which was synthesized by (Fe_2þ-H_2O_2)-initiated graft copolymerization of acrylamide onto banana stalks, followed by succinic anhydride functionalization (Shibi et al. [23]). Synthetic wastewater samples were

TABLE 6.2 Composition of cellulose in wood [13, 22].

Source	% Cellulose
Leaf fibers	40–50
Sisal fibers	55–73
Blast fibers	Flax 70–75
Hemp	75–80
Jute	60–65
Ramie	70–75
Kenaf	47–57 in canes
Bamboo	40–55
Bagasse	33–45
In cereal straw	Barley 48
Oat	44–53
Rice	43–49
Rye	50–54
Wheat	49–54
Cotton seed	90–99

treated with the adsorbent to determine its efficiency in removing Pb(II) and Cd (II) ions from industrial wastewaters. The maximum uptake of Pb(II) and Cd(II) from their aqueous solutions was found to be 99.8% and 90.1% respectively, at an initial concentration of 25 mg/L and pH = 6.5. The adsorbed Pb(II) and Cd (II) ions were efficiently desorbed by 0.2 M HCl and PGBS-COOH was reused after regeneration. The study examined the effectiveness of a new adsorbent prepared from banana stalks, one of the main available lignocellulosic agro-wastes, in removing Pb(II) and Cd(II) ions from aqueous solutions. The same adsorbent was used to eliminate Hg(II) from a water solution (Shibi [24]). The maximum adsorption capacity of the adsorbent for Hg(II) was found to be 138.0 mg/g. The nanocelluloses, elemental nano-sized constituents of plant fibers, have developed an extra character compare to conventional cellulose fibers due to their high surface area, high aspect ratio, and high Young's modulus of 145 GPa resulting from high crystallinity (Akil et al. [25]; Jin et al. [26]). Furthermore, as natural biomass materials, nanocelluloses are biodegradable, biocompatible, and renewable (Jin et al. [26]). The lateral size of cellulose molecule chains is around 0.3 nm, and these chains forms a number of bundles of elongated fibrils with nano-scale diameters. The cellulose chains are stabilized horizontally by hydrogen bonds between their hydroxyl groups (Zhang et al. [27]). Surface modification of the nanocellulose was achieved by adding specific groups such as carboxyl (Donia et al. [28]; Yu [29]), amine (Sun et al. [30]; Singh et al. [31]), ammonium (Lue et al. [32]), and xanthate (Pillai et al. [33]) on the surface of cellulose. Cellulose nanocrystal is an abundant source of nanomaterial that are extracted from many renewable resources (Dufresne [34]). To develop an efficient and sustainable multifunctional smart devices because of its physical properties, biodegradability, biocompatibility, and low cytotoxicity, making it an excellent candidate for a variety of applications such as water purification, drug release, paper, food package, personnel care, light weight composites for aerospace, automotive, building constructions, biomedical, and energy applications (Li et al. [35], Jorfi Foster et al. [36], Rojo et al. [37]). The physical, chemical and mechanical properties of natural fibers are influenced by several factors such as chemical composition, internal structure, cell dimension, and microbial angle, which are wide-ranging based on different plant species, as well as different parts of the plant (Ahmad et al. [38]). Table 6.3 presents the investigations and uses of cellulose-based nanostructured photocatalyst hybrids due to the nanocellulose hydrophilic nature, insolubility in neutral water pH conditions, and presence of abundant OH functional groups on their surface leading to high potential for water filtration membranes. The development of economic and ecofriendly process is necessary from a global demand of products (Ahmad et al. [38]). But the projections for nanocellulose-based hybrid adsorbents in wastewater treatment have never been discussed. Novel practical challenges encountered for implementation (treatment duration, upscaling, and material toxicology) for some cases and different parameters for process development must be more focused. Additionally, the fabrication of nanocellulose-based adsorbents is mostly developed on CNF, and limited when it comes to CNC (Shak et al. [4]).

TABLE 6.3 The method of cellulose nanofibers isolation from different cellulose sources.

Method of isolation	Cellulose sources	Average size
Mechanical treatment	Wood powder	Diameter: 15 nm
Acid hydrolysis	White and naturally colored	Diameter: 6–18 nm
	Cotton fibers	Length: 85–225 nm
Homogenization	Microcrystalline cellulose	Diameter: 28–100 nm
Enzymatic hydrolysis	Bleached sulfite softwood	Diameter: 5–6 nm
Assisted mechanical shearing	Cellulose pulp	
Ultra sound treatment	Bamboo fiber	Diameter: 30–80 nm
		Length: >1 mm
Electroplating	Fibrous cellulose and Surgical cotton batting	Diameter: 250–750 nm
TEMPO-Mediated Oxidation	Hardwood bleached Kraft pulp	Diameter: 3–4 nm
		Length: A few microns
High pressure homogenization Grinding		
Cryocrushing		
High-intense ultra sonication		
Steam explosion process	Pine apple leaves	
Ball milling	Hard wood	Diameter: 10–15 nm
Pretreatment	Acacia pod's	Diameter: 7–10 nm

Table 6.3 presents the applications mostly including sewage or by other wastes related to human and animal (FAO [39]). However, it is of the utmost importance to remove such contaminants from water so that it is drinkable.

6.2 Characterization of nanocellulose (cellulose nanocrystals and cellulose nanofibrils)

Nanocellulose/nanocrystals and nanofibers characterization plays a pivotal role for the assessment of a fiber dealumination process. A plethora of morphological systems, such as scanning electron microscope (SEM), transmission electron microscope (TEM), atomic force microscopy (AFM), and field emission SEM/TEM, are used to study the length distribution, surface modification, and optical properties of nanocellulose. Apart from these instruments, other instruments include ^{13}C NMR (nuclear magnetic resonance), wide angle X-ray scattering, as well as Raman and Fourier transform infrared (FTIR) spectroscopy, which are used to characterize the dimensions, crystallinity, and chemical structure of nanocellulose. In general, the diameter of CNF is from 20 to 200 nm, while the diameters of individual CNF remains around 320 nm (Sassi and Chanzy [40]).

6.3 Treatment of contaminated water with nanocellulose/ nanocellulose based nanohybrid composites

Nanocellulose fibers have a long rod-like elongated nano-fibrillated morphology with an average grain size of 6 nm. Hookanen et al. [22] reported on the preparation of nanocellulose fibers with high removal efficiency of 9.7 mg/g Cd (II), 9.4 mg/g Pb(II), and 8.6 mg/g Ni (II) ions. From the regeneration studies, the results showed that the nanocellulose fibers could be successively used for up to three cycles. Nanocellulose based nanohybrid composites can be found in environmental applications. Aminopropyl triethoxysilane removal of Ni (Hookanen et al. [22]), phosphorylated nanocellulose removal of Ag(I), Cu (I), F(II) [21], carboxylate-modified cellulose nanocrystal (CNC) for the removal of crystal white violet, methylene blue, malachite green, and basic fuchsin (Qiao et al. [41]), and sulfonated cellulose for Fe(II), Pb(II), and Cu (II) (Dong et al. [42]). But nanocellulose-amino interconnected by maleic acid supported on magnetite was used for the removal of arsenic (Taleb et al. [43]). ZnO/cellulose nanocomposite for the removal of photocatalytic degradation of methylene blue. Lafeatshe et al. [44] and Gao et al. [45] have reported on cellulose fiber-supported zinc phthalocyanine for photocatalytic degradation of basic green. Recently, focus was devoted on membrane technology for waste water treatment to remove urea (Cruz-Tato et al. [46]) and carbon dioxide capture (Kim et al. [47]). Finally, the removal of heavy metals using ion exchange of Cd(II) ions and oil adsorption on nanocellulose was reported (Chitpong and Husson [48]).

6.4 Removal of oil from waste water

Oil pollution is related to oil spills from large tankers. However, there are additional sources of oil pollution that collectively discharge more oil into water than the major oil spills. There are several adverse effects occurring when large amount of oil is accumulated on the surface of water. The most common case is the spreading of the oil over a large water surface area since the oil spreads very rapidly, with lighter oils, like gasoline, spreading faster than heavy crude oils. The presence of wind and warm temperatures will also cause the oil to spread faster over the water's surface. Oil pollution usually damages the ecosystems, including plants and animals, and contaminate water for drinking and other purposes (Rahman et al. [6]). One of the most important drawbacks of oil spill in ecosystem is affecting the birds. When the feathers of birds and marine animals become coated with oil and when the animal furs are covered with oil, they can no longer insulate themselves against the cold water, and the birds have difficulty flying. However, plants growing in or near the water can be harmed by oil pollution. In addition, oil spill can block the sunlight that plants need for photosynthesis, which kills the plants that are grown in water (Korhonen et al. [49]). Another exciting work was reported by synthesizing sponge-like nanocellulose-based carbon aerogels via a heat treatment of cellulose microfibril aerogel used for oil absorption (Meng et al. [50]). It was shown that this particular aerogel was useful as a highly porous oil absorbent (99% of porosity) with an ultra-light density (0.01 g/cm^3), hydrophobic properties (144° static contact angle), a fast absorption rate, and extensive reusability. Their research indicated that carbon aerogels heat treated at 700°C showed higher oil absorption capacity for various types of oils. This was attributed to the fact that a 3D-network structure was formed by the entanglement of carbonized cellulose fibrils and their large surface energy. Meng et al. [50] also recommended that the sponge-like nanocellulose based carbon aerogels would be perfect for oil adsorption and applicable for waste water treatment. Zhang et al. [27] reported the fabrication of a highly efficient cellulose-based aerogel exhibiting high mechanical strength via a novel method. The synthesized aerogel structure was a perfect 3D skeleton and interconnected pores similar to a honeycomb, exhibiting outstanding oil/water selectivity. Furthermore, these aerogels exhibited stable super-hydrophobic and superoleophilic properties, even under a strong mechanical abrasion. The excellent activity of cellulose aerogels was also shown which can be reused up to 30 cycles (Zhang et al. [27]). Wang et al. [51] reported an ultra-lightweight, elastic, cost-effective, and highly recyclable super absorbent from microfibrillated cellulose fibers for oil spillage cleanup. They developed a new method via a simple micro-fibrillation treatment and freeze drying to produce a low-cost, environmentally friendly, and highly recyclable super absorbent from renewable cellulose fibers. Wang et al. [51] demonstrated that the hydrophobic superabsorbent could selectively absorb oil from an oil-water mixture, and showed an ultra-high absorption capacity of 88–228 g/g, which was comparable to those of other novel carbon-based super absorbents. A vital property was the superabsorbent capacity showing excellent flexibility which could be repeatedly squeezed without structure failure (more than 30 times) (Wang et al. [51]). Jin et al. [26] reported that super hydrophobic and super oleophobic nanocellulose aerogels consist of fibrillary networks and aggregates with structures at different-length scales supporting substantial load on the water surface, and also on oils. They showed that the load carrying capacity of the aerogels might be at least one order of magnitude larger than the weight of the aerogel itself. They suggested that this would open up a new platform for carriers or for coatings that have large load-bearing capability on various organic and aqueous liquids, including oil-polluted water (Jin et al. [26]). Among these various methods, adsorption is a greater and widely used method owing to its accessibility, environmentally benign nature, and high efficiency. In fact, there are several chemical or physical sorbent materials, including activated carbons (Rakić [52]; Carrales-Alvarado et al. [53]) multiwalled carbon nanotubes (Fei [9]), activated sludge (Xu et al. [54]), bamboo charcoal (Liao et al. [55]), and graphene oxide (Ghadim et al. [56]), which have been applied for removing antibiotics from aqueous solutions. Nevertheless, they suffer from low removal capacities and difficult separation leading to secondary environmental pollution and unsatisfactory recycling ability significantly limiting these adsorbents in practical applications.

6.4.1 Removal of drugs with cellulose nanohybrid fibrils

The worldwide presence of antibiotics in water resources has raised alarms for potentially negative effects on aquatic ecosystems such as fish egg production inhibition and sex reversal of males, as well as human health (Richardson et al. [57]; Johnson et al. [58]). Therefore, it is urgent to put forward some rational and feasible recommendations for antibiotic pollution control. In recent times, antibiotics can be removed by a variety of methods such as adsorption (Zhang et al. [59]), catalytic degradation (Zhao et al. [60]), photo-catalytic degradation (Porras et al. [61]), advanced oxidation, and biodegradation (Lu et al. [32]). Therefore, it is urgent to develop new adsorbents with high reusability and efficiency to remove antibiotics in water. But graphene oxide has a single atomic layer of sp^2 hybridized carbon organized in a honeycomb structure, while graphene oxide (GO) nanosheets have unusual properties such as excellent mechanical properties. Both materials have been applied for several applications (Zhu et al. [62]; Park et al. [63]). Over the las few years, a great deal of work was reported showing that GO might be a favorable material to adsorb pollutants from water due to its extreme hydrophilicity, ultrahigh theoretical surface area, and abundant surface oxygen-containing groups (Wang et al. [64]). In previous studies, GO-based materials were used as an efficient adsorbent for antibiotics like tetracyclines (Pierre et al. [65]), dyes like methylene blue and basic red (Rakic et al. [52]) and triphenylmethane (Sun et al. [66]), as well as heavy metal ions such as copper, zinc, cadmium, and lead (Tan et al. [67]). Nevertheless, less research was devoted to study the potential application for removal of multiple antibiotics in water (Alexis et al. [68]). Quifang et al. [69] reported the fabrication of cellulose nanofibril/graphene oxide hybrid (CNF/GO) aerogel via a one-step ultrasonication method for adsorptive removal of 21 categories of antibiotics in water. The prepared CNF/GO aerogel has an interconnected 3D network microstructure, in which GO nanosheets with a 2D structure were well bonded with CNF through hydrogen bonds. The aerogel displayed superior adsorption capacity toward the antibiotics. Hence, the removal efficiencies (R%) of the antibiotics were above 69% and the sequence of six categories of antibiotics according to the adsorption efficiency was as follows: tetracyclines > quinolones > sulfonamides > chloramphenicols > β-lactams > macrolides. The adsorption mechanism was suggested to be electrostatic attraction, p-π interaction, π-π interaction, and hydrogen bonds. Hence the adsorption capacities of CNF/GO aerogel were as follows: 418.7 mg/g for chloramphenicol, 291.8 mg/g for β-lactams, 227.3 mg/g for macrolides, 128.3 mg/g for quinolones, and 230.7 mg/g for tetracyclines as calculated by Langmuir isotherms. In addition, the regenerated aerogels could still be repeatedly used after ten cycles without obvious degradation (Quifang et al. [69]).

6.4.2 Separation processes and wastewater treatment

Cellulose-based functional materials are frequently applied in separations, mainly in water treatment. Bio-polymers such as chitosan, chitin, and lignin are well known to adsorb heavy metal ions from aqueous solutions. A cellulose-based functional material evaluated the excellent adsorption capacity for water treatment methods (Li et al. [70]; Karim et al. [71]; Voisin et al. [72]). Ever since nanotechnology has emerged, new technologies have been applied as a great potential for the development of economic and efficient pollution prevention, treatment, and cleanup. Hence, two applications for cellulose nanostructures in this area have created interest as active absorbent materials for contaminants and as stabilizers for other active particles (Carpenter et al. [68]). Usually, wastewater contains both soluble compounds and small solid colloidal materials ranging in particle size from 0.001 to 10 μm. These colloidal particles increase the turbidity, color, and chemical oxygen demand (COD) of water. They also have a tendency to adsorb various ions from the neighboring medium which imparts an electrostatic charge to the colloids relative to the bulk of the adjacent water in addition to dissociation (Wang et al. [64]; Carpenter, et al. [68]). The smaller size of colloidal particles (<1 μm) implies that the sedimentation rates are very slow compared to diffusion and hence a colloid suspension remains stable as long as repulsive interactions between the particles preventing their aggregation due to collisions are constant (Wang et al. [64]). Thus, the colloids must be grouped together into forming larger aggregates of particles in order to remove them and reduce the turbidity of municipal and industrial wastewater that are viable options in commercial applications.

6.4.3 Cellulose nanomaterials in membranes for waste water treatment

Cellulose nanomaterials are alternative adsorbent owing to their high surface area-to-volume ratio, low cost, high natural abundance, and intrinsic environmental inertness. Furthermore, NC are easily functionalize on their surface allowing for the modification of chemical moieties that may increase the binding efficiency of pollutants to CN. Carboxylation of CN is the most studied method for increasing their sorptive capacity (Quifang et al. [69]). Of particular relevance to membranes was the option that the mechanical properties of the CN-polymer composites may be sensitive to fluctuations in humidity. For PEI (polyethylenimine)-CNC films for example, the Young's modulus at a relative humidity of 30%, 42%, and 64% was found to be 16, 12, and 3.5 GPa, respectively. Water sorption starts mainly at the hydroxyl groups on the CN surface and consequently decreases some of the cellulose properties, including their mechanical properties as shown in Table 6.5 (Quifang et al. [69]). If the bulk polymer is highly hydrophilic, such as starch-based polymers, NC were shown to decrease the water vapor sorption. NC may increase the hydrophilicity of the membranes to improve water diffusion, but this may decrease their mechanical properties. Using hydrophilic polymers prevents the loss of NC mechanical properties, but this may also reduce the water diffusion through the membrane. As a result of this trade-off, the increased hydrophilicity and increased mechanical strength associated with integrating NC into polymer membranes must be balanced (Moon et al. [73]). Cellulose nanomaterials have enormous application in nano remediation approaches in an inert manner where it functions as a scaffold or particle-stabilizer for reactive nanoparticles. Nanoparticles are often reformed with polymers to prevent aggregation and to help in transport (Liu et al. [74]). Nevertheless, these surface-bound stabilizers coat the reactive particle surface and inhibit the sorption and degradation of the target compounds. Aggregation issues plague the use of iron oxide nanoparticles as adsorbent materials for arsenic removal. Yu et al. [17] reported a simple one-pot solvothermal method for growing aminated iron oxide particles onto a CNF matrix for use in arsenic remediation (Yu et al. [17]). The CNF matrix prevented particle aggregation and enabled for increased modification of the magnetite particles with amines. As a result, these materials exhibited substantially higher arsenic removal (36.49 mg/g As(V)) than any previously published iron oxide-based adsorbents (Alexis et al. [68]).

Poly(acrylonitrile) (PAN) and poly(3-hydroxybutyrate) (PHB) polymer-CN composite membranes have been considered in a variety of membrane processes (micro-filtration, ultrafiltration, hemodialysis, nanofiltration, and membrane distillation) and taken together to validate a wide range of membrane characteristics. With the addition of carbon nanotubes, the incorporation of CN within polymer matrices can modify the membrane properties even at very low content (1 and 2% wt.). The most outstanding property is the high increases in membrane tensile strength achieved (42% and 36%) from these small concentrations.

Additional beneficial properties includes the modifications of the membrane surface hydrophilicity, greater permeability, greater selectivity, and greater resistance to biofouling. In contrast to CNT (carbon nanotubes), CN are highly biocompatible and environmentally benign, which is of importance for the biomedical and environmental applications of such nanocomposite membranes (Alexis et al. [68]).

6.4.4 TiO_2 photocatalysts for waste water treatment

The rapid industrialization of the developed countries and developing countries led to the growth of the manufacturing industries and also the generation of waste and wastewater from various industrial processes (Ravelli et al. [75]). In recent years, researchers showed an increased interest in developing hybrid nanocomposites based on titanium dioxide for the treatment of wastewater (Bhattacharya et al. [76], CERAM [77]). Titanium dioxide is considered as the most favorable material for photocatalytic applications as it displays a variety of attractive characteristics when it comes to its high photo stability nature, strong oxidizing agent, and redox selective agent, but mainly for its light absorbing capability, its electronic structure arrangement, and its cost effectiveness (Ravelli et al. [75]; Ohtani et al. [78]). The three different forms of TiO_2 are: anatase, brookite, and rutile (Burdett et al. [79]). However, anatase and rutile are the most common forms to achieve nanomaterials for photocatalytic applications (Hwang et al. [80]; Slimen et al. [81]). Anatase and rutile correspond to a tetragonal crystalline system, while brookite exhibits an orthorhombic crystalline system (Dambournet et al. [82]). Nano-TiO_2 offers a larger active surface area than micro-TiO_2, which makes it a very useful photocatalytic material owing to the greater absorption rate of UV light (Landsiedel et al. [83]). Unquestionably, the anatase form is the most highly active form for photocatalytic efficiency due to its effectiveness to generate electron-hole pairs (Sclafani and Herrmann [84]). Kobayakawa et al. [85] attributed the higher photocatalytic activity of anatase to the degree of hydroxylation and tetragonal crystalline structure (Fig. 6.1).

6.4.5 Methods for the synthesis of TiO_2

There are several preparation methods for the synthesis of TiO_2, such as precipitation (Li and Demopoulos [86]), hydrothermal and solvothermal reaction

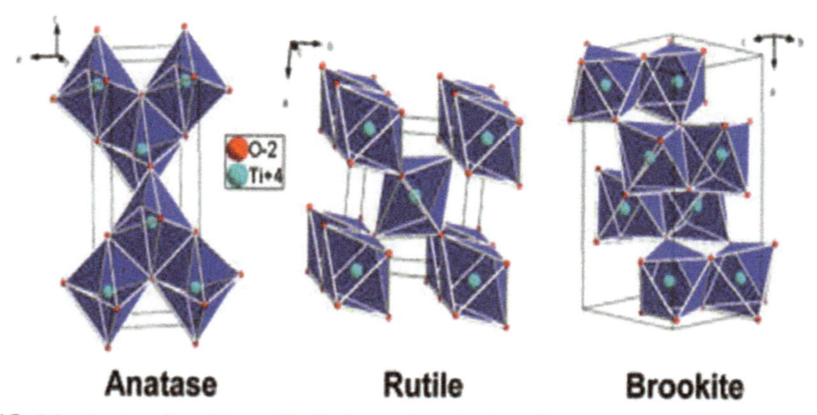

FIG. 6.1 Anatase, Brookite and Rutile forms of the titanium dioxide [82].

(Li and Demopoulos [86]; Zhu et al. [62]) and dip coating (Patil et al. [87]), as well as chemical solvent deposition (CSD) and chemical vapor deposition (CVD) (Kim et al. [88]); plasma treatment to inverse micelle (Arimitsu et al. [89]), multi-gelation (Inoue et al. [90]), ultrasonic irradiation (Peng et al. [91]); and continuous reaction method (Kim and Kim [92]). However, the sol-gel method is the most popular because it is performed at low synthesis temperature, but needs high purity and very crystalline nano powder, as well as a good control of the TiO_2 stoichiometry over the process under an homogeneous synthesis (Akpan and Hameed [93]). Some of the properties of the anatase and rutile phases of TiO_2 are given in Table 6.4. In the ultraviolet region (wavelength < 387 nm), TiO_2 displays greater reactivity and stability. To modify the photocatalytic activity, improve the surface chemistry and higher suitability for solar energy applications, this material is modified by doping with various transition metals such as: Ag, Au, Co, Cr, Cu, Fe, Nb, Ni, Mo, Mn, Pt, and V (Anpo [96]).

TABLE 6.4 Properties of anatase and rutile phases of TiO_2.

Property	Anatase phase	Rutile phase	References
Crystal structure	Tetragonal	Tetragonal	Fisher and Egerton [14]
Atoms per unit cell (Z)	4	2	Peters and Vill [2], Burdett et al. [79]
Lattice parameters (nm)	a = 0.3785 c = 0.9514	a = 0.4594 c = 0.29589	Peters and Vill [2], Burdett et al. [79]
Unit cell volume (nm^3)	0.1363	0.0624	Peters and Vill [2], Burdett et al. [79]
Density ($kg\ m^{-3}$)	3894	4250	Peters and Vill [2], Burdett et al. [79]
Refractive index	2.54, 2.49	2.79, 2.903	Fisher and Egerton [14], CERAM [77]
Calculated indirect band gap (nm) (eV)	 345.5–383.9 3.23–3.59	 382.7–410.1 3.02–3.24	 Beltrán et al. [94] Daude et al. [95]
Experimental band gap (nm) (eV)	 ~387 ~3.2	 ~413 ~3.0	 Beltrán et al. [94] Daude et al. [95]

6.4.6 Application of TiO_2-composite material in the wastewater treatment

Several industries, such as textile, cosmetics, plastics, paper, food, and leather, produce dyes that are common pollutants in wastewater (Rafatullah et al. [97]). The presence of these dyes makes the wastewater potentially harmful even at low concentrations. Researchers are still assertive in their efforts to find a unique treatment for the degradation of these hazardous dyes in wastewater. Photocatalytic material, such as TiO_2, can be extensively used in the efficient mineralization of almost all organic compounds and the total destruction of pathogenic micro-organisms in wastewater (Lee et al. [98]). Recently, the application of photocatalytic materials with adsorption capacity have gained interest for dye removal. This process can function in two different ways: (i) activated charcoal (AC) is used for the adsorption of the dye contaminants, which is a non-destructive process, and (ii) application of TiO_2 as a photocatalyst for the degradation of the contaminants, called advanced oxidation process (AOP). By combining both processes the disadvantages of using TiO_2, such as less adsorption capability, agglomeration, difficulty in separation, and recycling, can be overcome (Bhattacharyya et al. [76]). Many researchers paid attention in developing the TiO_2-composite material in the wastewater treatment comprising of TiO_2-zeolite (Chong et al. [99]), TiO_2-SiO_2 nanocomposite (Dong et al. [100]), TiO_2-Al_2O_3 (Wang et al. [101]), TiO_2-clay (Belessi et al. [102]), TiO_2-bentonite (Rosetto et al. [103]), TiO_2-carbon nanotubes (Lee et al. [104]), TiO_2-graphene (Muthirulan et al. [105]), biomineralized N-doped CNT/TiO_2 (Lee et al. [104]) and TiO_2-AC (Asilturk and Sener [106]). Overall, TiO_2-AC materials have attracted the attention of many researchers in the degradation of wastewater containing dyes due to the tunable pore structure, higher adsorption capability, and higher active surface area from activated carbon (Rahman et al. [6]). Wang et al. [107], Slimen et al. [81], and Xing et al. [108] have all explored TiO_2-AC composites as photocatalysts in the degradation of methyl orange, methylene blue, and rhodamine B dye containing wastewaters (Chun-Chang et al. [109]). Another study reported by Qiao et al. [41] explored the selective photodegradation of rhodamine B dye containing wastewaters with zirconia based TiO_2 composites (TiO_2-ZrO_2). Lian et al. [10] reported on the biodegradation of three sulfonamide antibiotics, namely sulfamethoxazole (SMX), sulfadimethoxine (SDM), and sulfamonomethoxine (SMM), in an activated sludge system. Jamil et al. [110] investigated the photocatalytic degradation of methyl orange from wastewater using AC impregnated TiO_2. The efficiency of TiO_2-SrO core-shell nanowires as photocatalytic degradation of methylene blue dye was studied by Wang et al. [3]. Many researchers considered the application of TiO_2 composites in the purification of contaminated water. Inactivation of the bacteria *Escherichia coli* from the water was studied by Sethi and Sakthivel [111] via the photocatalytic application of TiO_2-ZnO based composites. Mills and Le Hunte [112] investigated a water purification method by integrating the different membrane matrix with the incorporation of TiO_2 material. Lee and Yang [113] explored TiO_2-graphene oxide composites for the elimination of heavy metals such as Pb^{2+}, Cd^{2+}, and Zn^{2+} from wastewater. Zhao et al. [60] probed the application of TiO_2-composites with a combination of nickel oxide and boron (Ni_2O_3/TiO_2-xBx) for the efficient mineralization of organic compounds contained in wastewaters. Their method included light radiation for 4 h leading to the degradation of 80% COD (chemical oxygen demand), 70% TOC (total organic carbon) and the conversion of chloride into Cl^- ions. Bandara et al. [114] showed that the use of UV light and TiO_2-MgO composites with varying MgO content can produce an efficient degradation of a variety of organic compounds. Senevirathna et al. [115] reported that organic compounds are removed by photocatalytic degradation with TiO_2-Cu_2O composite (Table 6.5).

6.4.7 Photocatalytic reactions using TiO_2/TiO_2-composite

Photocatalytic reactions occur by irradiating a photocatalytic surface of a semiconductor, such as TiO_2 in one of the forms such as nanoparticles, colloids, and solution containing dissolved ions or molecules. This photocatalytic semiconductor is known to have an electronic band structure consisting of the valence band (VB, highest occupied energy band) and the conduction band (CB, lowest occupies energy band). These valence and conduction bands are disconnected by a region without electron, called the band gap (Daude et al. [95]) (Fig. 6.2).

The difference in the energies between VB and CB is called the band gap energy (Eg) (Hoffmann et al. [116]). The photocatalytic reaction initiation takes place through irradiating the semiconductor by a photon with an energy ($h\nu$) equivalent or larger than the bandgap energy. An electron from the semiconductor will get excited and promoted from the VB into the CB, leading to the formation of a hole and electron (h/e^-) pair on the VB and CB, respectively. The recombination of the hole and electron pair leads to the dissipation of the given energy as heat. The photo-oxidation takes place through the charge transfer from the h/e^- pair to the reactant species adsorbed on the surface of the semiconductor (Akpan et al. [93]). In a similar way, the photocatalytic reaction using the TiO_2-composites is facilitated by the reaction between the adsorbed O_2 or H_2O molecule on the titania surface and the photo-generated charge carrier from the h/e^- pair upon the interaction of the TiO_2 with the radiation in the UV-region (~290–380 nm) (Scheme 6.1). The e^-in the CB enables the photo-reduction of electron acceptors and holes leading to the photo-oxidation of electron donors (Herrmann [84]). As the h/e^- pair is generated on the surface of the TiO_2-composites, the hydroxyl radicals ($^{\bullet}OH$) and hydrogen ions (H^+) are formed upon the capture of holes by the H_2O molecules. The electrons contribute in the formation of H_2O_2, that further produces more of $^{\bullet}OH$ radicals.

TABLE 6.5 Summary of the stress, strain, Young's modulus and percent change of membranes formed from various cellulose nanomaterial-polymer composites.

Membrane type	Material	Stress (Mpa)	% Change in stress	Strain %	Young's modulus (MPa)	% Change in Young's modulus	No of replicates
MD electro spun fibers	PVDP-HFP	12.6	17.5		72	3	
	PVDP-HFP + 2 wt. % CNC	17.2	36.5	16.4	105	45.8	
UF	PES	5.1		9	56.7		NA
	PES + 1 wt. % CNF	7.3	42.2	11.6	62.5	10.2	
NF	PES + 4 wt. % CNF	5.5	7.8	10.3	53.4	−5.8	
	PVA	57		NA	NA	NA	5
	PVA + 5 wt. % CNC	75.2	31.9				
	CNF	129		10.3	1250		10
	CNF + PHB	142	10.1	9.2	150	24	
	CNF	35.9		5	681.7		NA
	CNF + CdSe	31.9	−11.1	6.2	325.9	−52.2	
UF	CA	4.6		6	77.2		3
	CA + CNF	6.1	31.5	11.4	53.8	−30.3	

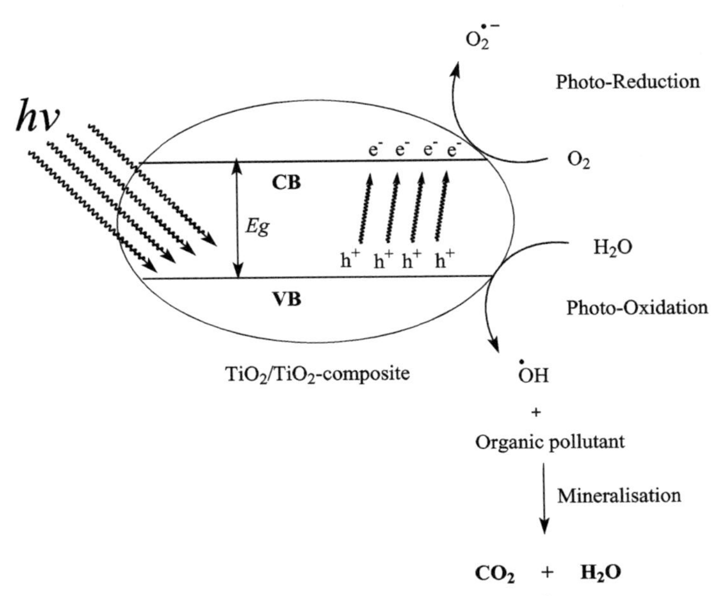

FIG. 6.2 Mechanism for general photocatalytic oxidation and reactions.

$$TiO_2 / TiO_2\text{-composites} + h\nu \leftrightarrow h^+ + e^- \quad \text{.......................} (i)$$

$$h^+ + H_2O_{(adsorbed)} \leftrightarrow H^+ + e^- \quad \text{.......................} (ii)$$

$$O_{2(adsorbed)} + e^- \leftrightarrow O_2^{\bullet -} \quad \text{.......................} (iii)$$

$$H^+ + H_2O_{(adsorbed)} \leftrightarrow {}^{\bullet}OH_{(adsorbed)} + H^+_{(adsorbed)} \quad \text{.......................} (iv)$$

$$O_2^{\bullet -} + H^+ \leftrightarrow HO_2^{\bullet} \quad \text{.......................} (v)$$

$$HO_2^{\bullet} + HO_2^{\bullet} \leftrightarrow H_2O_2 + O_2 \quad \text{.......................} (vi)$$

$$H_2O_{2(adsorbed)} \leftrightarrow {}^{\bullet}OH_{(adsorbed)} + {}^{\bullet}OH_{(adsorbed)} \quad \text{.......................} (vii)$$

$${}^{\bullet}OH + \text{Organic pollutant} \leftrightarrow \text{Reaction Intermediates} \quad \text{.......................} (viii)$$

$$\text{Reaction Intermediates } {}^{\bullet}OH \leftrightarrow CO_2 + H_2O \quad \text{.......................} (ix)$$

$$\text{Organic pollutant}_{(adsorbed)} + {}^{\bullet}OH_{(adsorbed)} \leftrightarrow \text{Reaction Intermediates} \quad \text{.......................} (x)$$

$$\text{Reaction Intermediates } {}^{\bullet}OH \leftrightarrow CO_2 + H_2O \quad \text{.......................} (xi)$$

SCHEME 6.1 General mechanism involved in the photocatalytic reaction.

These $^{\bullet}OH$ radicals react with pollutant species to produce the intermediates, which further get oxidized to CO_2 and H_2O, called complete mineralization. The simple general mechanism of the photo-oxidation of the organic pollutant is presented in Scheme 6.1 with equations (i)–(xi).

6.5 Conclusions

This chapter presented the key components of cellulose nanofibers/fibrils/TiO_2 for wastewater treatment. In recent years, the discovery of cellulose nanomaterial represent an important milestone in the field of materials science and engineering. It is well known that cellulose nanomaterials received remarkable attention for various applications, including bio-based food packaging, optoelectronics, and tissue regeneration to name a few. Few research studies have shown that cellulose, which is abundant in nature, can be physically and chemically modified to create novel materials due to large surface area, environmentally friendly nature, and functionality making it a prospective candidate for waste treatment. Hence, cellulose-based materials have natural binding capacity for heavy metals. Nevertheless, modifications and pretreatments have been found to improve the adsorption capacity of cellulose-based biopolymer. Nanofibrils/cellulose based TiO_2 materials are sustainable for the removal of different contaminants such as heavy metals, radionuclides, antibiotics, oil, dyes, pulp, microorganisms, waste plastics, and residual waste water during wastewater treatment in under developed and developing countries as they arise from various waste water sources: industries and domestic. A focus on the preparation and application of nanocellulose hybrid/fibrils and TiO_2 was presented. Based on the current literature, these materials are believed to play a pivotal role in the future of waste water treatment for large scale industries and water limited countries worldwide.

Acknowledgments

The author's acknowledges the University of Namibia (Namibia) and the University of Zululand (South Africa) for encouragement in this project.

Funding

This research received no external funding.

Conflict of interest

The authors declare no conflict of interest.

References

[1] Hlongwane GN, Sekoai PT, Meyyappan M, Moothi K. Simultaneous removal of pollutants from water using nanoparticles: a shift from single pollutant control to multiple pollutant control. Sci Total Environ 2019;656:808–33. Available from: https://doi.org/10.1016/j.scitotenv.2018.11.257.

[2] Peters G, Vill V. Index of modern inorganic compounds. In: Subvolume A. Landolt-Bornstein numerical data and functional relationships in science and technology. Berlin: Springer Verlag; 1989.

[3] Wang W, Yang J, Gong Y, Hong H. Tunable synthesis of TiO_2/SrO core/shell nanowire arrays with enhanced photocatalytic activity. Mater Res Bull 2013;48:21–4.
[4] Shak KPY, Yean LP, Shee KM. Nanocellulose: recent advances and its prospects in environmental remediation. Beilstein J Nanotechnol 2018;9:2479–98.
[5] Macedo E, Santos MSF, Maldonado-Hodar FJ, Alves A, Madeira LM. Insights on carbonaceous materials tailoring for effective removal of the anticancer drug 5-fluorouracil from contaminated waters. Ind Eng Chem Res 2018;57:3932–40.
[6] Rahman A, Heita J, Likius D, Uahengo V, Johannes S, Bhaskaruni SVHS, et al. Chemical preparation of activated carbon from Acacia erioloba seed pods using H_2SO_4 as impregnating agent for water treatment: an environmentally benevolent approach. J Clean Prod 2019;117689:1–21.
[7] Hager DG. Activated carbon used for large scale water treatment. Environ Sci Technol 1967;1:287–91.
[8] Sun Z, Chai L, Shu Y, Li Q, Liu M, Qiu D. Chemical bond between chloride ions and surface carboxyl groups on activated carbon. Colloid Surf A 2017;530:53–9.
[9] Fei Y, Jie M, Sheng H. Adsorption of tetracycline from aqueous solutions onto multi-walled carbon nanotubes with different oxygen contents. Sci Rep 2013;4:1370–9.
[10] Lian C. Fate of sulfonamide antibiotics in contact with activated sludge—sorption and biodegradation. Water Res 2012;46:1301–8.
[11] Legrini O, Oliveros E, Braun AM. Photochemical processes for water treatment. Chem Rev 1993;93:671–98. Available from: https://doi.org/10.1021/cr00018a003.
[12] Poyatos JM, Munio MM, Almecija MC, Torres JC, Hontoria E, Osorio F. Advanced oxidation processes for wastewater treatment: state of the art. Water Air Soil Pollut 2010;205:187–204. Available from: https://doi.org/10.1007/s11270-009-0065-1.
[13] Hokkanen S, Repo E, Suopajcarvi T, Liimatainen H, Niinimaa J, Sillanpcaca M. Adsorption of Ni (II), Cu (II) and Cd (II) from aqueous solutions by amino modified nanostructured microfibrillated cellulose. Cellulose 2014;21:1471–87.
[14] Fisher J, Egerton TA. Titanium compounds, inorganic. In: Kirk-Othmer encyclopedia of chemical technology. New York: John Wiley and Son's; 2012. https://doi.org/10.1002/0471238961.0914151805070518.a01.pub3.
[15] Henriques B, Goncalves G, Emami N, Pereira E, Mercedes V, Marques PAAP. Optimized graphene oxide foam with enhanced performance and high selectivity for mercury removal from water. J Hazard Mater 2016;301:453–61.
[16] Minnesota Department of Health. Drinking water standards for contaminants: microbiological, radiological, and inorganic contaminants; 2005. http://www.health.state.mn.us/divs/eh/water/factsheet/com/ioc.html.
[17] Yu X, Tong S, Ge M, Wu L, Zuo J, Cao C, Song W. Adsorption of heavy metal ions from aqueous solution by carboxylated cellulose nanocrystals. J Environ Sci 2013;25:933–43.
[18] Ortelli S, Blosi M, Albonetti S, Vaccari A, Dondi M, Costa AL. TiO2 based nanophotocatalysis immobilized on cellulose substrates. Photobiol J Photochem A Chem 2014;276:58–64.
[19] Marinho BA, Cristóvão RO, Djellabi RVJP, Loureiro JM, Boaventura RAR, Vilar S. Photocatalytic reduction acetate of Cr(VI) over TiO2-coated cellulose monolithic structures using solar light. Appl Catal B Environ 2017;203:18–30.
[20] Wei H, Rodriguez K, Renneckar S, Vikesland P. J. Environmental science and engineering applications of nanocellulose-based nanocomposites. Environ Sci Nano 2014;1:302–16.
[21] Liu Q, Li J, Zhao Y, Zhou Y, Li C. CdS nanoparticle-functionalized cotton natural cellulose electrospun nanofibers for visible light photocatalysis. Mater Lett 2015;138:89–91.

[22] Hokkanen S, Repo E, Johansson Westholm L, Lou S, Sainio T, et al. Adsorption of Ni^{2+}, Cd^{2+}, PO_4^{3-} and NO_3^- from aqueous solutions by nanostructured microfibrillated cellulose modified with carbonated hydroxyapatite. Chem Eng J 2014;252:64–74.
[23] Shibi IG, Anirudhan TS. Polymer-grafted banana (Musa paradisiaca) stalk as an adsorbent for the removal of lead(II) and cadmium(II) ions from aqueous solutions: kinetic and equilibrium studies. J Chem Technol Biotechnol 2006;81:433–44.
[24] Shibi IG, Anirudhan TS. Synthesis, characterisation, and application as a mercury (II) sorbent of banana stalk-polyacrylamide grafted copolymer bearing carboxyl groups. Ind Eng Chem Res 2002;41:5341–52.
[25] Akil HM, Omar MF, Mazuki AAM, Safiee S, Ishak ZAM, Bakar A. Kenaf fibre reinforced composites: a review. Mater Des 2011;32:4107–21.
[26] Jin H, Kettunen M, Laiho A, Pynnconen H, Paltakari J, Marmur A, et al. Superhydrophobic and superoleophobic nanocellulose aerogel membranes as bioinspired cargo carriers on water and oil. Langmuir 2011;27:1930–4.
[27] Zhang H, Li Y, Zexiang LU, Chen L, Huang L, Fan M. Versatile fabrication of superhydrophobic and ultralight cellulose based aerogel for oil spillage cleanup. Phys Chem Chem Phys 2016;18:28297–306.
[28] Donia AM, Atia AA, Abouzayed FI. Preparation and characterization of nanomagnetic cellulose with fast kinetic properties towards the adsorption of some metal ions. Chem Eng J 2012;191:22–30.
[29] Yu X, Tong S, Ge M, Zuo J, Cao C, Song W. One-step synthesis of magnetic composites of cellulose@iron oxide nanoparticles for arsenic removal. J Mater Chem A 2013;1:959–65.
[30] Sun X, Yang L, Li Q, Zhao J, Li X, Wang X, et al. Amino-functionalized magnetic cellulose nanocomposite as adsorbent for removal of Cr (VI): synthesis and adsorption studies. Chem Eng J 2014;241:175–83.
[31] Singh K, Arora JK, Sinha JMT, Srivastava S. Functionalization of nanocrystalline cellulose for decontamination of Cr(III) and Cr(VI) from aqueous system: computational modeling, approach. Clean Technol Environ Policy 2014;16:1179–91.
[32] Lu M, Xu Y, Guan X, Wei D. Preliminary research on Cr (VI) removal by bacterial cellulose. J Wuhan Univ Technol Mater Sci Ed 2012;25:572–5.
[33] Pillai SS, Deepa B, Abraham E, Girija N, Geetha P, Jacob L, et al. Biosorption of Cd(II) from aqueous solution using xanthated nano banana cellulose: equilibrium and kinetic studies. Ecotoxicol Environ Saf 2013;98:352–60.
[34] Dufresne A. Nanocellulose: from nature to high performance tailored materials. France: Walter De Gruyter GmbH & Co. KG; 2012. p. 1–460.
[35] Li F, Mascheroni E, Piergiovanni L. The potential of nanocellulose in the Packaging field: a review. Technol Sci 2015;28:475–508. https://doi.org/10.1002/pts.2121.
[36] Jorfi M, Foster EJ. Recent advances in nanocellulose for biomedical applications. J Appl Polym Sci 2015;132:1–19. https://doi.org/10.1002/app.
[37] Rojo E, Peresin MS, Sampson WWOJ, Hoeger IC, Vartiainen J, Laine J, et al. Comprehensive elucidation of the effect of residual lignin on the physical, barrier, mechanical and surface properties of nanocellulose films. Green Chem 2015;17:1853–66.
[38] Ahmad R, Hamid R, Osman SA. Effect of fibre treatment on the physical and mechanical properties of kenaf fibre reinforced blended cementitious composites. KSCE J Civ Eng 2019;23:4022–35.
[39] FAO. Potential pollutants, their sources and their impacts. In: Sciortino JA, Ravikumar R, editors. Fishery harbour manual on the prevention of pollution. Chennai, India: Bay of Bengal Programme for Fisheries Management (BOBP/MAG/22); 1999. p. 1–200.

[40] Sassi JF, Chanzy H. Ultrastructural aspects of the acetylation of cellulose. Cellulose 1995;2:111–27.
[41] Qiao H, Zhou Y, Yu F, Wang E, Min Y, Huang Q, et al. Effective removal of cationic dyes using carboxylate-functionalized cellulose nanocrystals. Chemosphere 2015;141:297–303.
[42] Dong C, Zhang F, Pang Z, Yang G. Efficient and selective adsorption of multimetal ions using sulfonated cellulose as adsorbent. Carbohydr Polym 2016;151:230–6.
[43] Taleb K, Rusˇmirovic J, Rancic M, Nikolic J, Drmanic S, Velickovic Z, Marinkovic A. Efficient pollutants removal by amino-modified nanocellulose impregnated with iron oxide. J Serb Chem Soc 2016;81:1199–213.
[44] Lefatshe K, Muiva CM, Kebaabetswe LP. Extraction of nanocellulose and in-situ casting of ZnO/cellulose nanocomposite with enhanced photocatalytic and antibacterial activity. Carbohydr Polym 2017;164:301–8.
[45] Gao M, Li N, Lu W, Chen W. Role of cellulose fibers in enhancing photosensitized oxidation of basic green 1 with massive dyeing auxiliaries. Appl Catal B 2017;147:805–12.
[46] Cruz-Tato P, Ortiz-Quiles EO, Vega-Figueroa K, Santiago-Martoral L, Flynn M, Díaz-Vázquez LM, et al. Metalized nanocellulose composites as a feasible material for membrane supports: design and applications for water treatment. Environ Sci Technol 2017;51:4585–95.
[47] Kim TJ, Li B, Hagg MB. Novel fixed-site–carrier polyvinylamine membrane for carbon dioxide capture. J Polym Sci B 2004;23:4326–36.
[48] Chitpong N, Husson SM. High-capacity, nanofiber-based ion-exchange membranes for the selective recovery of heavy metals from impaired waters. Sep Purif Technol 2017;179:94–103.
[49] Korhonen JT, Kettunen M, Ras RHA, Ikkala O. Hydrophobic nanocellulose aerogels as floating, sustainable, reusable, and recyclable oil absorbents. ACS Appl Mater Interfaces 2011;3:1813–6.
[50] Meng Y, Young TM, Liu P, Contescu CI, Huang B, Wang S. Ultralight carbon aerogel from nanocellulose as a highly selective oil absorption material. Cellulose 2015;22:435–47.
[51] Wang J, Chen B. Adsorption and coadsorption of organic pollutants and a heavy metal by graphene oxide and reduced graphene materials. Chem Eng J 2015;281:379–88.
[52] Rakić V, Rac V, Krmar M, Otman O, Auroux A. The adsorption of pharmaceutically active compounds from aqueous solutions onto activated carbons. J Hazard Mater 2015;282:141–9.
[53] Carrales-Alvarado DH, Ocampo-Pérez R, Leyva-Ramos R, Rivera-Utrilla J. Removal of the antibiotic metronidazole by adsorption on various carbon materials from aqueous phase. J Colloid Interface Sci 2014;436:276–85.
[54] Xu J, Sheng GP, Ma Y, Wang LF, Yu HQ. Roles of extracellular polymeric substances (EPS) in the migration and removal of sulfamethazine in activated sludge system. Water Res 2013;47:5298–306.
[55] Liao P. Adsorption of tetracycline and chloramphenicol in aqueous solutions by bamboo charcoal: a batch and fixed-bed column study. Chem Eng J 2013;228:496–505.
[56] Ghadim EE. Adsorption properties of tetracycline onto graphene oxide: equilibrium, kinetic and thermodynamic studies. PLoS One 2013;8:79254–64.
[57] Richardson SD, Ternes TA. Water analysis: emerging contaminants and current issues. Anal Chem 2014;86:2813–48.
[58] Johnson AC, Keller V, Dumont E, Sumpter JP. Assessing the concentrations and risks of toxicity from the antibiotics ciprofloxacin, sulfamethoxazole, trimethoprim and erythromycin in European rivers. Sci Total Environ 2015;511:747–55.
[59] Zhang D, Yin J, Zhao J, Zhu H, Wang C. Adsorption and removal of tetracycline from water by petroleum coke-derived highly porous activated carbon. J Environ Chem Eng 2015;3:1504–12.

[60] Zhao W, Ma W, Chen C. Efficient degradation of toxic organic pollutants with Ni_2O_3/TiO_2-xBx under visible irradiation. J Am Chem Soc 2004;126:4782–3.

[61] Porras J. Role of humic substances in the degradation pathways and residual antibacterial activity during the photodecomposition of the antibiotic ciprofloxacin in water. Water Res 2016;94:1–9.

[62] Zhu J, Deng Z, Chen F. Hydrothermal doping method for preparation of Cr^{3+}-TiO_2 photocatalysts with concentration gradient distribution of Cr^{3+}. Appl Catal B 2006;62:329–35.

[63] Park S. Graphene oxide papers modified by divalent ions—enhancing mechanical properties via chemical cross-linking. ACS Nano 2008;2:572–8.

[64] Wang S, Peng X, Zhong L, Tan J, Jing S, Cao X, et al. An ultralight, elastic, cost-effective, and highly recyclable superabsorbent from microfibrillated cellulose fibers for oil spillage cleanup. J Mater Chem A 2015;3:8772–81.

[65] Pierre AC, Pajonk GM. Chemistry of aerogels and their applications. Chem Rev 2003;102:4243–65.

[66] Sun H, Cao L, Lu L. Magnetite/reduced graphene oxide nanocomposites: one step solvothermal synthesis and use as a novel platform for removal of dye pollutants. Nano Res 2011;4:550–62.

[67] Tan P. Adsorption of Cu^{2+}, Cd^{2+} and Ni^{2+} from aqueous single metal solutions on graphene oxide membranes. J Hazard Mater 2015;297:251–60.

[68] Alexis WC, Charles François de L, Mark RW. Cellulose nanomaterials in water treatment technologies. Environ Sci Technol 2015;49:5277–87.

[69] Yao Q, Fan B, Xiong Y, Jin C, Sun Q, Sheng C. 3D assembly based on 2D structure of cellulose nanofibril/graphene oxide hybrid aerogel for adsorptive removal of antibiotics in Water. Sci Rep 2017;45914:1–13. https://doi.org/10.1038/srep45914.

[70] Li YY, Wang B, Ma MG, Wang B. Review of recent development on preparation, properties, and applications of cellulose-based functional materials. Int J Polym Sci 2018;, 8973643.

[71] Karim Z, Mathew AP, Grahn M, Mouzon J, Oksman K. Nanoporous membranes with cellulose nanocrystals as functional entity in chitosan: removal of dyes from water. Carbohydr Polym 2014;112:668–76.

[72] Voisin H, Bergström L, Liu P, Mathew A. Nanocellulose-based materials for water purification. Nanomaterials 2017;7–57.

[73] Moon RJ, Martini A, Nairn J, Simonsen J, Youngblood J. Cellulose nanomaterials review: structure, properties and nanocomposites. Chem Soc Rev 2011;40:3941–94.

[74] Liu P, Borrell PF, Božic M, Kokol V, Oksman K, Mathew AP. Nanocelluloses and their phosphorylated derivatives for selective adsorption of Ag from industrial effluents. J Hazard Mater 2015;294:177–85.

[75] Ravelli D, Dondi D, Fagnoni M, Albini A. Photocatalysis. A multi-faceted concept for green chemistry. Chem Soc Rev 2009;38:1999–2011.

[76] Bhattacharyya A, Kawi S, Ray MB. Photocatalytic degradation of orange II by TiO_2 catalysts supported on adsorbents. Catal Today 2004;98:431–9.

[77] CERAM. Titanium dioxide—titania, AZoM version 2.0. AZoM Pty. Ltd; 2002. https://www.azom.com/article.aspx?ArticleID=1179. accessed on 1 June 2020.

[78] Ohtani B, Ogawa Y, Nishimoto S. Photocatalytic activity of amorphous-anatase mixture of titanium (IV) oxide particles suspended in aqueous solutions. J Phys Chem B 1997;101:3746–52.

[79] Burdett JK, Hughbanks T, Miller GJ, Richardson Jr JW, Smith JV. Structural-electronic relationships in inorganic solids: powder neutron diffraction studies of the rutile and anatase polymorphs of titanium dioxide at 15 and 295 K. J Am Chem Soc 1987;109:3639–41.

[80] Hwang YK, Park SS, Lim JH, Won YS, Huh S. Preparation of anatase/rutile mixed-phase titania nanoparticles for dye-sensitized solar cells. J Nanosci Nanotechnol 2013;13:2255–61.
[81] Slimen H, Houas A, Nogier JP. Elaboration of stable anatase TiO_2 through activated carbon addition with high photocatalytic activity under visible light. J Photochem Photobiol A Chem 2011;221:13–21.
[82] Dambournet D, Belharouak I, Amine K. Tailored preparation methods of TiO_2 anatase, rutile, brookite: mechanism of formation and electrochemical properties. Chem Mater 2010;22:1173–9.
[83] Landsiedel R, Kapp MD, Schulz M, Wiench K, Oesch F. Genotoxicity investigations on nanomaterials: methods, preparation and characterization of test material, potential artifacts and limitations-many questions, some answers. Mutat Res 2009;68:241–58.
[84] Herrmann J-M. Heterogeneous photocatalysis: fundamentals and applications to the removal of various types of aqueous pollutants. Catal Today 1999;53:115–29.
[85] Kobayakawa K, Nakazawa Y, Ikeda M. Influence of the density of surface hydroxylgroups on TiO_2 photocatalytic activities. Phys Chem Chem Phys 1990;94:1439–43.
[86] Li Y, Demopoulos GP. Precipitation of nanosized titanium dioxide from aqueous titanium (IV) chloride solutions by neutralization with MgO. Hydrometallurgy 2008;90:26–33.
[87] Patil KR, Sathaye SD, Khollam YB. Preparation of TiO_2 thin films by modified spincoating method using an aqueous precursor. Mater Lett 2003;57:1775–80.
[88] Kim BH, Lee JY, Choa YH. Preparation of TiO_2 thin film by liquid sprayed mist CVD method. Mater Sci Eng B Adv 2004;107:289–94.
[89] Arimitsu N, Nakajima A, Kameshima Y. Preparation of cobalt-titanium dioxide nanocomposite films by combining inverse micelle method and plasma treatment. Mater Lett 2007;61:2173–7.
[90] Inoue S, Muto A, Kudou H, Ono T. Preparation of novel titania support by applying the multigelation method for ultra-deep HDS of diesel oil. Appl Catal A 2004;269:7–12.
[91] Peng F, Cai L, Yu H. Synthesis and characterization of substitutional and interstitial nitrogen-doped titanium dioxides with visible light photocatalytic activity. J Solid State Chem 2008;181:130–6.
[92] Kim KD, Kim HT. Synthesis of titanium dioxide nanoparticles using a continuous reaction method. Colloid Surf A 2002;207:263–9.
[93] Akpan UG, Hameed BH. The advancements in sol-gel method of doped-TiO_2 photocatalysts. Appl Catal A 2010;375:1–11.
[94] Beltrán A, Andrés J, Sambrano JR, Longo E. Density funcional theory study on the structural and electronic properties of low índex rutile surfaces for $TiO_2/SnO_2/TiO_2$ and $SnO_2/TiO_2/SnO_2$ composite systems. J Phys Chem A 2008;112(38):8943–52.
[95] Daude N, Gout C, Jouanin C. Electronic band structure of titanium dioxide. Phys Rev B 1977;15:3229–35.
[96] Anpo M. Use of visible light. Second-generation titanium oxide photocatalysts prepared by the application of an advanced metal ion-implantation method. Pure Appl Chem 2000;72:1787–92.
[97] Rafatullah M, Sulaiman O, Hashim R, Ahmad A. Adsorption of methylene blue on low-cost adsorbents: a review. J Hazard Mater 2010;177:70–80.
[98] Lee S, Nishida K, Otaki M, Ohgaki S. Photocatalytic inactivation of phage Qβ by immobilized titanium dioxide mediated photocatalyst. Water Sci Technol 1997;35:101–6.
[99] Chong MN, Tneu ZY, Poh PE, Jin B, Aryal R. Synthesis, characterization and application of TiO_2-zeolite nanocomposites for the advanced treatment of industrial dye wastewater. J Taiwan Inst Chem Eng 2015;50:288–96.

[100] Dong WY, Lee CW, Lu XC. Synchronous role of coupled adsorption and photocatalytic oxidation on ordered mesoporous anatase TiO_2–SiO_2 nanocomposites generating excellent degradation activity of RhB dye. Appl Catal B Environ 2010;95:197–207.

[101] Wang XD, Shi F, Huang W, Fan CM. Synthesis of high quality TiO_2 membranes on alumina supports and their photocatalytic activity. Thin Solid Films 2012;520:2488–92.

[102] Belessi V, Lambropoulou D, Konstantinou I. Structure and photocatalytic performance of TiO_2/clay nanocomposites for the degradation of dimethachlor. Appl Catal B Environ 2007;73:292–9.

[103] Rossetto E, Petkowicz DI, dos Santos JHZ, Pergher SBC, Penha FG. Bentonites impregnated with TiO_2 for photodegradation of methylene blue. Appl Clay Sci 2010;48:602–6.

[104] Lee WJ, Lee JM, Kochuveedu ST. Biomineralized N-doped CNT/TiO_2 core/shell nanowires for visible light photocatalysis. ACS Nano 2012;6:935–43.

[105] Muthirulan P, Nirmala Devi CK, Meenakshi SM. Fabrication and characterization of efficient hybrid photocatalysts based on titania and graphene for acid orange seven dye degradation under UV irradiation. Adv Mater Lett 2014;5:163–71.

[106] Asilturk M, Sener S. TiO_2-activated carbon photocatalysts: preparation, characterization and photocatalytic activities. Chem Eng J 2012;180:354–63.

[107] Wang XJ, Liu YF, Hu ZH, Chen YJ, Liu W, Zhao GH. Degradation of methyl orange by composite photocatalysts nano-TiO_2 immobilized on activated carbons of different porosities. J Hazard Mater 2009;169:1061–7.

[108] Xing B, Shi C, Zhang C, Yi G, Chen L, Guo H, et al. Preparation of TiO_2/activated carbon composites for photocatalytic degradation of RhB under UV light irradiation. J Nanomater 2016;8393648:1–10.

[109] Ou C-C, Yang C-S, Lin S-H. Selective photo-degradation of Rhodamine B over zirconia incorporated titania nanoparticles: a quantitative approach. Cat Sci Technol 2011;1:295–307.

[110] Jamil TS, Ghaly MY, Fathy NA, Abd El-Halim TA, Osterlund L. Enhancement of TiO_2 behavior on photocatalytic oxidation of MO dye using TiO_2/AC under visible irradiation and sunlight radiation. Sep Purif Technol 2012;98:270–9.

[111] Sethi D, Sakthivel R. ZnO/TiO_2 composites for photocatalytic inactivation of *Escherichia coli*. J Photochem Photobiol B 2017;168:117–23.

[112] Mills A, Hunte SL. An overview of semiconductor photocatalysis. J Photochem Photobiol A 1997;108:1–35.

[113] Lee YC, Yang JW. Self-assembled flower-like TiO_2 on exfoliated graphite oxide for heavy metal removal. J Ind Eng Chem 2012;18:1178–85.

[114] Bandara J, Hadapangoda CC, Jayasekera WG. TiO_2/MgO composite photocatalyst: the role of MgO in photoinduced charge carrier separation. Appl Catal B 2004;50:83–8.

[115] Senevirathna MKI, PKDDP P, Tennakone K. Water photoreduction with Cu_2O quantum dots on TiO_2 nano-particles. J Photochem Photobiol A 2005;171:257–9.

[116] Hoffmann MR, Martin ST, Choi W, Bahnemann DW. Environmental applications of semiconductor photocatalysis. Chem Rev 1995;95:69–96.

Chapter 7

Hybrid nanocomposites based on cellulose nanocrystals/nanofibrils and zinc oxides: Energy applications

Kalsoom Jan
Department of Plastic Engineering, University of Massachusetts Lowell, Lowell, MA, United States

7.1 Cellulose and derivatives from renewable sources

Keeping in view the rapid growth of nano-technology, fossil fuels, and environmental crises, there is a need for sustainable, biodegradable, and recyclable materials to be used for energy applications. In this regard, cellulose (linear series of glucose molecules) has been explored to be the renewable efficient biomaterial for energy and other diverse range of applications.

Cellulose has attained significant importance in the field of energy due to its potential tunable properties such as biodegradability, availability, recyclability, renewability, low thermal expansion coefficient, high surface area, lower density, strong mechanical and physicochemical properties. Cellulose has been reported to be extracted from various natural sources such as forest residues, cotton stalk, fruits wastes, corn husk including hardwood and softwood, bamboo, sugar beet, cotton, potato tubers, soybean stock, banana rachis [1], marine biomass, and rice straw [2, 3], non-woody [4] (water hyacinth, hemp, cotton, flax, potato tuber cells, sugar beet, soybean, ficus barkcloth, jute, pea hull, tomato peel [5]), wood (pine, douglas, birch) agriculture raw material (maize and wheat straw, rice, barley and coconut husk, tomato, lemon, pineapple peel [6]) and animals (exoskeletons of crabs) [7]. The greater source of nanocellulosic materials is cotton.

7.2 Types of cellulose

Generally, nano-cellulose materials are categorized into three main types: cellulose nanocrystals (CNC), cellulose nanofibrils (CNF), and bacterial

Cellulose Nanocrystal/Nanoparticles Hybrid Nanocomposites. https://doi.org/10.1016/B978-0-12-822906-4.00005-0

nano-cellulose (BNC) [8]. CNC and CNF have been used in various applications including electronics, devices, automobiles, technical films, textiles, medical equipment, and aeronautics.

7.2.1 Cellulose nanofibrils (CNF)

The natural cellulose fibers with a diameter of 20–60 nm with several micrometers in length can be expressed as CNF with high volume and large surface area. For CNF, different chemical and enzymatic pretreatments have been proposed such as phosphorylation, enzymatic hydrolysis, carboxymethylation, and 2,2,6,6-tetramethylpiperidine-1-oxyl (TEMPO) oxidation [66]. TEMPO oxidation is the most commonly used pretreatment for CNF extraction [9]. The source of CNF controls its basic properties such as the degree of crystallinity, morphological, and structural properties. Commonly available CNF sources include rice straw, sugar beet pulp, banana, etc. CNF from the primary fiber cells is longer and thinner as compared to the one obtained from the secondary one [10].

7.2.2 Cellulose nanocrystals (CNC)

Cellulose nanocrystals are isolated by the biosynthesis process from the cellulose, showing dimensions of 3–50 nm thickness with 100 nm-several micrometers in length [11]. CNC has been considered to show excellent crystallinity and higher rigidity than CNF. The different forms of CNC have been reported such as nano-rods, nano-whiskers, and rod-like cellulose crystals, which can be obtained by the acid hydrolysis of cellulose fibers. Generally, the extraction methods, such as alkali and bleaching treatment, have been applied to obtain the CNC. The nature of the source and the acid type (commonly used sulphuric acid at 64%wt.) have a direct effect on the resulting CNC.

7.2.3 Bacterial nanocellulose (BNC)

BNC is obtained from the bacteria species such as *acetobacter xylinium* using carbon and nitrogen in the aqueous culture [12]. BNC is reported to have a higher Young's modulus (78 GPa), higher molecular weight (8000 Da), and excellent water-holding capacity [13].

7.3 Metal oxide-based cellulose nanohybrid composites

The cellulose and its metal-based derivatives have been reported to be used in functionalizing materials as well. Recently, the cellulose designation with metal or metal-oxide (ZnO, TiO_2, MgO, ZnO, CuO, Ag, and Fe_3O_4) has drawn a great deal of attention. Nanocellulose surface modification is facilitated by the interaction of metallic particles and hydroxyl groups of cellulose to be adsorbed, as cellulose is rich in hydroxyl groups [14].

7.3.1 Zinc-oxide based cellulose hybrid nanocomposite

Among all the metals or metal oxides, zinc oxide (ZnO) has shown relatively better potential applications in the paint, plastics, electronics, and paper industry since it has higher chemical stability, lower production cost, reactive oxygen species, and photocatalytic properties [15, 16]. Cellulose-ZnO compounds have been reported to exhibit excellent electrical conductivity, thermal stability, and UV barrier enhancer. Moreover, they have been considered to be effective as antibacterial, anticancer, antioxidant, antidiabetic, and wound healer as well [15]. In the case of ZnO, the modification can be done by two approaches such as nano-cellulosic surface modification by oxidation or cationization and by using nano-cellulose as a template to make a hybrid material with ZnO imparting functional properties.

ZnO has greatly attracted the researcher as being environmentally friendly, lower production cost, abundant in nature with non-toxicity, which exhibits better photocatalytic activity and higher quantum efficiency. ZnO exhibits high polarity which enables its uniform dispersion in nano-cellulosic composites. Therefore, better hybrid nanocomposite ZnO and cellulose proper synthesis methods should be monitored carefully. Cellulose hybrid composites have been used extensively in cosmetics, medicines, recyclable electronics, and 3D-printing. All these applications of nanocellulose/ZnO are associated with their excellent properties, including nontoxicity, lightweight, dimensional stability, better thermal conductivity, higher thermal stability, environmentally friendly, recyclability, reusability, and biodegradability [17].

7.3.2 Synthesis methods and surface modification

ZnO exists into different morphological patterns such as a sphere, nanorod, mulberry-like, nano-flower, dumbbell shape, rice flake, nano-flake, and ring; which could be produced depending on the synthesis technique, considering the physical and chemical parameters used. Synthesis methods of ZnO particles are adopted according to the requirement of the specific application such as hydrothermal, solvothermal, microwave decomposition, wet chemical route, and simple precipitation methods [18–21]. The solvothermal and hydrothermal method exhibits a high degree of purity when high pressure and high temperatures are provided [22]. The mechano-chemical technique is applied on a large scale with low production cost, but there is a risk of a higher concentration of impurities presence [23]. The precipitation method is low-cost and simple but sometimes results in agglomerated particles [24].

The nanocellulose/ZnO hybrid composite can be prepared by the surface modification of the nanocellulose, mixing of the nanocellulose/ZnO with and without external reducing species. A simple mixing of ZnO with CNC or CNF can be done, but during the drying process, an uneven distribution of ZnO can be obtained due to aggregates.

7.3.3 Cellulose/ZnO energy and sensing properties

Polymer composites based on metal oxides can have extensive applications such as photocatalysis, photodetector, and in sensors due to the possibility to modify the nature of their surface (ZnO films). Cellulose hybrid composites based on ZnO can significantly enhance the sensing activity. ZnO has shown better biocompatibility than reported TiO_2 as an inorganic photocatalytic material. Many researches have been done on the ZnO nanocomposites as UV sensors. The resultant polymer exhibits excellent electrical photoconductivity changes upon exposure to UV light. The cellulose addition to such materials made it more flexible to be used as a wearable sensor [25]. Cellulose/ZnO nanocomposites can be synthesized via a two-step chemical method; by ZnO seeding and by growing the ZnO nanorods on the cellulose surface in an aqueous media with different temperatures [26].

The presence of cellulose enhances the porosity of the ZnO surface leading to improved UV radiation absorption. Therefore, cellulose-ZnO is highly recommended for ultra-high sensitive UV sensors [26]. Cellulose/ZnO hybrid composites can also be used as multifunctional composites with improved dielectric, thermal, mechanical, sensing, and energy storage capacity.

ZnO nanocomposites with nano-cellulose are expected to revolutionize nanotechnology with potential in the field of mechanical, electronic, and chemical sectors. Nanocellulose/ZnO exhibits UV sensing and photocatalytic properties with flexibility, so it has a future in flexible and wearable electronics with lighter weight and non-toxicity. There is a need for more investigation on the industrialization of nano cellulose/ZnO nanocomposites for more useful human needs.

7.4 Cellulose-based composites for energy applications

7.4.1 State of art

Day by day, the world's demand for energy is increasing. Fossil fuels are not only insufficient for energy needs but also cause a hazardous effect on the earth by increasing pollution and global warming. To overcome this challenge, there is a need to develop an energy system that is not only environmentally friendly and effective but also utilized renewable sources for energy. There is a need for energy conversion or storage devices based on renewable sources such as solar cells, rechargeable batteries, and supercapacitors to fulfill future energy needs. Nowadays, the researchers are prioritizing renewable cellulose sources based materials to use for energy conversion or storage with easy processing, higher efficiency, and lower pollutant emission. Although cellulose is not considered to be electrically or thermally conductive, the literature suggests that it is possible to obtain cellulosic nanocomposites with good electrical properties. For example, Van den Berg et al. [27] reported conductive nano-cellulosic surfaces by using pi-conjugated polymers such as a poly(p-phenylene ethynylene) and polyaniline derivative with quaternary ammonium side chains.

7.4.2 Cellulose-based material for energy conversion

Several materials (silicon-based, inorganic, organic, and hybrid) have been used for the conversion of solar energy into electric energy using photovoltaic (PV) cells. Inorganic materials, compared to organic ones for PV, give higher conversion efficiency as 10%–30% and 2.5%, respectively. Inorganic and organic materials prioritized the usage of renewable resources for PV and PV devices. Cellulose-based material is promising in terms of sustainability, renewability, and abundance. In particular, CNC have a high surface area making them more suitable for higher efficiency conversion devices.

7.4.2.1 Organic photovoltaics (OPV)

The replacement of conventional conducting substrates by cellulose for photovoltaic devices has the advantage of good mechanical flexibility and low weight. CNC and CNF have been used as substrates in organic solar cells, which are reported to be characterized by X-ray diffraction (XRD), atomic force microscopy (AFM), and differential scanning calorimetry (DSC). The resulting substrate exhibits better homogeneity and crystallinity, but lower surface roughness. The power conversion efficiency (PCE) values of inverted OPV cells based on CNC and CNF are about 0.5%–3%. Cellulose and silver-based solar cells have been reported with a conversion efficiency of about 2.7% [28]. But more recently developed solar cells with CNC as a substrate using film transfer lamination were found to have a conversion efficiency of about 4% [29].

Nevertheless, using cellulose for solar cells needs more improvement for efficient performance. Other energy applications for cellulose-based nanocomposites are for energy harvester, display devices, actuators, and paper transistors [30]. Cellulose nanocrystals based electroactive paper, which works based on the ion-migration and piezoelectric effect, holds the potential capability to be used in sensors, actuators, biomimetic robots, and other haptic technology.

7.4.2.2 Nanocellulose-based paper substrate for solar cell development

CNF as an insulator has been used in solar cell development as a transparent substrate. CNF exhibits high optical haze with low transmittance combined with high optical transparency and high optical haze, both highly desirable for solar cell substrates. The carboxymethylated nano-cellulose has been fabricated to study their optical application. The results revealed that their PCE (power conversion efficiency) was 0.2% which a Jsc (short circuit current) and Voc (open circuit voltage) of 2.4 mA cm^{-2} and 0.4 V, respectively [31].

Organic solar cells based on CNC substrates have been reported to yield higher efficiencies and rectification. Zhou et al. [28] reported that CNC-based solar cells using polyethylenimine (PEI)-modified Ag electrode exhibited an efficiency of about 4%. In this study, the dry lamination transfer process was introduced to minimize the cellulose degradation from aqueous-based processes. Moreover, good diode rectification of the device was achieved. More recently, the nano-cellulose paper was improved using a tetramethylpiperidine-1-oxy (TEMPO)-oxidized cellulose as a building block, which exhibits excellent optical and ultrahigh haze of 96% and 60% respectively, which was associated to their higher packing density.

Generally, the nano cellulose paper offered an enhanced photon absorption with a greater extent of collecting ambient light, make it superior with higher efficiency than any glass substrates.

7.4.2.3 CNF-templated mesoporous structure as solar cell electrodes

In solar design, high photocurrent density can be achieved by the CNF large surface and extensive network-like structure. The intrinsic insulating property of CNF was not used much for the development of scaffold for solar cell electrodes as compared to the substrate. The main strategy is to use CNF as a template to fabricate the mesoporous structure of any functional material which has photoactive and conductive characteristics. An example is the CNF-Templated TiO_2 nanofibers providing better photon interaction and charge transport [29, 32].

The nanofibers can be directly cast on a transparent conducting glass substrate at the photoelectrode of a dye-sensitized solar cell (DSSC). For better overall PV performance of the nano-cellulose templated solar cells, the surface area density must be improved without losing the porosity. Typicaly, the nanofiber-based device typically showed a higher open-circuit voltage as compared to the nanoparticle-based DSSC.

7.4.2.4 Cellulose in photoelectrochemical (PEC) cell development

Photoelectrochemical (PEC) water splitting can be used to directly convert solar energy into hydrogen fuels. The structural merits of 3D nano-cellulose can also improve the PEC [33]. The hydrophilic property of cellulose has an advantage in the PEC water splitting reactions. Li et al. [34] studied the CNF-templated TiO_2 capillary nanostructures of PEC systems for water splitting outside the electrolyte body. In the study, a mesoporous CNF film was produced by using atomic layer deposition for the fabrication of an anatase TiO_2 nanofibrous 3D structure. The CNF strip, attached to the TiO_2 nanostructure, acted as the photoanode, while keeping it outside of the electrolyte facing the light source [35].

By using the capillary force, the electrolyte has been supplied to the CNF film channels. The results revealed a photocurrent density (Jph) of 0.87

mAcm^{-2}. The study also revealed that the PEC performance with the capillary design was attributed to lower light scattering from minimal electrolyte coating and higher charge transport kinetics [36].

The surface reaction-limited pulsed chemical vapor deposition (SPCVD) techniques can be used for synthesizing the high-density TiO_2 branches within the mesoporous CNF templates. The SPCVD has been used under high temperature (600 °C) with ZnO overcoating. A ZnO layer with $TiCl_4$ vapor interaction resulted in the conversion to a polycrystalline TiO_2 thin films followed by the Kirkendall effect (phenomenon which results from the difference in intrinsic diffusivities of chemical constituents of substitution solid solutions (non-reciprocal diffusion) [37]. The final ZnO based heterostructure offered longer optical paths with high quality of charge transport potential and a larger electrolyte-semiconductor interface area. The resulting composite reported an efficiency as PEC photoanode about 200% which is higher than those of CNF-templated TiO_2. This study concluded that the addition of metal oxide can highly improve the efficiency by charge transport for solar cell development [38].

7.5 Cellulose for energy storage

Due to its structural advantage, cellulose can be used in energy storage systems. Due to the porous nature of cellulose, liquid electrolytes and ionic species can move between the electrode surface. Thus many conducting materials, such as metal oxides, metal nanowires, graphene, and other polymers, can be used to incorporate with the electrodes on the cellulose surface.

Flexible energy storage devices based on CNT-cellulose nanocrystals and RTIL (room temperature ionic liquid) in the form of nanocomposites sheets can be used in configuring energy-storage devices like supercapacitors and Li-ion batteries. The nanocrystals, nanofibers, and their derivatives can be highly interesting or demanding because of their simple integration process to use in Li-ion batteries (LIB) as separators, electrolytes, and electrodes. For LIB, the nano-fibrillated cellulose can served as binding the electrode materials and acting as a separator [39].

CNF can be used more in batteries as electrolytes due to multiple advantages such as superior mechanical strength with higher porosity for permeable electrolytes. The elastic modulus of a single cellulose nanofibril has been reported up to 140 GPa [40]. For LIB applications, the micro-fibrillated cellulose has been used as a reinforcement for LIB electrolytes with ionic conductivity of about 103 S/cm [41], and nano-fibrillated cellulose composite with a liquid electrolyte with electrical high ionic conductivity of 5×10^{-5} S/cm [42].

7.5.1 Cellulose in sodium-ion battery (SIB)

Cellulose can be used in another storage system such as a sodium-ion battery (SIB) as an electrolyte. The cellulose mesoporous structure is helpful for the transport of the ions via the fiber surface. The wood fiber as an electrolyte with an initial capacity of 399 mAh/g has been reported for SIB [43, 44]. The electrochemical behavior of the cellulose-based composite electrolyte can be evaluated based on the resulting currents from the cyclic voltammetry measurements. In the case of SIB (using cellulose), there is a need to investigate the Na-ion based batteries at a commercialized production scale and lower cost. There is also a need to develop a paper-based battery with a longer cell lifetime and with ultrathin recyclable flexible paper-based battery [45]. If chemical modification techniques are applied to NC, they can modify the functionality of the hybrid material to a greater extent for applications such as paper batteries, supercapacitors, paper displays, solar cells, etc. [30].

Cellulose has been used as a substrate material as well in polypyrrole (pPy) based electrodes for energy storage. To increase the charge storage capacity, a pPy coating on cellulose was performed by chemical polymerization and the composite (pPy/cellulose) was doped with anthraquinone-2,6-disulfonic acid (pPy[AQS]) via electropolymerization. The voltammogram showed increased charge capacity from 125 C/g to 250 C/g after the electropolymerization [46].

7.5.2 Cellulose-based lithium-ion batteries (LIB)

The conventional LIB electrodes mostly use lithium intercalation materials in both the anodes and cathodes. For the current collector and structural support, the electrode material is usually coated onto the metal foils. The lithium-ion storage material, or any other active material, is being used in conjugation with binding and electronic conductivity enhancing materials such as carbon black (CB). The researchers are using a renewable resource (cellulose) as the substrate for LIB electrodes instead of metal foil current collectors [47]. The process of coating the current collector and electrode materials onto the cellulose paper sheet requires great care. Hu et al. [48] were the first to report the simple Meyer rod coating method to coat CNT and silver nanowires on the xerox papers with sheet resistance as low as 1 ohm per square.

Generally, the cellulose paper substrate, compared to plastic sheets, highly improves the film adhesion and helps in simplifying the coating process at a low cost. Moreover, the porous structure of the cellulose paper facilitates a more conformal coating of the conductive materials because a strong capillary force is available leading to a large contacting surface area between the materials (nanotubes/nanowires and cellulose paper after the solvent is being absorbed and dried out). The cellulose-based LIB has advantages over conventional LIB because they are thin, flexible, and can be used in diverse applications. The main applications are for batteries and, bendable electronics by helping in filling the void spaces in the irregularly shaped batteries such as in the chassis of electric vehicles [49, 50].

The lamination process has been reported to integrate all the components of the LIB into a single paper sheet. The lamination process assembles the CNT current collector layer and active material layer by a slurry ($Li_4Ti_5O_{12}$ (LTO, anode material)/$LiCoO_2$ (LCO, cathode material)) on stainless steel, followed by transferring this double-layer structure to a cellulose paper. A coating with polyvinylidene fluoride (PVDF) can be used as glue. The anode and cathode double-layer structure can be peeled off and from the cellulose paper can be glued, which serves as a separator membrane and also provide mechanical support. Compared to other processes, the lamination process provides better mechanical integrity of the electrode materials and helps in the prevention of leaks through the paper separator. A similar configuration of lamination can be applied to zinc-air batteries [51]. In another study, the flexible zinc-air batteries by printing zinc/carbon/polymer composite anode on one side whereas, the poly(3,4-ethylenedioxythiophene) on the other side of the cellulose paper has been successfully fabricated [52].

7.5.2.1 Cellulose-based binders for LIB

For LIB, the anode and cathode for the electrode material are based on lithium-ion phosphate, sulpfur, TiO_2, silicon, and graphite, while cellulose can be used as a binder. The binders typically in LIBs greatly affect the electrode's mechanical and electrochemical performances. Many binders, most commonly PVDF, require a solvent like *N*-methyl-pyrrolidone (NMP) which is highly toxic. furthermore, PVDF and NMP are both expensive and difficult to reuse/recycle, and can sometimes lead to explosion hazards.

Keeping in mind the cost, environmental safety, and towards safer and greener LIB, linear polymeric cellulose derivatives with substitution by anionic carboxymethyl group such as water-soluble carboxymethyl cellulose (CMC) is highly used and recommended in literature. The literature also suggests that CMC as a binder is safer, cheaper, and can be used as an effective alternative to PVDF with enhanced electrode performance [53]. CMC can be stiff with a low strain at fracture but provides strong bonding with the electrode substrate and electrode materials, which is associated with the high number of carboxymethyl groups on the cellulose chains. Sodium CMC has been used with styrene-butadiene rubber with a specific weight ratio. This blend provides good adhesion between the electrode material and copper substrate. By using a Si-based cathode or graphite-based cathode, the improved lifetime of LIB by about 10 and 100 times was obtained, respectively. Significant improvement in the cyclability has been also observed when sulphur was used as the cathode material [54, 55].

Natural cellulose can also be used as a binder without any modification. However, cellulose is not soluble in water, so ionic liquids can be used as a solvent.

7.5.2.2 Cellulose-based separators for LIB

It has been estimated that the cost of the separator in batteries, especially in LIBS, can be about 20% of the cost of the LIB. Typical separators used in

LIB have limitations because of the lack of a thermal shutdown mechanism causing the closure of the pores and at high temperature terminates the ionic flow [56].

The hydrophilic nature of cellulose papers can be taken into serious account as the water can cause the degradation of lithium salts. However, Jabbour et al. [57] and Leijonmarck et al. [58] suggested that this problem can be minimized by prolonged thermal treatment of cellulose papers. The low cost of cellulose has put forward efforts to replace the existing LIB separators based on polyethylene(PE)/polypropylene (PP) for low power applications [59, 60]. Rice paper has been used as a separator in LIB with cellulose fibers of about 5–40 μm in diameter. The performance was compared to the commercial one and a lower ionic resistance was observed [61, 62].

7.5.2.3 Cellulose-based electrolyte for LIB

Ethylene carbonate and linear carbonates are most commonly used as electrolytes in LIB. But these compounds have drawbacks by being highly volatile and flammable. Water and ionic liquid have also been used as electrolytes, but ionic liquids are limiting the LIB lifetime by leakage. This leakage problem was addressed by the modification of the polymer electrolytes in which the liquid electrolyte are trapped. Keeping in mind these issues, cellulose can be used as a filler material in polymer electrolytes for mechanical reinforcement (mechanical strength).

Nair et al. [63] studied natural cellulose in methacrylic-based thermoset gel polymer electrolyte. By grating poly(ethylene glycol) methyl ether methacrylate under UV, the cellulose was modified to exhibit superior mechanical and electrochemical properties. Whereas, the tunicin cellulose whisker reinforced cross-linked POE (poly-oxyethylene) as an electrolyte with higher ionic conductivity and long-term temperature stability compared to poly(oxyethylene) without cellulose, was investigated by Samir et al. [64, 65], the high mechanical properties of the resultant composites were enhanced by the formation of the cellulosic network through the whisker hydrogen bonds.

7.5.3 Supercapacitors

Similar to LIB, cellulose papers can be used in supercapacitor substrates to give flexibility to the devices compared to the conventional ones, as well as to meet the better requirements for design and power.

Cellulose paper/CNT/MnO_2/CNT-based supercapacitor electrodes have been developed [66]. Many comparisons revealed that cellulose gives better performance compared to cellulose paper-based electrodes and Al_2O_3 coated polyester textile. The cellulose showed better ionic and electron conductivity with excellent capacitance retention up to 50k cycles. In capacitance, the cellulose has the advantage of a higher surface area for higher performance.

Bacterial nano-cellulose (BNC) with smaller sizes (20–100 nm) has drawn high attention for supercapacitor electrodes due to better mechanical strength, smaller size, and chemical stability [12, 67]. Kim et al. [68] made BNC paper from *gluconacetobacter xylinum* to produce the CNT/BNC composite papers which showed an excellent specific capacity of about 20 mF/cm^2. They are shown less than 0.5% reduction over 5k cycles with a current density of 10 A/g. The bending properties were also retained for about 200 bending cycles.

Cellulose can be used as a binder and separator in supercapacitors. Conventionally, active carbon-based double-layer capacitors have been used. Varzi et al. [69] studied cellulose fibers binders that enabled superior performance retention under float tests at high voltage compared to PVDF. The study showed a more significant performance in the case of cellulose fibers. Cellulose/ZnO were found to be multifunctional composite materials that not only improved the dielectric, mechanical, sensing, and thermal properties, but significantly improved the energy storage (supercapacitor activity) even at very low ZnO concentration as well. The ZnO and nano-cellulose prove to be a sustainable component which can revolutionize the entire energy field providing exceptional applications in mechanical, electronic, and chemical sectors [70].

7.5.3.1 Nanocellulose as substrate materials for paper supercapacitors

Nano-cellulose has been used as a robust 1D substrate serving as an excellent electrolyte reservoir and mechanical buffer [71]. The main challenge regarding cellulose is the large pores between the cellulose units. Thus a paper, nontransparent, has difficulty being used in optoelectric devices, but chemical and physical methods, like 2,2,6,6-tetramethylpiperidine-1-oxyl radical (TEMPO) oxidation, can be done to improve the pores scale between the cellulose units down to the nanometer level [72].

7.5.4 Cellulose as electrodes for pseudo-capacitors

For the higher performance of pseudo-capacitors, advanced binder-free electrode materials with high mass loading are required. Bacterial cellulose polypyrrole nanofibers with multi-walled carbon nanotubes by vacuum filtering methods at a mass loading of 7–12 mg/cm^2 were reported to be highly conductive electrodes for pseudo-capacitors. The high capacitance of 2.43 F/cm^2 at a mass of 11.23 mg/cm^2 has been achieved under standard conditions with these cellulose-based electrodes combined with excellent cycling stability (94.5% retention after 5000 cycles) [73].

The structural configuration, such as porosity, stability, and mechanical properties of cellulose, is attractive for flexible supercapacitor paper-based, portable, and wearable electronics. These characteristics make cellulose an excellent substrate material. Different strategies, such as using cellulose hybrid composites based on active carbon, graphene, conductive polymers (polypyrrole and polyaniline), and metal oxides (ZnO, MnO_2, and V_2O_5) have been used for flexible capacitor by direct printing, filtration, surface coating, and conformal coating [74].

7.5.5 Cellulose nanomaterials for nanogenerator developments

The triboelectric nanogenerator (TENG) technology originated in 2012, which converts environmental mechanical energy to electrical energy. The TENG proved to be more desirable in terms of high efficiency, lighter weight, higher power density, lower cost, and manufacturability. But cellulose-based piezoelectric nanogenerators (PENG) technology can also be used for mechanical energy harvesting.

7.5.5.1 Cellulose nanostructure-based triboelectric nanogenerators

Natural cellulose, being a triboelectric positive material, can be used in generating triboelectric output when paired with strong triboelectric negative materials for TENG device fabrication.

Bacteria cellulose has been used to obtain transparent cellulose films. In general, BC has a similar dielectric and triboelectric polarization as wood cellulose leading to a similar level of triboelectric output. Besides these, pure cellulose films and cellulose composites have been developed for TENG applications. Chandrasekhar et al. [75] fabricated polydimethylsiloxane (PDMS)/cellulose composite films for TENG with improved flexibility and mechanical integrity. The cellulose composite film paired with an Al film was assembled for TENG leading to better performance (open circuit voltage = 28 V, short circuit = 2.8 μA, instantaneous peak power = 576 μW) mainly due to internal polarization from the cellulose nanocrystals.

7.5.5.2 Cellulose-based piezoelectric nanogenerators

Piezoelectric nanogenerators (PENG) is an applied principle based on harvesting mechanical energy using piezoelectric nanomaterials [76]. Zhang et al. [77] reported a piezoelectric composite paper based on ferroelectric $BaTiO_3$ (BTO) nanoparticles. Crystalline cellulose is considered a polymeric piezoelectric material. The piezoelectric polarity of the crystalline cellulose is mainly coming from the glucose unit alignment through the glycosidic linkages along the C2 monoclinic lattice [75].

It has been observed that cellulose paper films (xanthate cellulose) morphology consists of ordered and disordered regions, which exhibits better mechanical deflection when a small electrical potential is provided. But this paper was observed to be highly sensitive to humidity. Nevertheless, oriented CNC films, due to the high ordered structure, help in the generation of appreciable piezoelectricity. Levente et al. prepared ordered CNC films with a combination of shear and electric field [78]. Integrating the bulk-related piezoelectric and surface-related triboelectric effect in cellulose nanostructure can be an effective strategy to enhance the mechanical energy harvesting capability.

7.6 Summary

Cellulose-based materials and their derivatives have been identified for potential energy applications. The advanced synthesis, manufacturing, and characterization techniques of the cellulose-based material for energy purpose were evaluated with different strategies leading to better performance. This chapter provided the platform and advantage of using cellulose-based derivatives coupled with zinc oxide for energy material technology. New approaches for the functionalization and modification to assemble these cellulose nanomaterial with different crystal structures and scales were highlighted. Cellulose in energy conversion and storage was also highly explained for application like rechargeable metal batteries compared with cellulose-based state-of-art battery technology. Furthermore, the interfacial interaction between cellulose and metal-oxide (ZnO) effect on the physical and mechanical properties of these materials was investigated. Different strategies to improve the energy application performances have been discussed. Although a great deal of literature can be found on experimental works, no significant computational/theoretical investigation has been done on the energy evaluation capacity of cellulose-based composites, providing a large room for research in these areas.

References

[1] Hiasa S, et al. Isolation of cellulose nanofibrils from mandarin (Citrus unshiu) peel waste. Ind Crop Prod 2014;62:280–5.
[2] Shamskar KR, Heidari H, Rashidi A. Preparation and evaluation of nanocrystalline cellulose aerogels from raw cotton and cotton stalk. Ind Crop Prod 2016;93:203–11.
[3] Yang X, et al. Two-dimensional magnetic field sensor based on silicon magnetic sensitive transistors with differential structure. Micromachines 2017;8(4):95.
[4] Alila S, et al. Non-woody plants as raw materials for production of microfibrillated cellulose (MFC): a comparative study. Ind Crop Prod 2013;41:250–9.
[5] Wang B, Sain M. Isolation of nanofibers from soybean source and their reinforcing capability on synthetic polymers. Compos Sci Technol 2007;67(11-12):2521–7.
[6] Alemdar A, Sain M. Isolation and characterization of nanofibers from agricultural residues—wheat straw and soy hulls. Bioresour Technol 2008;99(6):1664–71.
[7] Ifuku S. Chitin and chitosan nanofibers: preparation and chemical modifications. Molecules 2014;19(11):18367–80.
[8] Habibi Y, Lucia LA, Rojas OJ. Cellulose nanocrystals: chemistry, self-assembly, and applications. Chem Rev 2010;110(6):3479–500.
[9] Oun AA, Rhim J-W. Characterization of nanocelluloses isolated from Ushar (Calotropis procera) seed fiber: effect of isolation method. Mater Lett 2016;168:146–50.

[10] Desmaisons J, et al. A new quality index for benchmarking of different cellulose nanofibrils. Carbohydr Polym 2017;174:318–29.
[11] Hubbe MA, et al. Cellulosic nanocomposites: a review. Bioresources 2008;3(3):929–80.
[12] Klemm D, et al. Nanocelluloses: a new family of nature-based materials. Angew Chem Int Ed 2011;50(24):5438–66.
[13] Yano H, et al. Optically transparent composites reinforced with networks of bacterial nanofibers. Adv Mater 2005;17(2):153–5.
[14] Lizundia E, et al. Increased functional properties and thermal stability of flexible cellulose nanocrystal/ZnO films. Carbohydr Polym 2016;136:250–8.
[15] Alavi M. Modifications of microcrystalline cellulose (MCC), nanofibrillated cellulose (NFC), and nanocrystalline cellulose (NCC) for antimicrobial and wound healing applications. E-Polymers 2019;19(1):103–19.
[16] Wang H, et al. Preparation and characterization of multilayer films composed of chitosan, sodium alginate and carboxymethyl chitosan-ZnO nanoparticles. Food Chem 2019;283:397–403.
[17] Abitbol T, et al. Nanocellulose, a tiny fiber with huge applications. Curr Opin Biotechnol 2016;39:76–88.
[18] Talebian N, Amininezhad SM, Doudi M. Controllable synthesis of ZnO nanoparticles and their morphology-dependent antibacterial and optical properties. J Photochem Photobiol B Biol 2013;120:66–73.
[19] Stanković A, Dimitrijević S, Uskoković D. Influence of size scale and morphology on antibacterial properties of ZnO powders hydrothemally synthesized using different surface stabilizing agents. Colloids Surf B: Biointerfaces 2013;102:21–8.
[20] Ma J, et al. Synthesis of large-scale uniform mulberry-like ZnO particles with microwave hydrothermal method and its antibacterial property. Ceram Int 2013;39(3):2803–10.
[21] Kumar S, Gautam S, Sharma A. Antimutagenic and antioxidant properties of plumbagin and other naphthoquinones. Mutat Res 2013;755(1):30–41.
[22] Rai P, Kwak W-K, Yu Y-T. Solvothermal synthesis of ZnO nanostructures and their morphology-dependent gas-sensing properties. ACS Appl Mater Interfaces 2013;5(8):3026–32.
[23] Ao W, et al. Mechanochemical synthesis of zinc oxide nanocrystalline. Powder Technol 2006;168(3):148–51.
[24] Kołodziejczak-Radzimska A, Markiewicz E, Jesionowski T. Structural characterisation of ZnO particles obtained by the emulsion precipitation method. J Nanomater 2012;2012:.
[25] Mun S, et al. Flexible cellulose and ZnO hybrid nanocomposite and its UV sensing characteristics. Sci Technol Adv Mater 2017;18(1):437–46.
[26] Sahoo K, Biswas A, Nayak J. Effect of synthesis temperature on the UV sensing properties of ZnO-cellulose nanocomposite powder. Sensors Actuators A Phys 2017;267:99–105.
[27] Van den Berg O, Capadona JR, Weder C. Preparation of homogeneous dispersions of tunicate cellulose whiskers in organic solvents. Biomacromolecules 2007;8(4):1353–7.
[28] Zhou Y, et al. Recyclable organic solar cells on cellulose nanocrystal substrates. Sci Rep 2013;3(1):1–5.
[29] Zhou Y, et al. Efficient recyclable organic solar cells on cellulose nanocrystal substrates with a conducting polymer top electrode deposited by film-transfer lamination. Org Electron 2014;15(3):661–6.
[30] Kim J-H, et al. Review of nanocellulose for sustainable future materials. Int J Precision Eng Manuf Green Technol 2015;2(2):197–213.
[31] Hu L, et al. Transparent and conductive paper from nanocellulose fibers. Energy Environ Sci 2013;6(2):513–8.
[32] Iqbal S, Ahmad S. Recent development in hybrid conducting polymers: synthesis, applications and future prospects. J Ind Eng Chem 2018;60:53–84.

[33] Zhengying W, et al. Molybdenum disulfide based composites and their photocatalytic degradation and hydrogen evolution properties. Progr Chem 2019;31(8):1086.
[34] Li Z, et al. Cellulose nanofiber-templated three-dimension TiO2 hierarchical nanowire network for photoelectrochemical photoanode. Nanotechnology 2014;25(50):504005.
[35] Ke D, et al. CdS/regenerated cellulose nanocomposite films for highly efficient photocatalytic H2 production under visible light irradiation. J Phys Chem C 2009;113(36):16021–6.
[36] Li Z, et al. Highly efficient capillary photoelectrochemical water splitting using cellulose nanofiber-templated TiO2 photoanodes. Adv Mater 2014;26(14):2262–7.
[37] Yin Y, et al. Formation of hollow nanocrystals through the nanoscale Kirkendall effect. Science 2004;304(5671):711–4.
[38] Shi J, Wang X. Hierarchical TiO 2–Si nanowire architecture with photoelectrochemical activity under visible light illumination. Energy Environ Sci 2012;5(7):7918–22.
[39] Pushparaj VL, et al. Flexible energy storage devices based on nanocomposite paper. Proc Natl Acad Sci 2007;104(34):13574–7.
[40] Iwamoto S, et al. Elastic modulus of single cellulose microfibrils from tunicate measured by atomic force microscopy. Biomacromolecules 2009;10(9):2571–6.
[41] Chiappone A, et al. Microfibrillated cellulose as reinforcement for Li-ion battery polymer electrolytes with excellent mechanical stability. J Power Sources 2011;196(23):10280–8.
[42] Willgert M, et al. Cellulose nanofibril reinforced composite electrolytes for lithium ion battery applications. J Mater Chem A 2014;2(33):13556–64.
[43] Zhang T, et al. Recent advances of cellulose-based materials and their promising application in sodium-ion batteries and capacitors. Small 2018;14(47):1802444.
[44] Mukherjee S, et al. Electrode materials for high-performance sodium-ion batteries. Materials 2019;12(12):1952.
[45] Zhu H, et al. Tin anode for sodium-ion batteries using natural wood fiber as a mechanical buffer and electrolyte reservoir. Nano Lett 2013;13(7):3093–100.
[46] Kim SY, Hong J, Palmore GTR. Polypyrrole decorated cellulose for energy storage applications. Synth Met 2012;162(15-16):1478–81.
[47] Qi W, et al. Nanostructured anode materials for lithium-ion batteries: principle, recent progress and future perspectives. J Mater Chem A 2017;5(37):19521–40.
[48] Hu L, et al. Highly conductive paper for energy-storage devices. Proc Natl Acad Sci 2009;106(51):21490–4.
[49] Jabbour L, et al. Cellulose-based Li-ion batteries: a review. Cellulose 2013;20(4):1523–45.
[50] Zhang H, et al. Preparation of cellulose-based lithium ion battery membrane enhanced with alkali-treated polysulfonamide fibers and cellulose nanofibers. J Membr Sci 2019;591:117346.
[51] Mahmood N, Hou Y. Electrode Nanostructures in Lithium-Based Batteries. Adv Sci 2014;1(1):1400012.
[52] Hilder M, Winther-Jensen B, Clark N. based, printed zinc–air battery. J Power Sources 2009;194(2):1135–41.
[53] Lux S, et al. Low cost, environmentally benign binders for lithium-ion batteries. J Electrochem Soc 2010;157(3):A320.
[54] He M, et al. Enhanced cyclability for sulfur cathode achieved by a water-soluble binder. J Phys Chem C 2011;115(31):15703–9.
[55] Liu W-R, et al. Enhanced cycle life of Si anode for Li-ion batteries by using modified elastomeric binder. Electrochem Solid-State Lett 2004;8(2):A100.
[56] Chung Y, Yoo S, Kim C. Enhancement of meltdown temperature of the polyethylene lithium-ion battery separator via surface coating with polymers having high thermal resistance. Ind Eng Chem Res 2009;48(9):4346–51.

[57] Jabbour L, et al. Flexible cellulose/LiFePO 4 paper-cathodes: toward eco-friendly all-paper Li-ion batteries. Cellulose 2013;20(1):571–82.
[58] Leijonmarck S, et al. Single-paper flexible Li-ion battery cells through a paper-making process based on nano-fibrillated cellulose. J Mater Chem A 2013;1(15):4671–7.
[59] Xue X, et al. Hybridizing energy conversion and storage in a mechanical-to-electrochemical process for self-charging power cell. Nano Lett 2012;12(9):5048–54.
[60] Xu X, et al. Cellulose nanocrystals vs. cellulose nanofibrils: a comparative study on their microstructures and effects as polymer reinforcing agents. ACS Appl Mater Interfaces 2013;5(8):2999–3009.
[61] Zhang L, et al. Rice paper as a separator membrane in lithium-ion batteries. J Power Sources 2012;204:149–54.
[62] Li J, et al. Development of separators for lithium ion battery applied in vehicles. Sci Sin Chim 2014;44(7):1116–24.
[63] Nair JR, et al. Novel cellulose reinforcement for polymer electrolyte membranes with outstanding mechanical properties. Electrochim Acta 2011;57:104–11.
[64] Azizi Samir MAS, Alloin F, Dufresne A. High performance nanocomposite polymer electrolytes. Compos Interfaces 2006;13(4-6):545–59.
[65] Samir MASA, et al. Cellulose nanocrystals reinforced poly (oxyethylene). Polymer 2004;45 (12):4149–57.
[66] Gui Z, et al. Natural cellulose fiber as substrate for supercapacitor. ACS Nano 2013;7 (7):6037–46.
[67] Gan S, et al. Highly porous regenerated cellulose hydrogel and aerogel prepared from hydrothermal synthesized cellulose carbamate. PLoS One 2017;12(3)e0173743.
[68] Kang YJ, et al. All-solid-state flexible supercapacitors fabricated with bacterial nanocellulose papers, carbon nanotubes, and triblock-copolymer ion gels. ACS Nano 2012;6(7):6400–6.
[69] Varzi A, Balducci A, Passerini S. Natural cellulose: a green alternative binder for high voltage electrochemical double layer capacitors containing ionic liquid-based electrolytes. J Electrochem Soc 2014;161(3):A368.
[70] Yu Y, et al. Evolution of hollow TiO2 nanostructures via the Kirkendall effect driven by cation exchange with enhanced photoelectrochemical performance. Nano Lett 2014;14(5):2528–35.
[71] Chen C, Hu L. Nanocellulose toward advanced energy storage devices: structure and electrochemistry. Acc Chem Res 2018;51(12):3154–65.
[72] Luo Y, et al. The cellulose nanofibers for optoelectronic conversion and energy storage. J Nanomater 2014;2014:.
[73] Li S, et al. Freestanding bacterial cellulose–polypyrrole nanofibres paper electrodes for advanced energy storage devices. Nano Energy 2014;9:309–17.
[74] Wang Z, et al. Surface modified nanocellulose fibers yield conducting polymer-based flexible supercapacitors with enhanced capacitances. ACS Nano 2015;9(7):7563–71.
[75] Chandrasekhar A, et al. A microcrystalline cellulose ingrained polydimethylsiloxane triboelectric nanogenerator as a self-powered locomotion detector. J Mater Chem C 2017;5 (7):1810–5.
[76] Uddin AI, Chung G-S. Wide-ranging impact-competent self-powered active sensor using a stacked corrugated-core sandwich-structured robust triboelectric nanogenerator. Sensors Actuators B Chem 2017;245:1–10.
[77] Zhang G, et al. Novel piezoelectric paper-based flexible nanogenerators composed of batio3 nanoparticles and bacterial cellulose. Adv Sci 2016;3(2):1500257.
[78] Way AE, et al. pH-responsive cellulose nanocrystal gels and nanocomposites. ACS Macro Lett 2012;1(8):1001–6.

Chapter 8

Cellulose nanocrystal (CNC): Inorganic hybrid nanocomposites

Marya Raji[a], Hamid Essabir[a,b], Hala Bensalah[c], Kamal Guerraoui[c], Rachid Bouhfid[a], and Abou el Kacem Qaiss[a]
[a]*Moroccan Foundation for Advanced Science, Innovation and Research (MAScIR), Composites and Nanocomposites Center, Rabat Design Center, Rabat, Morocco,* [b]*Mechanic, Materials, and Composites (MMC), Laboratory of Energy Engineering, Materials and Systems, National School of Applied Sciences of Agadir, Ibn Zohr University, Agadir, Morocco,* [c]*Equipe de Mécanique et des Matériaux (E.M.M), Centre de Recherche en Enérgie (C.R.E.), Faculty of Sciences, Mohammed V University in Rabat, Rabat, Morocco*

8.1 Introduction

Hybrid nanocomposites are the result of the combination of various phases where at least two reinforcements are integrated into a matrix to improve the composite's physical, mechanical and chemical properties, as well as durability and thermal stability [1–3]. Currently, this kind of material based on the combination of two different materials gained increasing attention in both academic research and in industry, as a way to overcome their environmental limitations including aggregation, low thermal stability, toxicity, poor processability, high cost, etc. [4]. Several kinds of nanomaterials have shown increased structural or functional properties when incorporated into organic-inorganic nanocomposites. For example, graphene [5–7], nano-clays [8–15] ferrite [16], natural fibers [11, 17–24] and carbon nanotubes [17], as compared to each constituent alone. In the last years, hybridization of inorganic and organic components gained more attention because of their high request in many application fields as like optical, catalytic, biomedical and electrochemical devices typically using inorganic materials like metal ions, metal cations (or metal clusters), salts, oxides, sulfides, nonmetallic elements and their derivatives [25]. Although organic materials including organic groups or molecules, ligands, biomolecules, pharmaceutical molecules and polymers have been used, biomolecules like natural fibers have attracted a great deal of attention because of their interesting

Cellulose Nanocrystal/Nanoparticles Hybrid Nanocomposites. https://doi.org/10.1016/B978-0-12-822906-4.00002-5

properties attributed to their high cellulose content associated to its rigid rod-shaped nanostructure, good aqueous colloidal stability, high surface area, favorable surface modification, excellent biodegradability, biocompatibility and outstanding mechanical properties as high Young's modulus in the longitudinal (100 GPa) or transversal (10–50 GPa) direction.

Cellulose nanoparticles are the most renewable, abundant and biodegradable materials available from several biomass sources such as trees, tunicate, plants and bacteria [26]. They can also achieve other innovative properties like magnetic, electrical and optical properties by combination with inorganic nanoparticles. Their chemical structure consists of hydroxyl groups on linear chains of β-D-glucopyranose, but suffers from their low grafting ratios and limited functional groups. However, due to their high surface energy, large surface area, water absorption, reductive surface, functional group and due to environmental and human health concerns, there is increasing interest in using cellulose nanocrystals as supports for the growth of several inorganic nanoparticles [27, 28]. In general, inorganic nanoparticles are characterized by high surface energy leading to a mutual attraction of the nanoparticles through van der Waals forces or chemical bonding leading to aggregates and agglomerates. These complex phenomena affect their properties and applications. In order to overcome these problems, high-surface supporting materials like cellulose nanocrystals are used to adsorb and produce inorganic nanoparticles at small dimensions [29, 30]. The overall properties of the produced inorganic nanoparticles are generally depending on their size (1–100 nm), shape, surface area, crystallinity and stability which are significantly related to their synthesis method and conditions. Several methods are used for the production of inorganic nanoparticles by using cellulose nanocrystals as a template including sol-gel synthesis, co-precipitation, hydrothermal method and microemulsion techniques [31].

8.2 Cellulose nanocrystals

8.2.1 General overview on the chemistry and properties of cellulose

Cellulose nanocrystals (CNC) are generally referred as cellulose whiskers with nanosized particles that were firstly extracted by Rånby and Ribi from a wide variety of renewable cellulosic sources (trees, annual plants, algae, bacteria, tunicates and biomass) [32]. Such products are characterized by high crystallinity and their aqueous suspensions displaying a colloidal behavior that is of high interest to numerous industrial areas because of their availability, high aspect ratio, rod-like morphology, lightweight, low cost and renewability [33, 34]. Compared to bulk cellulose, these nanocrystals lead to chemical modification and transparency of their nanocomposites. The nanocomposites based on this organic compound received high interest in the last years owing to their extraordinary mechanical performances, such as high tensile strength and Young's modulus, at relatively low concentrations [35]. The fraction of crystalline region in cellulose fibers generally depend on the plant sources. As mentioned before, the main sources of cellulose nanocrystals are plants, algae, bacteria and dome sea animals. Their chemical structure formed by condensation, consists of monomers joined together by glycosidic oxygen bridges which is the same in all sources and is defined as a macromolecule of anhydroglucose units (AGU) covalently linked via β-1,4-glycosidic bonds of variable length through acetal functions between the equatorial hydroxyl group of the carbon atom number 4 and the carbon atom number 1. However, the dimensions of cellulose nanocrystals depend on their origin and the conditions of the preparation method like time, purity, temperature, etc. [36–38]. In general, cellulose nanocrystals are stiff rod-like particles with diameters in the range of 2–20 nm and 100–600 nm in length comprising of cellulose chain segments in an almost perfect crystalline structure. Table 8.1 summarizes some of the cellulose nanocrystals sources with their dimensions depending on the preparation method, while Table 8.2 summarizes some examples of cellulose nanocrystals dimensions prepared by acid hydrolysis at different times (120, 150 and 180 min) from coconut husk fibers [41]. Furthermore, the type of acid and the preparation conditions may also influence the thermal stability of cellulose nanocrystals which is a critical parameter in the field of nanocomposites production. Table 8.3 summarizes some examples of the effect of the type of acid on the thermal stability of cellulose nanocrystals. Cellulose

TABLE 8.1 Overview of the dimensions of cellulose nanocrystals depending on the source and production method [39, 40].

Source	Preparation method	Length (nm)	Width (nm)	Aspect ratio (*L*/*D*)
Wood	H_2SO_4 hydrolysis	100–300	3–5	20–100
Sisal	H_2SO_4 hydrolysis	100–300	3–5	33–60
Ramie	H_2SO_4 hydrolysis	70–200	5–15	12–40
Acacia pulp	H_2SO_4 hydrolysis	100–250	5–15	6–50
Bacteria	HCl hydrolysis	160–420	15–25	7–23
Bacteria	H_2SO_4 hydrolysis	100–1000	10–50	2–50
Tunicates	Enzymatic Hydrolysis	1300	20	>70
Cotton	HCl hydrolysis	100–150	5–10	10–30
Coconut husk fibers	H_2SO_4 hydrolysis	58–477	5.3–6.6	35–44

TABLE 8.2 The effect of hydrolysis conditions on the cellulose nanocrystal's properties [1].

Hydrolysis time (min)	Length, L (nm)			Width, D (nm)	Average aspect ratio (L/D)
	min	max	average		
120	74	408	210±78	5.3±1.3	42±16
150	75	515	218±99	6.6±1.7	44±20
180	73	409	177±80	6.1±1.7	35±16

TABLE 8.3 The effect of acid type on the thermal stability of cellulose nanocrystals [42].

Sample code	Hydrolysis catalysis	Acid concentration (%)	T_{max} (°C)
UFC-ref	–	–	341.6
CNC-S 31%	H_2SO_4	31.1	310.8
CNC-S 43%		43.4	297.8
CNC-S 56%		56.2	218.8
MCC-P 60%	H_3PO_4	60.0	336.8
MCC-P 69%		69.0	310.7
MCC-P 74%		74.0	279.0
Sample denotation	Modifier	Molar ratio of SA:OH	T_{max} (°C)
CNC-SA 2:1	Succinic anhydride	2:1	361.0
CNC-SA 3:1		3:1	359.1
CNC-SA 5:1		5:1	362.4
CNC-SA 10:1		10:1	360.8

nanocrystals are also known for their reinforcing ability to enhance polymer nanocomposite performances at low concentrations due to their impressive mechanical properties including an elastic modulus between 120–143 GPa and a tensile modulus of 7500 MPa, which make them interesting candidates for the production of nanocomposites.

8.2.2 Extraction techniques of cellulose nanocrystals

In general, two main sources of nanocellulose are mostly used to produce cellulose nanocrystals: lignocellulosic material and bacteria. The extraction of cellulose nanoparticles from lignocellulosic sources usually requires a pre-treatment or pulping process in order to purify the micron-sized holocellulose fibers via complete or partial removal of the matrix materials including hemicellulose, lignin and impurities [24]. The alkaline pre-treatment, also named delignification, is one of the simplest and cheapest processes to partially remove lignins and other noncellulosic components [11, 43]. The bleaching process is another common process used to chemically purify and extract the cellulose fraction from lignocellulosic materials [44–48]. The extraction of cellulose nanocrystals can be performed via three routes: acid hydrolysis, enzymatic hydrolysis and mechanical treatment. The mechanical process imposes high shear forces in order to defibrillate the cellulose fibers and convert the micro-sized fibers into nano-sized scales. This processing is generally carried out through either high-pressure homogenization, grinding, cryocrushing, microfluidization and high-intensity sonification processing. The produced nanofibrillated cellulose via these methods is known for its efficiency and is negatively charged imparting higher mechanical properties owing to the high surface area since longer and highly entangled fibers (percolation networks) are produced [49, 50]. However, they require the use of high energy and long-time making them costly. On the other hand, strong acid hydrolysis is a promising approach for a simpler and eco-friendly production of cellulose microfibrils based on the transversal cleavage of noncrystalline fractions of cellulose microfibrils, leading to the so-called cellulose nanocrystals or (nano) whiskers. This process is based on decreasing the degree of polymerization of cellulose chains until a cut-off level is achieved which is known as the level-off degree of polymerization (LODP) [51]. Strong acid hydrolysis is the main method used to produce highly crystalline cellulose due to the high accessibility of hydroxyl groups present on the cellulose surface by sulfate groups of the acid during the hydrolysis process. This esterification process grafts the sulfate groups on the surface of cellulose chains causing electrostatic repulsion generated by the produced negative charge which is a crucial factor to obtain an excellent dispersion and stability of the cellulose nanocrystal's aqueous suspension. Another way to produce cellulose nanocrystals is by combining the enzymatic hydrolysis with mechanical shearing leading to cellulose nanomaterials with controlled dimensions down to the nanoscale with diameters of ~5-6 nm leading to better mechanical and thermal properties [38, 52].

8.3 Cellulose nanocrystals: Inorganic hybrid nanocomposites

Hybrid materials contain two or more different components having different functionalities which generally offer superior performance compared to any of the separate components in enhancing the mechanical, thermal, optical and electrical properties. The inorganic part can be ionic complexes comprising of cations and anions combined together via ionic bonding such as metal ions, metal clusters or particles, oxides, salts, sulfides, nonmetallic elements and their derivatives, while the organic components can be molecules, biomolecules, ligands, pharmaceutical substances, polymers, etc. [25, 53, 54].

8.3.1 Synthesis of cellulose-inorganic hybrid nanocomposites

The production of hybrid nanocomposites based on organic and inorganic fillers require a high dispersion-distribution of nanosized fillers inside the matrix as this may compromise the end properties of the nanocomposites. On this basis, synthesis techniques and manufacturing processes play a critical role allowing the production of the nanocomposite with high properties without agglomeration [55–57]. Many techniques have been developed to synthesize inorganic nanoparticles on cellulose nanocrystals surface for different applications and different properties including coprecipitation, sol-gel processing, micro-emulsion synthesis and hydrothermal processing [58–60]. In all cases, cellulose plays the role of a reducing agent, structure-directing agent and stabilizer via anchoring and stabilizing the inorganic particles with the hydroxyl groups on the surface of cellulose nanocrystals [61].

8.3.1.1 Coprecipitation process

The coprecipitation process, also called in situ reduction, is the most clean method to synthesize stable inorganic nanoparticles based on the precipitation of soluble metal ions from aqueous solutions via chemical reactions and subsequent thermal decomposition of the produced inorganic nanoparticles with controlled sizes and high properties [62, 63]. In this process, several reactions simultaneously occur starting by nucleation, growth, coarsening and finishing by agglomeration [64]. Coprecipitation is widely used in biomedical applications because of its ease of operation and low energy consumption without harmful materials and procedures. Generally, the coprecipitation technique of metal nanoparticles on cellulose surfaces can be performed via two ways: (i) chemical methods by the addition of acids, bases or complex-forming agents [65], or (ii) physical process under hydrothermal conditions like temperature change, pH, solvent evaporation and concentration of the reactant in aqueous media [57, 66]. As mentioned before, the precipitation formation of metal nanoparticles from a homogeneous liquid phase carried out under supersaturation conditions is composed of two steps: nucleation of the smallest elementary particles, which is controlled by the activities of the solutes and solubility product constant, followed by the growth or agglomeration of the new stable phase under the precipitation conditions (temperature and concentration gradients) [59]. This process may be affected by several precipitation factors such as the nature and rate of the gradients, temperature, pH, concentrations of the reactant, nanoparticles solubility, solvent evaporation, stirring rate and rate of the basic solution addition. To stabilize the produced inorganic nanoparticles via coprecipitation reactions, various chemical products, either introducing steric stabilization or electrostatic repulsion, can be used such as polyelectrolytes, surfactants, polymers and amphiphilic molecules [58].

8.3.1.2 *Sol-gel processing*

In 1640, Van Helmont precipitated for the first time SiO_2 upon acidifying an alkaline silicate solution. However, Graham in 1864 proposed the name of "sol-gel" for this process [67]. The sol-gel process is a wet technique typically used to produce an organic-inorganic nano-structured material. This process is performed in five steps starting with solvation using water or an organic solvent (acid or base), then hydrolyzing or alcoholizing the solute followed by condensation of the precursor leading to a colloidal suspension of inorganic networks in sol [59]. Simultaneously, the drying and gelling of the resulting hydroxyl groups in sol, considered as a solid macromolecule of an inorganic salt or metal alkoxide immersed in a solvent, are performed through solvent removal or by a chemical reaction to form a network in a continuous liquid-phase (gel) with a final post-treatment and transition into a solid oxide material. The basic reactions of the sol-gel process are summarized in Fig. 8.1. The advantages of this

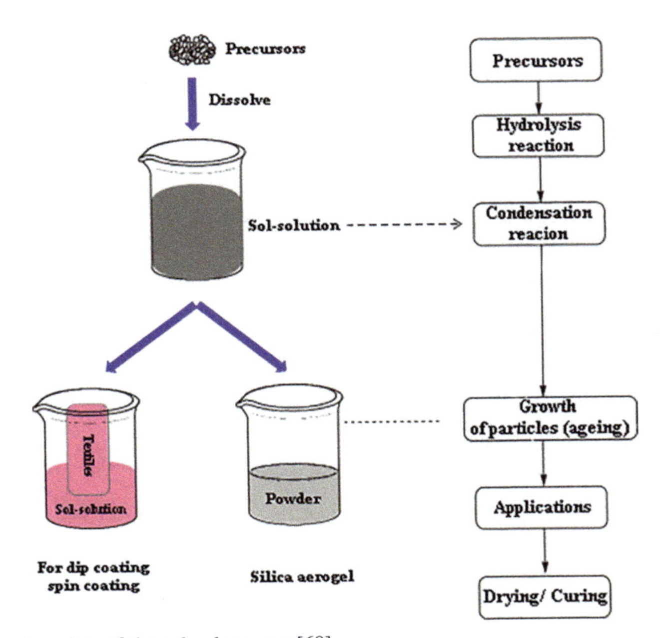

FIG. 8.1 Overview of the sol-gel process [68].

process are that it can be performed at ambient temperature and directly leads to nanosized particles. Various parameters may affect the quality of the produced nanoparticles: rate of hydrolysis and condensation, temperature, solution composition and pH of the reaction medium. Nowadays, the sol-gel technique is a common technology for the fabrication of ultrafine inorganic powders on cellulose fibers applied in various fields like superconducting materials, nonlinear optical materials, functional ceramic materials, porous glass materials, catalysts and enzyme carriers [67, 69].

8.3.1.3 Pickering emulsion synthesis

The microemulsion process was originally proposed in 1959 by Schulman et al. [70]. This process is a macroscopically homogeneous and thermodynamically stable isotropic dispersion of at least two immiscible components, a polar phase (water) and a nonpolar phase (oil), in the presence of an amphiphilic molecule (surfactant) imparting repulsive interactions resulting in a monolayer interface between both phases and the formation of a transparent solution [71]. The advantages of this process are the possibility of producing a large variety of nanosized, crystalline and high surface area particles by using simple conditions (ambient temperature and pressure) and equipment. There are many cationic, anionic and nonionic surfactants that can be used in this process to provide kinetic stability to water and oil dispersions in the form of macroemulsion with different physicochemical properties over a long lifetime. The choice of a suitable surfactant depending on the competitive adsorption of particles and surfactants to the free oil-water interface and its concentration are the main parameters in the formation of the transparent microemulsion [72–75]. Fig. 8.2 shows the particle and oil/water interfacial tension in Pickering emulsion.

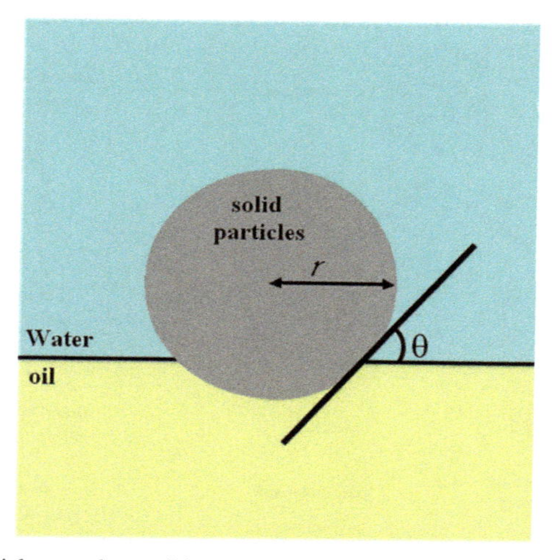

FIG. 8.2 Solid particle at a planar oil/water interface [74].

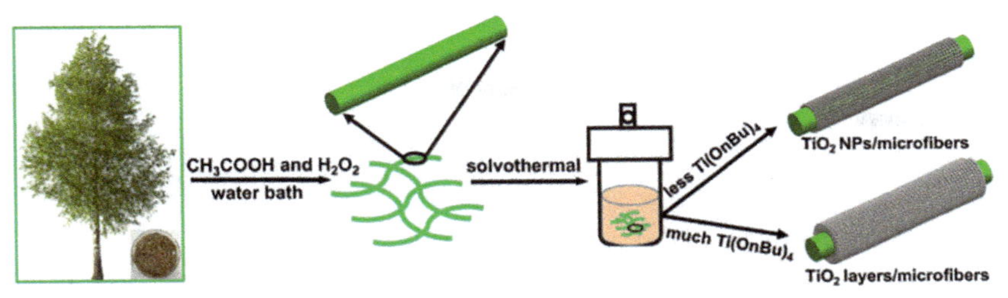

FIG. 8.3 Schematic representation of the steps to prepare TiO_2/cellulose microfibre composites [77].

8.3.1.4 *Hydrothermal/solvothermal processing*

The hydrothermal process or the solvothermal method are methods for the synthesis of cellulose-inorganic nanoparticles based on the reaction performed by overheating an aqueous solution as a precursor without using retention aids. It is generally performed in a sealed container like a steel autoclave at high temperature and high vapor pressure [55]. The choice of the chemical composition of the precursors and suitable solvent, temperature, volume and pressure are critical factors in producing inorganic nanoparticles with ultrafine sizes and shapes. This method results in a high coverage of nanoparticles on the cellulose surface and is very successful to create available active sites. However, it is a slow reaction even at any given temperature [76]. Fig. 8.3 presents an overview of all the steps involved.

8.3.2 Characterization of cellulose-inorganic hybrid nanocomposites

Nowadays, cellulose-inorganic hybrid nanocomposites have exciting characteristics and the combination of both types of particles leads to promising functional nanocomposites with exclusive properties in many applications. However, the use of cellulose suspensions as supports for inorganic nanoparticles to synthesize inorganic nanoparticles started in 2003 [60]. Cellulose nanocrystals are a perfect template including reducing agent, structure-directing agent and stabilizer to produce inorganic nanoparticles with different structures including mesoporous silica (Si, SiO_2, SiC) nanoparticles, silver (Ag) nanoparticles, calcium carbonate ($CaCO_3$) nanoparticles, gold (Au) nanoparticles, palladium (Pd) nanoparticles, nickel (Ni) nanocrystals, platinum (Pt) nanoparticles, selenium (Se) nanoparticles, titanium (Ti) core-shell and hollow nanoparticles TiO_2, nanocubes of cadmium sulfide (CdS), zinc sulfide (ZnS), lead sulfide (PbS), Au-Ag alloy nanoparticles and Ag-Pd alloy nanoparticles [27]. Fig. 8.4 presents some examples of cellulose-inorganic nanoparticles hybrid nanocomposites [57].

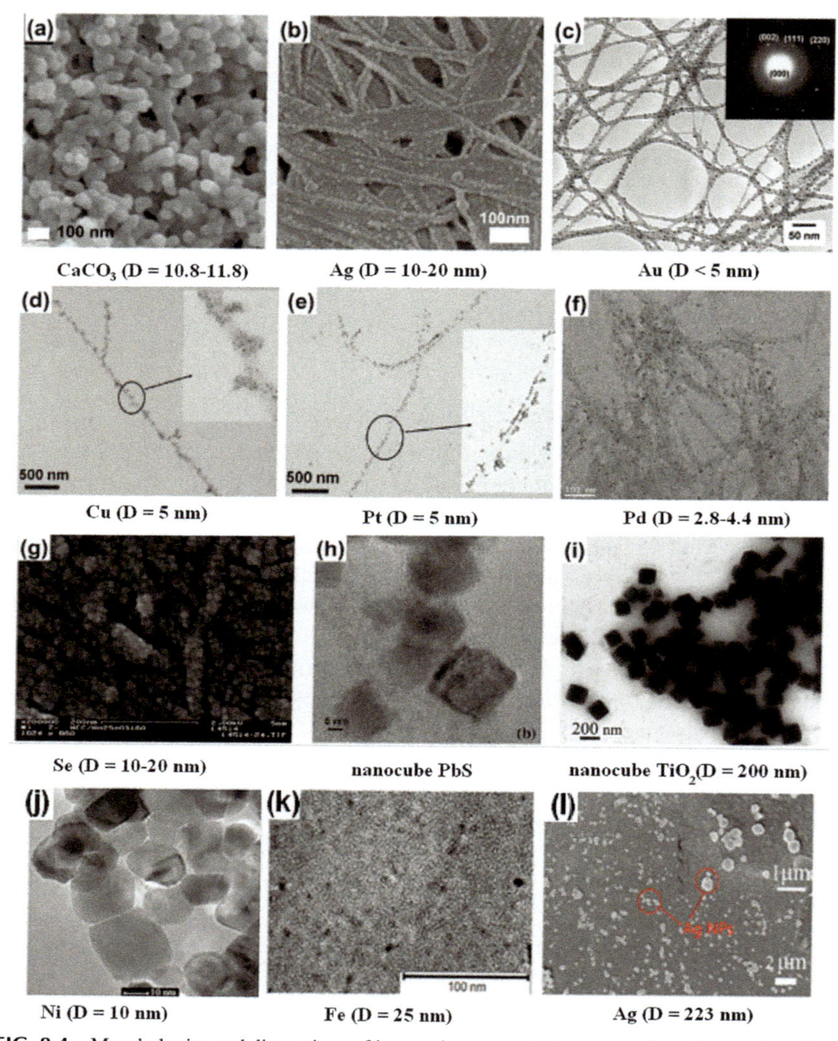

FIG. 8.4 Morphologies and dimensions of inorganic nanoparticle supported synthesis with cellulose nanocrystals based on: (A) $CaCO_3$ [61]; (B) Ag [61]; (C) Au [61]; (D) Cu [61]; (E) Pt [61]; (F) Pd [61]; (G) Se [61]; (H) PbS [61]; (I) TiO_2 [61]; (J) Ni [78]; (K) Fe [59]; Ag [79].

8.3.2.1 Cellulose-silica nanoparticles hybrid nanocomposites

Research effort on the development of cellulose-silicate nanoparticles hybrid nanocomposites with special performances consist of superhydrophobic and high thermal properties in many fields of applications. Xie et al. developed cellulose-silica nanoparticles hybrid nanocomposites via the sol-gel process using tetraethoxyorthosilicate (TEOS) as precursor, c-aminopropyltriethoxylsilane (APTES) as coupling agents and 2,4,6-tri[(2-epihydrin-3-bimethyl-ammonium)propyl]-1,3,5-triazine chloride (tri-EBAC) as crosslinking agent [54]. The produced film showed a good and smooth surface with enhanced thermal properties. Then, Xie et al. [80] used the sol-gel method to produce cellulose-silica nanoparticles hybrid nanocomposites with superhydrophobic properties for oil/water separation. The special structure of the prepared membrane was achieved by using TEOS and hexadecyltrimethoxysilane (HDTMS) with superhydrophobicity and superoleophilicity leading to water-in-oil and oil-in-water emulsion separation.

8.3.2.2 Cellulose-gold nanoparticles hybrid nanocomposites

Hybrid nanocomposites based on cellulose and gold nanoparticles having a wide range of interesting characteristics for several applications like for antioxidant materials, surface-enhanced Raman scattering (SERS), water purification, sensors, biomedical applications and conductive hybrid films were prepared. Rie et al. [56] presented an overview of various preparation methods for cellulose-gold hybrid nanocomposites and summarized the different areas of applications. For example, cellulose-gold nanoparticles hybrid nanocomposites for water treatment were prepared by Weng et al. [56] for the detection and recycling of heavy arsenic metal in water by cellulose-gold nanoparticles ester membrane coupled with laser-induced desorption/ionization mass spectrometry (LDI-MS). The prepared LDI-MS-based assay was used for the detection of AsO_2 using Au NP-modified membranes in alkaline solution leading to aggregation of the Au NP. Another application of cellulose-gold nanoparticles hybrid nanocomposite is for protein separation, You et al. [81] reported that quaternized cellulose-AuNP (QC-Au) coated capillaries allowed the separation of glycoproteins with high resolution. The hybrid nanocomposites can also be used in milk and tear samples. The results showed that the QC-Au coating has potential for use in the analysis of biological samples because of its high detection sensitivity for lysozymes. Furthermore, the cellulose-gold nanoparticles hybrid nanocomposites also showed tunable optical, mechanical and electrical properties. The conductive films by Liu et al. [82] were prepared in two steps. Firstly, the spin-coating of viscous solutions of cellulose and *N*-methylmorpholine-*N*-oxide/dimethyl sulfoxide and their immersion in aqueous Au NP to regenerate the metal. Schlesinger et al. [83] prepared gold nanoparticles into mesoporous photonic cellulose (MPC) to produce hybrid nanocomposite films. The size of the monodisperse gold nanoparticles increased with increasing the metal salt solution concentration and then used to prepare the mesoporous chiral nematic structure of the mesoporous photonic cellulose to stabilize the dimensions [83].

8.3.2.3 Cellulose-silver nanoparticles hybrid nanocomposites

Cellulose nanocrystals have been combined with silver nanoparticles for biosensor applications in the detection of catechol, for the preparation of antibacterial fibers, for catalytic materials and hydrogen storage. The cellulose-silver nanoparticles hybrid nanocomposites were prepared via in situ chemical reduction and chemical binding methods. In both techniques, the cellulose nanocrystals were used as stable templates and were doped with silver nanoparticles to produce nanosized fiber dimensions (1.5 nm) with anti-bacterial properties including high resistance against the growth of *Escherichia coli* (*E. coli*) and *Staphylococcus aureus* (*S. aureus*) [79, 84–86]. In the work of Abdel-Halim et al., [86] hydroxypropyl cellulose samples having a molar substitution of 0.4 or higher have been used in the preparation of silver nanoparticles through the reduction of silver nitrate. UV-vis spectra of the prepared silver nanoparticles provided full reduction of silver ions to silver nanoparticles generally performed at pH 12.5 under 90 °C for 90 min [86]. In another work, presented by Drogat et al., cellulose nanocrystals were used as a template for silver nanoparticles to produce hybrid nanocomposites for numerous medical applications such as fighting infections, more especially as components of wound dressings and antifouling coatings [84]. The method is based on periodate oxidation of cellulose nanocrystals to generate an aldehyde functions that can be used to reduce Ag+ into Ag^o in mild alkaline conditions. The prepared cellulose-silver nanoparticles have been verified for their antibacterial properties against strains of *Escherichia coli* (Gram negative) and *Staphylococcus aureus* (Gram positive) and have shown great potential. The combination of organic-inorganic hybrid nanocomposites also find application in water purification via metal ions removal from water. Attarad et al. prepared two kinds of the organic-inorganic hybrid nanocomposite (zinc, silver) based on cellulose-silver nanoparticles through the co-precipitation method and was exposed to water containing Hg^{2+}, Cr^{3+}, Co^{2+}, Pb^{2+} and Ni^{2+} at 25 °C (pH 5.5) for 400 min, separately [85]. The results showed that cellulose-silver nanoparticles hybrid nanocomposites can be efficiently used for competitive adsorption of heavy metal ions and their efficiency were in the sequence of $Hg^{2+}>Ni^{2+}>Cr^{3+}>Co^{2+}>Pb^{2+}$ from wastewater.

8.3.2.4 Cellulose-palladium nanoparticles hybrid nanocomposites

Palladium nanoparticles, like all metal nanoparticles, are characterized by small size and large surface to volume ratio. When combined with cellulose nanocrystals, they show several interesting properties [27, 87]. Wu et al. [87] developed cellulose-palladium nanoparticles hybrid nanocomposites via a green process in which the cellulose nanocrystals played a dual role as a supporting matrix and a reductant. The prepared hybrid nanocomposites showed a high catalytic activity for the reduction of methylene blue and 4-nitrophenol and may used for catalysis, sensors and other potential applications [87]. Another green process was proposed by Rezayat et al. [27] to produce catalytically active palladium nanoparticles supported on cellulose nanocrystals. The production of hybrid nanocomposites was performed using subcritical and supercritical CO_2 in one-step from Pd(hexafluoroacetylacetonate)2(Pd(hfac)2) [27]. The resulting product was shown to be catalytically active in palladium-mediated cross-coupling reactions of arylhalides and alkenes.

8.3.2.5 Cellulose-metal oxide nanoparticles hybrid nanocomposites

In recent years, a new approach for water purification was proposed by using cellulose-inorganic nanoparticles hybrid nanocomposites as antimicrobial agents with extreme toxicity to Gram-positive and Gram-negative bacteria at low concentration [60, 88]. In the work presented by Gouda et al. several metal oxide nanoparticles, including nano-nickel oxide (nNiO), nano-copper oxides (nCuO) and nano-iron oxides (nFe_3O_4), were deposited onto aminated cellulose (Acell)) using a co-precipitation method [88]. The purification of domestic wastewater was done using the synthesized metal oxide/aminated cellulose (nMO/Acell) nanocomposite as sterilizers. The results for the disk diffusion test evaluation of the prepared hybrid nanocomposites showed that cellulose-nFe_3O_4 had a higher antibacterial effect than cellulose-nCuO and cellulose-nNiO. Moreover, the purification of domestic wastewater using 1.0 mg of nFe_3O_4, nCuO and nNiO in 1 g of Acell was able to respectively kill 99.6%, 94.5% and 92.0% of the total and fecal coliforms within 10 min.

8.3.3 Cellulose-inorganic hybrid nanocomposites applications

Over the last decades, the applications for cellulose-inorganic nanoparticles hybrid nanocomposites have attracted a great deal of attention from academic and industrials researchers. As mentioned before, the combination of inorganic nanoparticles and cellulose nanocrystals provides a variety of performances that are exclusive and distinct as compared to those of their bulk or molecular counterparts [60]. In general, inorganic nanoparticles are thermodynamically unstable and have the tendency to aggregate leading to form a metal network. The use of cellulose as a template for inorganic nanoparticles led to several applications and Table 8.4 summarizes these works with their applications.

8.4 Conclusion

Nanocellulose (NC) is a versatile material, providing high mechanical properties, low density, high crystallinity, biodegradability and biocompatibility at an affordable price. Generally, NC acts as a reinforcing agent in different polymers or as a bio-polymer. Over the last years, cellulose nanocrystals started to play the role of an environmentally friendly support to enhance the efficiency of inorganic nanoparticles. This combination led to the development of new hybrid nanocomposites characterized by a very wide range of properties and which are useful for several applications. In this chapter, a review on cellulose nanocrystals methods of production and the effect of the processing conditions on

TABLE 8.4 Cellulose-inorganic nanoparticles hybrid nanocomposites applications.

Applications	Nanoparticles	Nanoparticlesize	Precursor	Type of cellulose used
Antibacterial activity	Ag NP	~7 nm	$AgNO_3$ (reduction with dopamine hydrochloride)	Polydopamine coated CNC
		6.3±3.1 nm	$AgNO_3$ (reduction by $NaBH_4$)	TEMPO-oxidized CNF
		spherical: 1–10 nm; dendritic: 5–10 μm	$AgNO_3$ (reduction by CNC: 100 °C, 12 h)	CNC
		5–14 nm	$AgNO_3$ (reduction by $NaBH_4$)	Bacteria CNF
		17.1±5.9 nm	$AgNO_3$ (reduction by bacterial CNF: 80 °C, 4 h)	Bacteria CNF
		10–15 nm	$AgNO_3$ (reduction by $NaBH_4$)	TEMPO-oxidized CNC
		20–45 nm	$AgNO_3$ (reduction by aldehyde groups on CNC surface)	Periodate-oxidized surface CNC
		8–15 nm	$AgNO_3$ (reduction by triethanolamine)	Bacteria CNF (triethanolamine as complexing agent)
		~30 nm	$AgNO_3$ (reduction by NH_2NH_2, NH_2OH, ascorbic acid)	Bacteria CNF (polyvinylpyrrolidone (PVP) and gelatin used as additional stabilizers)
		–	$AgNO_3$ (reduction by $NaBH_4$)	CNF (polymethacrylic acid (PMAA) used as stabilizer)
	AgCl NP	–	$AgNO_3$ and NaCl	Bacteria CNF
	Ag-ZnO NP	9-35 nm	$AgNO_3$ and $Zn(AcO)_2{\cdot}2H_2O$ (80–100 °C, 1–2 h, pH 10)	CNC

Catalytic reaction	Ag NP	1.3±0.3 nm	Ag wire and $AgNO_3$ (reduction by CNC)	CNC
		~10 nm	$AgNO_3$ (reduction by dopamine)	Polydopamine-coated CNC (β-cyclodextrin agent used as additional capping agent)
		~6.2 nm	$AgNO_3$ (reduction by $NaBH_4$)	NFC (poly(amidoamine), PAMAM, dendrimer)
		4.0±2.0 nm	$AgNO_3$ (reduction by $NaBH_4$)	TEMPO-oxidized CNF
		10–50 nm	$AgNO_3$ (reduction by CNC)	CNC
	Au NP	4.5±0.4, 5.6±0.6, 7.1±0.6 nm	$HAuCl_4$ (reduction by $NaBH_4$)	CNC modified into mesoporous photonic cellulose by co-templating with urea/formaldehyde resin
		2–4 nm	$HAuCl_4$ (reduction with and without $NaBH_4$)	PAMAM dendrimer-grafted CNC
		30.5±13.4 nm	$HAuCl_4$ (reduction by CNC)	CNC
		2-3 nm	$HAuCl_4$ (Reduction by –HS groups on the CNC surface)	HS-functionalized CNC
		2.95±0.06 nm	$HAuCl_4$ (reduction by $NaBH_4$)	Poly(diallyldimethyl ammonium chloride) (PDDA)-coated carboxylated CNC
		~9 nm	$HAuCl_4$ (reduction by poly (ethyleneimine)	Bacteria CNF (HEME proteins: horseradish peroxidase, hemoglobin and myoglobin immobilized onto CNF)
		~9 nm	$HAuCl_4$ (reduction by poly (ethyleneimine)	Bacteria CNF: poly(ethyleneimine) used as linking agent
		–	$HAuCl_4$ (reduction by $NaBH_4$)	CNF
		2–7 nm	$HAuCl_4$ (reduction by $NaBH_4$)	TEMPO-oxidized CNF
	Au-Pd NP	4–9 nm	$HAuCl_4 \cdot 3H_2O$ and $[Pd(NH_3)_4] \cdot Cl_2$ (reduction by $NaBH_4$)	TEMPO-oxidized CNF

Continued

TABLE 8.4 Cellulose-inorganic nanoparticles hybrid nanocomposites applications—cont'd

Applications	Nanoparticles	Nanoparticlesize	Precursor	Type of cellulose used
	CdS NP	10–20 nm	$CdCl_2$ (thermal treatment in presence of thiourea)	Bacteria CNF
	Cu NP	~5 nm	$CuCl_2$ (reduction by ascorbic acid)	TEMPO-oxidized CNF
	CuO NP	~7 nm	$CuSO_4$ (reduction by $NaBH_4$)	CNC
	Fe NP	–	$FeCl_3$ (reduction by H_2)	CNC from bamboo pulp
	Pd NP	1–7 nm	$PdCl_2$ (reduction by CNC)	CNC
			$PdCl_2$ (reduction by KBH_4)	Bacteria CNF
		3.6±0.8 nm	$PdCl_2$ (reduction by H_2)	CNC
	Pt NP	11–101 nm	H_2PtCl_6 (reduction by wood nanomaterial)	Wood nanomaterials
		~2 nm	H_2PtCl_6 (reduction by CNC)	CNC; carbon black as additional support
		3–4 nm	H_2PtCl_6 (reduction by $NaBH_4$ and HCHO)	Bacteria CNF
	Ru NP	8.0±2.0 nm	$RuCl_3$ (reduction by $NaBH_4$)	TEMPO-oxidized CNF
	TiO_2 NP	4.3–8.5 nm	$Ti(OBu)_4$ (thermal treatment)	Bacteria CNF

Biosensor	Ag NP	<10 nm	$AgNO_3$ (reduction by $NaBH_4$)	TEMPO-oxidized CNC
	Au NP	4.5 ± 0.4, 5.6 ± 0.6, 7.1 ± 0.6 nm	$HAuCl_4$ (reduction by $NaBH_4$)	CNC modified into mesoporous photonic cellulose by co-templating with urea/formaldehyde resin
Aerogels	Ag NP	3–4 nm	$AgNO_3$ (reduction by CNF)	TEMPO-oxidized CNF
Water treatment	Cu-Pd NP	Pd = ~3.7 nm Cu = ~4.0 nm	$PdCl_2$ and $CuCl_2$ (reduction by KBH_4)	Bacteria CNF
	TiO_2 NP	20–60 nm aggregates	Commercial TiO_2 NP	CNF
	Fe_3O_4	5.9–14.1 nm	$FeSO_4 \cdot 7H_2O$ (reaction in NaOH)	CNC (PDDA, PVP, SiO_2 and β-cyclodextrins used as additional stabilizers)

cellulose nanocrystals properties was given. Then, different preparation methods of organic-inorganic hybrid nanocomposites were presented and discussed. In the last part, an overview of the main examples of organic-inorganic hybrid nanocomposites was presented with their production and applications. Based on the information gathered, there is still potential to develop new materials and applications based on different particles and matrices.

References

[1] Bouhfid N, Raji M, Boujmal R, Essabir H, Bensalah M-O, Bouhfid R, el Kacem QA. Numerical modeling of hybrid composite materials. In: Jawaid M, Thariq M, editors. Modelling of damage processes in biocomposites, fibre-reinforced composites and hybrid composites. Woodhead Publ. Ser. Compos. Sci. Eng. Woodhead Publishing; 2019. p. 57–101.

[2] Raji M, Abdellaoui H, Essabir H, Kakou C-A, Bouhfid R, Qaiss AEK. Prediction of the cyclic durability of woven-hybrid composites. In: Durability and life prediction in biocomposites, fibre-reinforced composites and hybrid composites. Woodhead Publ. Ser. Compos. Sci. Eng.; 2019. p. 27–62.

[3] Essabir H, Raji M, Bouhfid R. Nanoclay and natural fibers based hybrid composites: mechanical, morphological, thermal and rheological properties. In: Nanoclay reinforced polymer composites; 2016. p. 29–49.

[4] Raji M, Essabir H, Essassi EM, Rodrigue D, Bouhfid R, Qaiss AEK. Morphological, thermal, mechanical, and rheological properties of high density polyethylene reinforced with Illite clay. Polym Compos 2016;39(5):1522–33. https://doi.org/10.1002/pc.24096.

[5] Raji M, Essabir H, Rodrigue D, Bouhfid R, Qaiss AEK. Influence of graphene oxide and graphene nanosheet on the properties of polyvinylidene fluoride nanocomposites. Polym Polym Compos 2017;39(5):1522–33. https://doi.org/10.1002/pc.

[6] Raji M, Zari N, Qaiss AEK, Bouhfid R. Chemical preparation and functionalization techniques of graphene and graphene oxide. In: Jawaid M, Bouhfid R, Qaiss AEK, editors. Micro Nano Technol. Elsevier; 2019. p. 1–20.

[7] Bouhfid N, Raji M, Bensalah M, Qaiss AEK. Functionalized graphene and thermoset matrices-based nanocomposites: mechanical and thermal properties. In: Jawaid M, Bouhfid R, Qaiss AEKD, editors. Micro Nano Technol. Elsevier; 2019. p. 47–64.

[8] Raji M, Qaiss AEK, Bouhfid R. Effects of bleaching and functionalization of kaolinite on the mechanical and thermal properties of polyamide 6 nanocomposites. RSC Adv 2020;10: 4916–26.

[9] Raji M, Essabir H, Essassi EM, Rodrigue D, Bouhfid R, Qaiss AEK. Morphological, thermal, mechanical, and rheological properties of high density polyethylene reinforced with Illite clay. Polym Polym Compos 2016;16:101–13.

[10] Raji M, Nekhlaoui S, Amrani I, Hassani E, Mokhtar E, Essabir H, Rodrigue D, Bouhfid R. Utilization of volcanic amorphous aluminosilicate rocks (perlite) as alternative materials in lightweight composites. Compos Part B 2019;165:47–54.

[11] Essabir H, Raji M, Bouhfid R, Qaiss AEK. Nanoclay and natural fibers based hybrid composites: mechanical, morphological, thermal and rheological properties. In: Nanoclay Reinf. Polym. Compos; 2016. p. 29–49.

[12] Mekhzoum M, Raji M, Rodrigue D, Qaiss AEK, Bouhfid R. The effect of benzothiazolium surfactant modified montmorillonite content on the properties of polyamide 6 nanocomposites. Appl Clay Sci 2020;185. https://doi.org/10.1016/j.clay.2019.105417, 105417.

[13] Zari N, Raji M, Mghari H, Bouhfid R. Nanoclay and polymer-based nanocomposites: Materials for energy efficiency. In: Polymer-based nanocomposites for energy and environmental applications. Publ. Woodhead; 2018. p. 75–103.

[14] Raji M, Essassi E, Essabir H, Rodrigue D, Qaiss AEK, Bouhfid R. Properties of nanocomposites based on different clays and polyamide 6/acrylonitrile butadiene styrene blends. In: Bio-based polymer nanocomposites. Cham: Springer International Publishing; 2019. p. 107–28.

[15] Raji M, Mohamed M, Rodrigue D, Qaiss AEK, Bouhfid R. Effect of silane functionalization on properties of polypropylene/clay nanocomposites. Compos Part B 2018;146:106–15. https://doi.org/10.1016/j.compositesb.2018.04.013.

[16] Essabir H, Raji M, Essassi EM, Rodrigue D, Bouhfid R, Qaiss AEK. Morphological, thermal, mechanical, electrical and magnetic properties of ABS/PA6/SBR blends with Fe3O4 nanoparticles. J Mater Sci Mater Electron 2017;28:17120–30. https://doi.org/10.1007/s10854-017-7639-2.

[17] Abdellaoui H, Raji M, Essabir H, Bouhfid R, Qaiss AEK. Mechanical behavior of carbon/natural fiber-based hybrid composites. In: Jawaid M, Thariq M, Saba N, editors. Mechanical and physical testing of biocomposites, fibre-reinforced composites and hybrid composites. Woodhead Publ. Ser. Compos. Sci. Eng. Woodhead Publishing; 2019. p. 103–22.

[18] Raji M, Essabir H, Bouhfid R, Qaiss AEK. Impact of chemical treatment and the manufacturing process on mechanical, thermal, and rheological properties of natural fibers-based composites. In: Handbook of composites from renewable materials. Hoboken, NJ, USA: John Wiley & Sons, Inc; 2017. p. 225–52.

[19] Essabir H, Raji M, Laaziz SA, Rodrigue D, Bouhfid R, Qaiss AEK. Thermo-mechanical performances of polypropylene biocomposites based on untreated, treated and compatibilized spent coffee grounds. Compos Part B Eng 2018;149:1–11. https://doi.org/10.1016/j.compositesb.2018.05.020.

[20] Semlali Aouragh Hassani FZ, Ouarhim W, Raji M, Mekhzoum MEM, Bensalah MO, Essabir H, Rodrigue D, Bouhfid R, Qaiss AEK. N-Silylated benzothiazolium dye as a coupling agent for polylactic acid/date palm fiber bio-composites. J Polym Environ 2019;27:2974–87.

[21] Abdellaoui H, Bensalah H, Raji M, Rodrigue D, Bouhfid R. Laminated epoxy biocomposites based on clay and jute fibers. J Bionic Eng 2017;14:1–11.

[22] Essabir H, Raji M, Bouhfid R. Recent advances in nanoclay/natural fibers hybrid composites. In: Nanoclay reinforced polymer composites; 2016. p. 29–49.

[23] Abdellaoui H, Raji M, Essabir H, Bouhfid R, Qaiss AEK. Nanofibrillated cellulose-based nanocomposites. In: Bio-based polymers and nanocomposites; 2019. p. 67–86.

[24] Sana A, Laaziz A, Raji M, Hilali E, Essabir H, Rodrigue D, Bouhfid R. Bio-composites based on polylactic acid and argan nut shell: production and properties. Int J Biol Macromol 2017;104:30–42.

[25] Ananikov VP. Organic-inorganic hybrid nanomaterials. Nanomaterials 2019;9:1197–203.

[26] Fang Q, Zhou X, Deng W, Zheng Z, Liu Z. Freestanding bacterial cellulose-graphene oxide composite membranes with high mechanical strength for selective ion permeation. Sci Rep 2016;6:1–11.

[27] Rezayat M, Blundell RK, Camp JE, Walsh DA, Thielemans W. Green one-step synthesis of catalytically active palladium nanoparticles supported on cellulose nanocrystals. ACS Sustain Chem Eng 2014;2:1241–50.

[28] Ojala J, Visanko M, Laitinen O, Österberg M, Sirviö JA, Liimatainen H. Emulsion stabilization with functionalized cellulose nanoparticles fabricated using deep eutectic solvents. Molecules 2018. https://doi.org/10.3390/molecules23112765.

[29] Yan W, Chen C, Wang L, Zhang D, Li AJ, Yao Z, Shi LY. Facile and green synthesis of cellulose nanocrystal-supported gold nanoparticles with superior catalytic activity. Carbohydr Polym 2016;140:66–73.

[30] Kaushik M, Moores A. Review: nanocelluloses as versatile supports for metal nanoparticles and their applications in catalysis. Green Chem 2016;18:622–37.

[31] Rathore DK, Prusty RK, Mohanty SC, Singh BP, Ray BC. In-situ elevated temperature flexural and creep response of inter-ply glass/carbon hybrid FRP composites. Mech Mater 2017;105:99–111.

[32] Lavoine N, Bergström L. Nanocellulose-based foams and aerogels: processing, properties, and applications. J Mater Chem A 2017;5:16105–17.

[33] Tang J, Sisler J, Grishkewich N, Tam KC. Functionalization of cellulose nanocrystals for advanced applications. J Colloid Interface Sci 2017;494:397–409.

[34] Xie J, Luo Y, Chen Y, Liu Y, Ma Y, Zheng Q, Yue P, Yang M. Redispersible Pickering emulsion powder stabilized by nanocrystalline cellulose combining with cellulosic derivatives. Carbohydr Polym 2019;213:128–37.

[35] Malha M, Nekhlaoui S, Essabir H, Benmoussa K, Bensalah M-O, Arrakhiz F-E, Bouhfid R, Qaiss A. Mechanical and thermal properties of compatibilized polypropylene reinforced by woven doum. J Appl Polym Sci 2013;130:4347–56. https://doi.org/10.1002/app.39619.

[36] Xie S, Zhang X, Walcott MP, Lin H. Applications of cellulose nanocrystals: a review. Eng Sci 2018;3–16.

[37] Giese M, Spengler M. Cellulose nanocrystals in nanoarchitectonics-towards photonic functional materials. Mol Syst Des Eng 2019;4:29–48.

[38] Jedvert K, Heinze T. Cellulose modification and shaping—a review. J Polym Eng 2017;37:845–60.

[39] Kalia S, Dufresne A, Cherian BM, Kaith BS, Avérous L, Njuguna J, Nassiopoulos E. - Cellulose-based bio- and nanocomposites: a review. Int J Polym Sci 2011. https://doi.org/10.1155/2011/837875.

[40] George J, Sabapathi SN. Cellulose nanocrystals: synthesis, functional properties, and applications. Nanotechnol Sci Appl 2015;8:45–54.

[41] Rosa MF, Medeiros ES, Malmonge JA, Gregorski KS, Wood DF, Mattoso LHC, Glenn G, Orts WJ, Imam SH. Cellulose nanowhiskers from coconut husk fibers: effect of preparation conditions on their thermal and morphological behavior. Carbohydr Polym 2010;81:83–92.

[42] Leszczyńska A, Radzik P, Haraźna K, Pielichowski K. Thermal stability of cellulose nanocrystals prepared by succinic anhydride assisted hydrolysis. Thermochim Acta 2018;663:145–56.

[43] Abdellaoui H, Bouhfid R, Qaiss AEK. Lignocellulosic fibres reinforced thermoset composites: preparation, characterization, mechanical and rheological properties; 2018. https://doi.org/10.1007/978-3-319-68696-7_5.

[44] Essabir H, Nekhlaoui S, Malha M, Bensalah MO, Arrakhiz FZ, Qaiss A, Bouhfid R. Biocomposites based on polypropylene reinforced with Almond Shells particles: mechanical and thermal properties. Mater Des 2013;51:225–30.

[45] Arrakhiz FZ, El Achaby M, Kakou AC, Vaudreuil S, Benmoussa K, Bouhfid R, Fassi-Fehri O, Qaiss A. Mechanical properties of high-density polyethylene reinforced with chemically modified coir fibers: impact of chemical treatments. Mater Des 2012;37:379–83.

[46] Arrakhiz FZ, El Achaby M, Benmoussa K, Bouhfid R, Essassi EM, Qaiss A. Evaluation of mechanical and thermal properties of Pine cone fibers reinforced compatibilized polypropylene. Mater Des 2012;40:528–35.

[47] Arrakhiz FZ, Malha M, Bouhfid R, Benmoussa K, Qaiss A. Tensile, flexural and torsional properties of chemically treated alfa, coir and bagasse reinforced polypropylene. Compos Part B Eng 2013;47:35–41.

[48] Essabir H, Bensalah MO, Rodrigue D, Bouhfid R, Qaiss A. Structural, mechanical and thermal properties of bio-based hybrid composites from waste coir residues: fibers and shell particles. Mech Mater 2016;93:134–44. https://doi.org/10.1016/j.mechmat.2015.10.018.

[49] Eichhorn SJ, Dufresne A, Aranguren M. Review: current international research into cellulose nanofibres and nanocomposites. J Mater Sci 2010;45:1–13. https://doi.org/10.1007/s10853-009-3874-0.

[50] Nagalakshmaiah M, Rajinipriya M, Sadaf Afrin MAA, Asad M, Karim Z. Cellulose nanocrystals-based nanocomposites. In: Bio-based polymers and nanocomposites; 2019. https://doi.org/10.1007/978-3-030-05825-8.

[51] Calvini P. The influence of levelling-off degree of polymerisation on the kinetics of cellulose degradation. Cellulose 2005;12:445–7.

[52] Jawaid M, Boufi S, Abdul Khalil HPS. Cellulose-reinforced nanofibre composites: production, properties and applications. Woodhead publishing series in composites science and engineering,; 2017.

[53] Duncan B, Landis RF, Jerri H, Normand V, Benczédi D, Ouali L, Rotello VM. Hybrid organic-inorganic colloidal composite "sponges" via internal crosslinking. Small 2015;11:1302–9.

[54] Xie K, Yu Y, Shi Y. Synthesis and characterization of cellulose/silica hybrid materials with chemical crosslinking. Carbohydr Polym 2009;78:799–805.

[55] Oprea M, Panaitescu DM. Nanocellulose hybrids with metal oxides nanoparticles for biomedical applications. Molecules 2020;25:4045–69.

[56] Rie J, Thielemans W. Cellulose-gold nanoparticle hybrid materials. Nanoscale 2017;9: 8525–54.

[57] Adel AM. Incorporation of nano-metal particles with paper matrices. Int J Chem 2016; 1:36–46.

[58] Islam MS, Chen L, Sisler J, Tam KC. Cellulose nanocrystal (CNC)-inorganic hybrid systems: synthesis, properties and applications. J Mater Chem B 2018;6:864–83.

[59] Liu S, Luo X, Zhou J. Magnetic responsive cellulose nanocomposites and their applications. In: Cellulose—medical, pharmaceutical and electronic applications; 2013. p. 105–24.

[60] Prusty K, Sahu D, Swain SK. Nanocellulose as a template for the production of advanced nanostructured material. In: Cellulose-reinforced nanofibre composites; 2017. p. 427–54. https://doi.org/10.1016/B978-0-08-100957-4.00019-X.

[61] Lin N, Huang J, Dufresne A. Preparation, properties and applications of polysaccharide nanocrystals in advanced functional nanomaterials: a review. Nanoscale 2012;4:3274–94.

[62] Lan Q, Liu C, Yang F, Liu S, Xu J, Sun D. Synthesis of bilayer oleic acid-coated Fe3O4 nanoparticles and their application in pH-responsive Pickering emulsions. J Colloid Interface Sci 2007;310:260–9.

[63] Wu W-B, Jing Y, Gong M-R, Zhou X-F, Dai H-Q. Preparation and properties of magnetic cellulose fiber composites. Bioresources 2011;6:3396–409.

[64] Wulandari IO, Santjojo DJDH, Shobirin RA, Sabarudin A. Characteristics and magnetic properties of chitosan-coated Fe3O4 nanoparticles prepared by ex-situ co-precipitation method. Rasayan J Chem 2017;10:1348–58.

[65] Philippova O, Barabanova A, Molchanov V, Khokhlov A. Magnetic polymer beads: recent trends and developments in synthetic design and applications. Eur Polym J 2011;47:542–59.

[66] Li L, Xiang S, Cao S, Zhang J, Ouyang G, Chen L, Su C-Y. A synthetic route to ultralight hierarchically micro/mesoporous Al(III)-carboxylate metal-organic aerogels. Nat Commun 2013;4:1774.

[67] Ciriminna R, Fidalgo A, Pandarus V, Beland F, Ilharco LM, Pagliaro M. The sol-gel route to advanced silica-based materials and recent applications. Chem Rev 2013;113:6592–620.

[68] Periyasamy AP, Venkataraman M, Kremenakova D. Progress in sol-gel technology for the coatings of fabrics. Materials 2020;13:1838–72.

[69] Zdravkov A, Listratenko M, Gorbachev S. Solvothermal sol-gel synthesis of TiO2 cellulose nanocrystalline composites. Cellulose 2021;1–26. https://doi.org/10.1007/s10570-020-03656-y.

[70] Tartaro G, Mateos H, Schirone D, Angelico R, Palazzo G. Microemulsion microstructure(s): a tutorial review. Nanomaterials 2020;10:1–40.

[71] Update N, Malik MA. Microemulsion method: a novel route to synthesize organic and inorganic nanomaterials. Arab J Chem 2014;5:397–417.

[72] Tang J, Quinlan PJ, Tam KC. Stimuli-responsive Pickering emulsions: recent advances and potential applications. Soft Matter 2015;11:3512–29.

[73] Udoetok I, Wilson LD, Headley JV. Stabilization of Pickering emulsions by iron oxide nanoparticles. Adv Mater Sci 2016;1:24–33.

[74] Fujisawa S, Togawa E, Kuroda K. Nanocellulose-stabilized Pickering emulsions and their applications. Sci Technol Adv Mater 2017;18:959–71.

[75] Chevalier Y, Bolzinger MA. Emulsions stabilized with solid nanoparticles: Pickering emulsions. Colloids Surfaces A Physicochem Eng Asp 2013;439:23–34.

[76] Lu AH, Salabas EL, Schith F. Magnetic nanoparticles: synthesis, protection, functionalization, and application. Angew Chem Int Ed Eng 2007;46:1222–44. https://doi.org/10.1002/anie.200602866.

[77] Chu S, Miao Y, Qian Y, Ke F, Chen P, Jiang C, Chen X. Synthesis of uniform layer of TiO2 nanoparticles coated on natural cellulose micrometer-sized fibers through a facile one-step solvothermal method. Cellulose 2019;26:4757–65.

[78] Foo YT, Chan JEM, Ngoh GC, Abdullah AZ, Horri BA, Salamatinia B. Synthesis and characterization of NiO and Ni nanoparticles using nanocrystalline cellulose (NCC) as a template. Ceram Int 2017;43:16331–9.

[79] Zhang Y, Chen L, Hu L, Yan Z. Characterization of cellulose/silver nanocomposites prepared by vegetable oil-based microemulsion method and their catalytic performance to 4-nitrophenol reduction. J Polym Environ 2019;27:2943–55.

[80] Xie A, Cui J, Chen Y, Lang J, Li C, Yan Y, Dai J. One-step facile fabrication of sustainable cellulose membrane with superhydrophobicity via a sol-gel strategy for efficient oil/water separation. Surf Coat Technol 2019;361:19–26.

[81] You J, Zhao L, Wang G, Zhou H, Zhou J, Zhang L. Quaternized cellulose-supported gold nanoparticles as capillary coatings to enhance protein separation by capillary electrophoresis. J Chromatogr A 2014;1343:160–6.

[82] Liu Z, Wang H, Huang P, et al. Highly stable and stretchable conductive films through thermal-radiation-assisted metal encapsulation. Adv Mater 2019;31:1–9.

[83] Schlesinger M, Giese M, Blusch LK, Hamad WY, Maclachlan MJ. Chiral nematic cellulose-gold nanoparticle composites from mesoporous photonic cellulose. Chem Commun 2015;51:530–3.

[84] Drogat N, Granet R, Sol V, Memmi A, Saad N, Klein Koerkamp C, Bressollier P, Krausz P. Antimicrobial silver nanoparticles generated on cellulose nanocrystals. J Nanopart Res 2011;13:1557–62.

[85] Ali A, Mannan A, Hussain I, Hussain I, Zia M. Effective removal of metal ions from aquous solution by silver and zinc nanoparticles functionalized cellulose: isotherm, kinetics and statistical supposition of process. Environ Nanotechnol Monitor Manag 2017;9:1–11.

[86] Abdel-halim ES, Al-deyab SS. Utilization of hydroxypropyl cellulose for green and efficient synthesis of silver nanoparticles. Carbohydr Polym 2011;86:1615–22.

[87] Wu X, Lu C, Zhang W, Yuan G, Xiong R, Zhang X. A novel reagentless approach for synthesizing cellulose nanocrystal-supported palladium nanoparticles with enhanced catalytic performance. J Mater Chem A 2013;1:8645–52.

[88] Gouda M, Al-Bokheet W, Al-Omair M. In-situ deposition of metal oxides nanoparticles in cellulose derivative and its utilization for wastewater disinfection. Polymers 2020;12:1834.

Chapter 9

Hybrid nanocomposites based on cellulose nanocrystals/nanofibrils with graphene and its derivatives: From preparation to applications

Lakkoji Satish[a], Ayonbala Baral[b], and Aneeya K. Samantara[c,d]

[a]*Department of Chemistry, Ravenshaw University, Cuttack, Odisha, India,* [b]*Faculty of Metallurgy & Energy Engineering, Kunming University of Science & Technology, Kunming, China,* [c]*National Institute of Science Education and Research (NISER), Khordha, Odisha, India,* [d]*Homi Bhabha National Institute (HBNI), Mumbai, India*

9.1 Introduction

Recent progress of nanotechnology could enable to produce new smart materials having a wide range of applications in various fields. The negative environmental impact of petroleum based polymers is the main concern of the researchers to develop sustainable natural products. Switching towards the alternative to petroleum based polymers forced the scientific community and industries to investigate potential methods of preparation and application of biopolymers. Nanocomposites received high attention in the last few decades for their unique properties due to the nanometric size effect compared to conventional composite materials. A nanocomposite is a multiphase solid material in which the matrix material is filled by one or more separate nanomaterials in order to improve the properties and performances. The matrix of nanocomposites can be made from polymers (nylon, polyepoxide, epoxy and polyethylimide, etc.), metals (iron, titanium, magnesium, etc.) and ceramics (alumina, porcelain, glass, etc.) [1, 2]. Up to now, several nanocomposites have been produced but inorganic/organic composites have been considered as a breakthrough for nanocomposite materials with significant characteristics as these materials carry both the characteristics of their inorganic-organic parent constituents. Polymer matrix nanocomposites (PMNC) are widely used in industry due

Cellulose Nanocrystal/Nanoparticles Hybrid Nanocomposites. https://doi.org/10.1016/B978-0-12-822906-4.00007-4

to their unique properties, ease of production, lightweight, and advanced applications [3].

Current research is focusing towards green chemistry and environmentally friendly materials. Biopolymers and their nanocomposites have been developed with improved functional properties. Biopolymers are an alternative to former existing polymers, but their properties still need to be improved before using them in different applications as these are costly and have inferior property compared to commercial thermoplastic polymers [4, 5]. A large variety of fillers, like nanotubes and silicates, are available for making hybrid structures with polymers enabling to increase the thermal, rheological, mechanical, and gas barrier properties [6–8]. Graphene and its derivatives have remarkable potential to improve the functional properties of polymers as compared to other fillers [9]. They have been used in a wide spectrum of applications such as photonic, biosensor, drug carrier, biomedical energy storage, polymer nanocomposite, and environmental pollution control due to their amazing features [10]. Likewise, cellulose nanomaterials have drawn scientific and technological interest rapidly for their excellent inherent physical and chemical properties like high tensile strength and elastic modulus, low density, high specific surface area, reactive surfaces, biodegradability, and renewability [11, 12].

Despite the rapid development in the synthesis of hybrid nanocomposites based on cellulose-graphene derivative, a cost-effective method for improved nanoscale cellulose remains scarce, thus limiting the potential implementation in practical use. This chapter encompasses the synthesis of cellulose nanocrystals/fibrils in combination with graphene and its derivatives from an engineering perspective, aiming to establish their feasibility and identify the current challenges. Also, potential applications of various cellulose nanocomposite materials based on graphene derivatives are discussed.

9.2 Cellulose based nanocrystals/nanofibrils

Materials from bio-based resources have attracted high research interest in recent years as a result of their very high potentials for fabricating several high-value products with a low impact on the environment. Effective utilization of various nature-based nanomaterials offers certain ecological advantages, extraordinary physicochemical properties, and high performance. Among various natural materials, cellulose holds a crucial position as being an abundant organic raw material. Cellulose is a fibrous, tough, water-insoluble substance playing a crucial role in preserving the structure of natural fibers. It is a linear natural polymer consisting of 1,4-anhydro-D-glucopyranose units as depicted in Fig. 9.1. Over the last few decades, it has attracted growing interest owing to its abundance and versatility when processing at the nanoscale in the form of cellulose nanomaterials. Using various reaction strategies, different types of nanomaterials can be extracted from cellulose owing to its hierarchical structure and semi-crystalline nature.

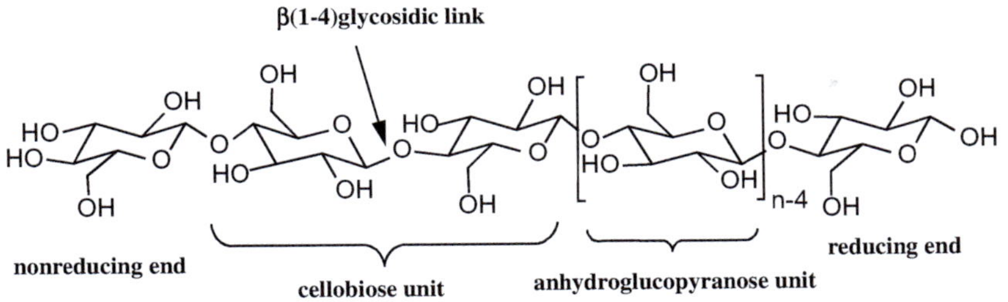

β(1-4)glycosidic link

nonreducing end

cellobiose unit

anhydroglucopyranose unit

reducing end

FIG. 9.1 Representation of a cellulose chain showing the anhydroglucose unit in the chair conformation along with atom numbering and the glycosidic link, as well as both the reducing and nonreducing ends of the polymer.

Fig. 9.2 illustrates the diverse hierarchical structure of cellulose nanomaterials [13]. Nanoscale cellulose can be divided into nanostructured materials and nanofibers resulting from the use of various isolation processes. Various terminologies of nanocellulose have been used in research including nanofibrillar cellulose, nanoscale fibrillated cellulose, cellulosic fibrillar fines, nanofibers, nanofibrils, fibril aggregates, microfibrils, cellulose nanowhiskers, cellulose crystallites, nanorods, and nanocrystalline cellulose, etc. Cellulose nanocrystals are highly crystalline needle-shaped nanometric or rod-like particles having at least one dimension less than 100 nm. They can be prepared from different starting materials like bacterial cellulose, microcrystalline cellulose, algal cellulose,

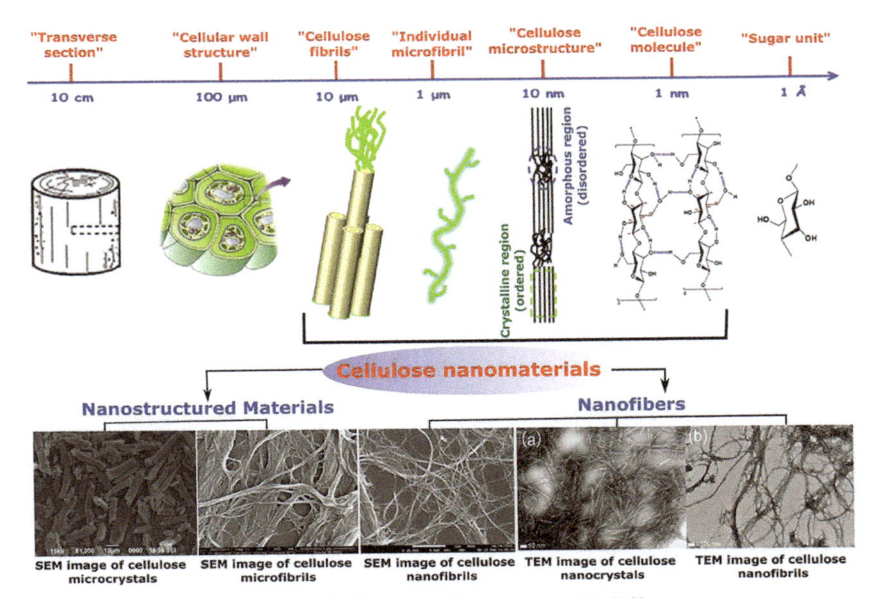

FIG. 9.2 Hierarchical structure of cellulose and its nanomaterials [13].

cotton linters, bast fibers and wood pulp etc. [14–19]. Cellulose nanocrystals also offer attractive biophysicochemical characteristics such as biocompatibility, biodegradability, optical transparency, sustainability, gas impermeability, renewability, low thermal expansion, lightweight, nontoxicity, stiffness, adaptable surface chemistry, and improved mechanical properties [20, 21]. The properties of nanocrystals strongly depends on the extraction process. The most frequently used technique is acid hydrolysis to prepare nanocrystals from various cellulose based materials where strong acids like sulphuric and hydrochloric acids are used [15, 17, 22, 23]. However few reports are available for the production of nanocrystals from other mineral and organic acids [24–26] Several other approaches have been reported for the fabrication of CNC, such as ionic liquid treatment, enzymatic hydrolysis, subcritical water hydrolysis, mechanical refining, oxidation method, and combined processes [27–33]. Recently, several reviews have been published on different aspects related to nanocrystals such as characterization, chemical modification of surfaces, isolation processes, and self-assembly of suspensions [14, 15, 34–39].

9.3 Graphene based composites

Graphene is a one-atom-thick single layer of graphite sheet made of polycyclic aromatic carbon dislocated from a 3D graphite structure that was first produced by mechanical exfoliation. The monolayer unique plane structure and geometry of graphene contribute towards its excellent properties like exceptional electron transport, mechanical properties, and high surface area. Graphene is the lightest, thinnest, and strongest material ever discovered which can be synthesized via two different methods: (1) top-down and (2) bottom-up [40]. The top-down method is extensively used which involves the exfoliation of the individual graphene sheet. This method limits its application due to the use of highly toxic reagents. However, the bottom-up approach for the synthesis of graphene is a completely different method in which graphene is directly generated on a surface. In this method, small carbon molecular blocks form a large or single-layer graphene structure through: (i) organic synthesis, (ii) chemical vapor deposition, and (iii) epitaxial growth or decomposition of silicon carbide SiC at high temperatures ($>1100°C$). The activity of graphene strongly depends on the structure, surface defect chemistry, and substrates [41]. Several derivatives of graphene are reported, but the most studied and interesting materials are: graphene oxide (GO), reduced GO (rGO), and graphene quantum dots (GQD) whose properties are completely different from the parent molecule. The molecular structure of graphene, graphene oxide, reduced graphene oxide, and graphene quantum dots are shown in Fig. 9.3.

GO, rGO and GQD have gained popularity for their unique characteristics such as good electrocatalytic activity, biocompatibility, controllable size, good signal amplification, and multiplexed detection ability. These distinguished properties of graphene derivatives are due to structural defects caused by the

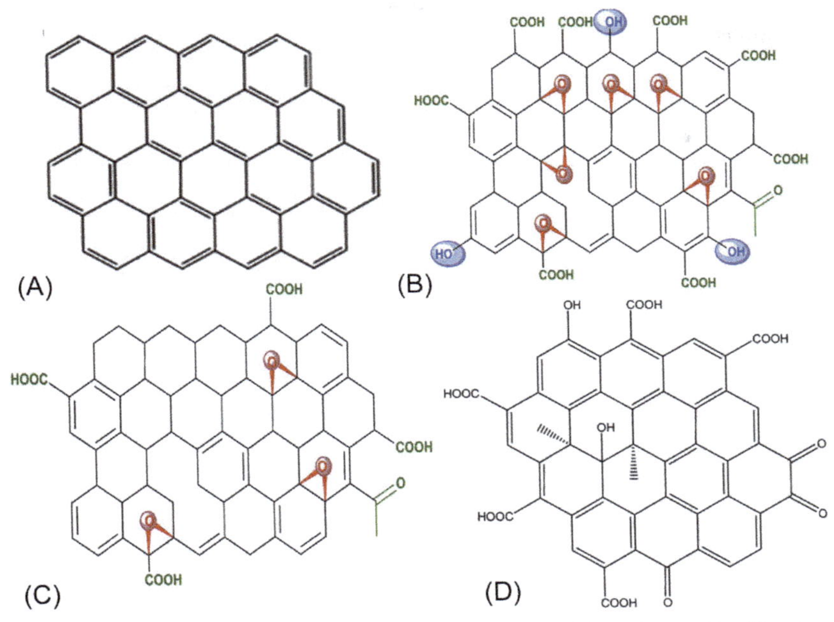

FIG. 9.3 Structure of graphene and its derivatives: (A) graphene, (B) graphene oxide, (C) reduced graphene oxide, and (D) graphene quantum dot. The combined figure is reproduced from several figures appearing in Refs. [42–44] with permission.

oxidation and/or reduction processes. The significant properties of graphene and its derivatives make them interesting in different applications including energy storage, electronics, biotechnology, and especially improv fiber-based composite materials [10]. Thus, graphene derivatives can be multifunctional reinforcements improving the properties of polymers, even at very low loading.

9.4 Nanocomposites of cellulose nanocrystals/nanofibrils with graphene and its derivatives

The synthesis of nanocomposites requires the dispersion of graphene derivatives in the polymer matrix. The main governing factors affecting the interactions between graphene and nanocellulose include polarity, molecular weight, hydrophobicity, nature of the reactive groups present on graphene derivatives, cellulose, and solvent. In general, the properties of nanocomposites mainly depend on the dispersion and the preparation method. There are three key methods available for the incorporation of graphene into polymer matrices: solution intercalation, melt intercalation, and in situ intercalative polymerization. Also, methods like wet spinning, electrospinning, freeze-drying, and drop-casting have been used in the preparation of nanocomposites.

9.5 Solution intercalation

Solution blending or solution mixing is the most effective technique for the synthesis of cellulose based nanocomposites by creating a strong interface between the cellulose and graphene filler. In this method, the colloidal suspension of graphene/graphene derivatives and cellulose are mixed by simple stirring, followed by the evaporation of the solvent medium present in the system helping the adsorbed polymer to reassemble and form a sandwich structure between cellulose and graphene/graphene derivatives, resulting in the formation of nanocomposites [45]. It is important to note that irrespective of the molecular weight (low/high) and structure (semi-crystalline and amorphous), any form of cellulose can be used for the preparation of graphene-based composites. There are several articles available in the literature on solution blending for the incorporation of graphene in a cellulose matrix. As cellulose is insoluble in water, different solvents have been used for the solution mixing of cellulose and graphene derivatives such as *N*-methylmorpholine-*N*-oxide (NMMO), 1-ethyl-3-methylimidazolium acetate (EMIMAc) monohydrate, *N*,*N*-dimethylacetamide-lithium chloride (DMACLiCl), etc. [9]. NMMO has strong hydrogen bonding capabilities between the cellulose matrix and GO filler, which was proved to be a good solvent for dispersing and exfoliating GO in the nanocomposite [46]. Moreover, it was shown to be a remarkable green solvent for cellulose. LiOH-urea was also found to be an efficient dispersing agent for GO in cellulose. Freeze-dried sponge-like GO and cellulose were dispersed into aqueous LiOH-urea (7–12 wt%) solution followed by freezing, thawing, and casting on a glass plate. In another study, aqueous NaOH-urea served as a solvent for synthesis of RC-RGO nanocomposite [47].

9.6 Melt intercalation

Melt intercalation is an efficient method for large-scale manufacturing capability and does not require any dangerous/toxic solvents. This method involves mixing and shearing of graphene/graphene derivatives into the molten state of cellulose to produce nanocomposites. Several reaction conditions, like rotor speed, reaction temperature, residence time, post extrusion processing conditions can affect the mechanical and thermal properties of nanocomposites. It was demonstrated that mechanical and thermal properties were enhanced with increase in rotor speed. However, reduced speeds can be compensated by addition of a plasticizer [9]. In some cases, high-temperature reactions could affect both cellulose and graphene, resulting in a decreased aspect ratio of the filler. Jeon et al. used this methodology to improve the thermal and mechanical properties of cellulose acetate propionate (CAP) by the incorporation of graphite nanoplatelets [48]. The thermal stability of the CAP-exfoliated graphite nanocomposite was improved due to graphene forming an interlocking network with the polymer matrix. Also, the dynamic storage modulus of CAP was enhanced by the incorporation of graphene.

9.7 In situ polymerization

In situ polymerization is another important method for the fabrication of cellulose-graphene nanocomposites. This methodology involves mixing of graphene/graphene derivative dispersion with neat monomers and subsequent polymerization by heat or radiation. During the polymerization reaction, the monomers combine to form a polymer and then this polymer binds to the nanomaterials. This helps nanomaterials to intercalate and disperse which increases the interlayer spacing and exfoliation level of the layered graphene structure. This method creates a covalent crosslink between the graphene filler and cellulose matrix. This technique can also be applied for noncovalent-bonded composites, for example, polypropylene (PP)-GO, polyethylene (PE)-graphite, and polymethyl-methacrylate (PMMA)-GO [49–51]. A number of reports are available in the literature where in situ polymerization was used for the incorporation of graphene/graphene derivatives into cellulose matrix. Zhang et al. applied this methodology to prepare a ternary graphene oxide (GO)/polyacrylamide (PAM)/ carboxymethyl cellulose sodium (CMC) nanocomposite hydrogel by the addition of GO and CMC to the polymerization mixture of acrylamide followed by ionic cross-linking with Al^{3+} [52].

9.8 Applications

Cellulose based nanocomposites with graphene and its derivatives have shown promising results in several industrial applications such as green energy, sensors, electronics, aerospace, electro-magnetic interference (EMI) shielding, and automotive industries [9, 53, 54]. These composites have also been utilized in biomedical applications for drug delivery, drug encapsulation, and scaffolding materials. Moreover, apart from the existing applications, cellulose-graphene nanocomposites can be applied in different fields. Table 9.1 presents some of the most significant cellulose-graphene based nanocomposites synthesized in the last 10 years and their respective use in different fields.

Several reports published on cellulose-graphene nanocomposites described significant mechanical and thermal properties improvement compared to cellulose based materials. This could be attributed to the presence of small amount of graphene nano filler, although the properties mainly depend on the graphene distribution within the cellulose matrix, as well as the interactions between graphene and cellulose. Huang et al. achieved a remarkable mechanical properties improvement using graphene oxide nanosheet (GONS)/regenerated cellulose (RC) nanocomposite films [47]. Fig. 9.4 shows the tensile strength and Young's modulus of the different composite samples. It was observed that 1.64 vol% GONS increased the tensile strength by 67% and Young's modulus by 68%

TABLE 9.1 Different techniques for the synthesis and characterization of cellulose-graphene based nanocomposites and their prospective applications in different fields.

No	Material composition	Characterization techniques	Application	Reference
1	Magnetic cellulose/ graphene oxide composite (MCGO)	XRD, FTIR, SEM, BET surface analysis	Adsorption of azo dye (methylene blue)	[55]
2	Graphene oxide/ Cellulose aerogels nanocomposite	SEM, TEM, Raman spectroscopy, XRD, TGA	Application in EMI shielding	[56]
3	Cellulose/ Graphene composite hydrogels (CGH)	FTIR, XPS, TGA, FESEM	High mechanical strength and good thermal stability composite	[57]
4	Carboxymethyl cellulose/ graphene oxide nanocomposite film	FTIR, XRD, FESEM, Raman spectroscopy, Dynamic mechanical analyser	High thermal stability, tensile strength, and storage modulus	[58]
5	Cellulose/ graphene composite	FTIR, XRD, Raman spectroscopy, XPS, TEM	Adsorption of triazine pesticides	[59]
6	Bacterial cellulose/ graphene oxide composites	SEM, XRD, FTIR, XPS	Cytotoxic assays show biocompatibility	[60]
7	Nanocrystalline cellulose- graphene oxide based nanocomposite	FTIR, Surface plasmon resonance	Ni^{2+} sensing	[61]
8	Graphene oxide- cellulose nanocrystals- cadmium ion nanocomposite	FTIR, XRD, Raman spectroscopy, XPS	High mechanical strength and electrical conductivity	[62]
9	Carboxymethyl cellulose/ graphene quantum dot nanocomposite hydrogel films	FTIR, UV-Vis spectroscopy, SEM	Cytotoxic assays show biocompatibility and application in drug delivery	[63]

TABLE 9.1 Different techniques for the synthesis and characterization of cellulose-graphene based nanocomposites and their prospective applications in different fields—cont'd

No	Material composition	Characterization techniques	Application	Reference
10	Porous cellulose/ Graphene oxide nanocomposite	FESEM, FTIR, Cyclic voltameter	Supercapacitor electrode	[64]
11	Poly(vinyl alcohol) nanocomposites with cellulose nanocrystal-stabilized graphene	FESEM, UV-Vis spectroscopy, TGA	Enhanced mechanical properties	[65]
12	Carboxymethyl cellulose/ graphene oxide bio-nanocomposite hydrogel beads	FTIR, XRD, SEM, TEM	Prolonged release for Doxorubicin and no significant toxicity against colon cancer cells	[66]
13	Nanofibrillated cellulose/ Nanographite composite film	SEM	Enhanced electrical and mechanical properties	[67]
14	Carbamated cellulose nanocomposite film reinforced with graphene oxide	XRD, FTIR, FESEM, TEM	High thermal stability and heat resistance	[68]
15	Graphene oxide-cellulose nanocrystal films	XRD, AFM	High electrical conductivity, ultrahigh tensile strength	[69]

AFM, atomic force microscopy; *BET*, Brunauer-Emmett-Teller; *FTIR*, Fourier Transform Infrared Spectroscopy; *SEM*, scanning electron microscope; *FESEM*, field emission scanning electron microscopy; *TEM*, transmission electron microscopy; *TGA*, thermogravimetric analysis; *XPS*, X-ray photoelectron spectroscopy; *XRD*, X-ray diffraction.

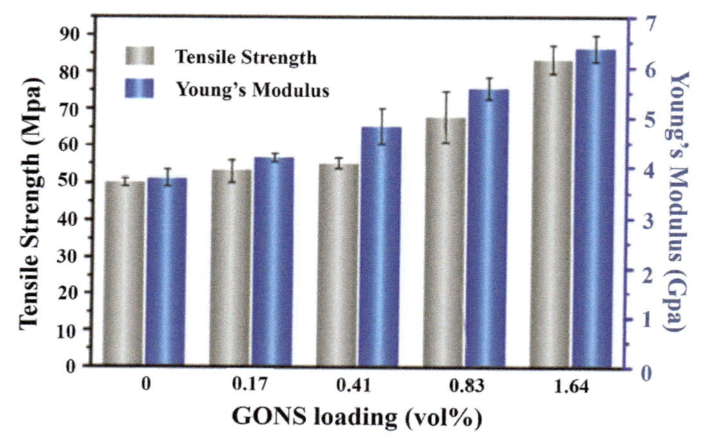

FIG. 9.4 Tensile strength and Young's modulus of RC and the cellulose nanocomposite films as a function of GONS loadings. *(Reprinted with permission from Huang H-D, Liu C-Y, Li D, Chen Y-H, Zhong G-J, Li Z-M. Ultra-low gas permeability and efficient reinforcement of cellulose nanocomposite films by well-aligned graphene oxide nanosheets. J Mater Chem A 2014;2(38):15853-15863.)*

relative to the RC film. In a similar study, Feng et al. presented a highly flexible nanocomposite film of bacterial cellulose (BC) and graphene oxide (GO) with enhanced mechanical properties [70]. The introduction of graphene nanoflakes into a cellulose matrix led to a remarkable improvement in the mechanical and thermal properties [71] Yadav et al. synthesized carboxymethylcellulose (CMC)/GO nanocomposite with excellent GO dispersion within the CMC matrix [58]. The tensile strength and Young's modulus of these materials were significantly improved upon incorporation of 1 wt% graphene oxide by 67% and 148% respectively, compared to neat CMC, and the dynamic mechanical analysis showed a high storage modulus up to 250 ^{0}C. Moreover, graphene-based materials have been used to improve the mechanical performance of regenerated cellulose using ionic liquids (e.g., *N*-methylmorpholine-*N*-oxide hydrate, *N*,*N*-dimethylacetamide/LiCl) and aqueous alkali/urea as solvent [46, 71–74]. Xu et al. prepared highly tough cellulose/rGO composite hydrogels by simply mixing cellulose and rGO at different ratios in BmimCl [57]. Fig. 9.5 shows the schematic representation of the synthesis of cellulose/rGO composite hydrogels. These hydrogels were found to have high mechanical strength and thermal stability compared to neat cellulose hydrogels. The results also depend on the doping ratio of rGO. In another study by Tian et al., a wet spinning method was applied to prepare regenerated cellulose/graphene composite fibers [75]. It is noteworthy that 50% increase in tensile strength was achieved by the addition of only 0.2 wt% graphene oxide and the onset decomposition temperature increased by 44°C compared to the neat regenerated cellulose. All these

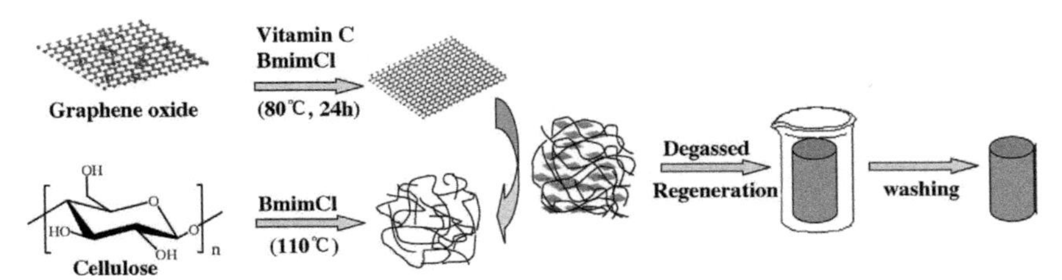

FIG. 9.5 Schematic diagram of the green synthesis of cellulose/graphene composite hydrogels (CGH) [57].

results suggested that a low graphene content was able to enhance the mechanical properties as well as the thermal stability.

Cellulose-graphene nanocomposites have also been used in the sorption of hazardous contaminants like heavy metal ions and pesticides. Shi et al. synthesized a novel bioadsorbent, magnetic cellulose/graphene oxide (MCGO) with an enhanced adsorption capacity for the removal of methylene blue [55]. The composite was found to be stable and easily recyclable. The rough surface nature of MCGO, as evident from SEM images (Fig. 9.6), improves the rate of adsorption and adsorption capacities.

Zhang et al. prepared a highly efficient sorbent cellulose/graphene composite (CGC). It has a rough and wrinkled surface with a lamellar structure which improves the adsorption rate compared to that of graphite carbons, primary secondary amine (PSA), graphite carbon black (GCB), cellulose, and graphene for triazine pesticides [59] (Fig. 9.7).

Cellulose-graphene nanocomposites have shown significant progress in sensing applications. In a recent study, Daniyal et al. described the synthesis and application of nanocrystalline cellulose-graphene oxide based nanocomposite (CTA-NCC/GO) towards Ni ion sensing [61]. Interestingly, it was observed that the composite was able to differentiate and detect Ni^{2+} in a range of 0.01–0.1 ppm. Enzyme based biosensors has received high attention in recent years for selective and sensitive detection. Palanisamy et al. fabricated a laccase biosensor utilizing laccase immobilized on graphene-cellulose microfibers (GR-CMF) composite for the detection of catechol (CC) [76]. The biosensor has great potential for selective detection of CC in the presence of potentially active biomolecules and phenolic compounds while exhibiting a low limit of detection (85 nM), high sensitivity (0.932 μ Aμ M^{-1} cm^{-2}), and fast response time (2 s).

A number of electrochemical applications can be envisioned with cellulose-graphene composites. Kafy et al. synthesized a porous structured cellulose/graphene oxide (GO) nanocomposite as a renewable electrode material for supercapacitor [64]. Enhanced current density in CV (cyclic voltammetry) curve, galvanostatic charging-discharging time, and specific capacitance make it a good choice for electrical double layer supercapacitor. Wang et al. presented

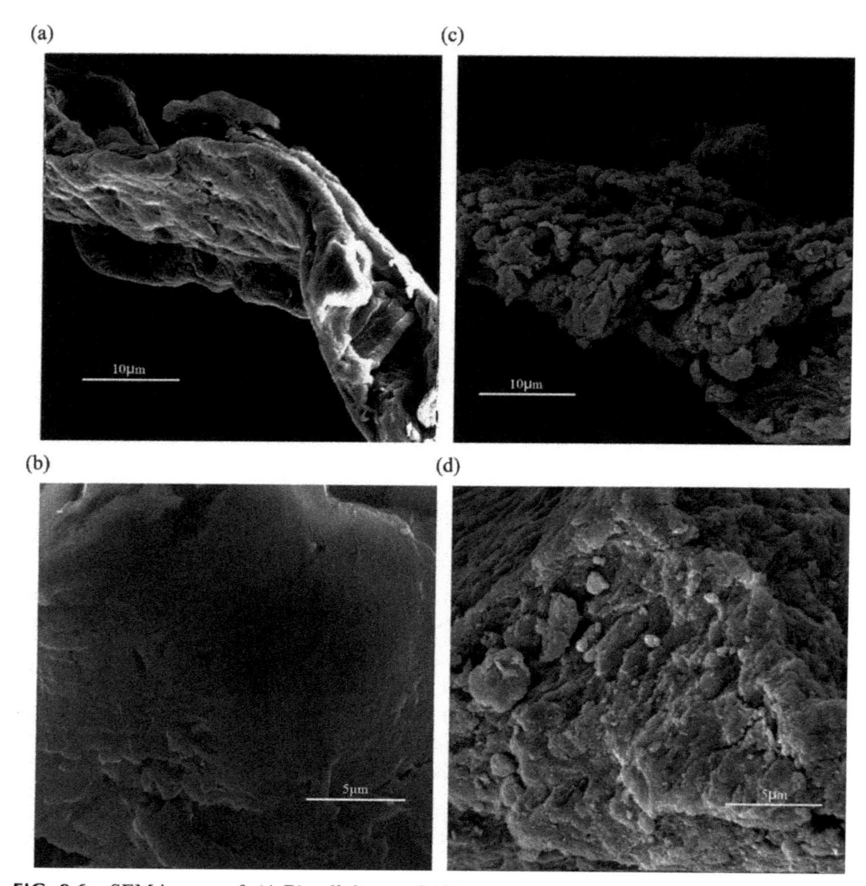

FIG. 9.6 SEM images of: (A,B) cellulose and (C,D) MCGO [55].

thin films composed of graphene nanoplatelets (GnP) and cellulose nanocrystals (CNC) with enhanced electrical and mechanical properties [77].

Liu et al. demonstrated the synthesis of hybrid composites based on graphene oxide and bacterial cellulose and their application towards supercapacitor electrodes [78]. The best composite was found to be highly efficient with high electrical conductivity (1320 S m^{-1}) and volumetric capacitance (278 F cm^{-3}). Finally, Sen et al. described the enhanced electroactive performance of cellulose-based composite actuators by adding graphene nanoplatelets (0.10, 0.25, and 0.50 wt%) into a cellulose matrix [79].

9.9 Conclusion

The use of graphene and its derivatives in cellulose nanocomposites makes the production of hybrid -applications possible. These materials are considered

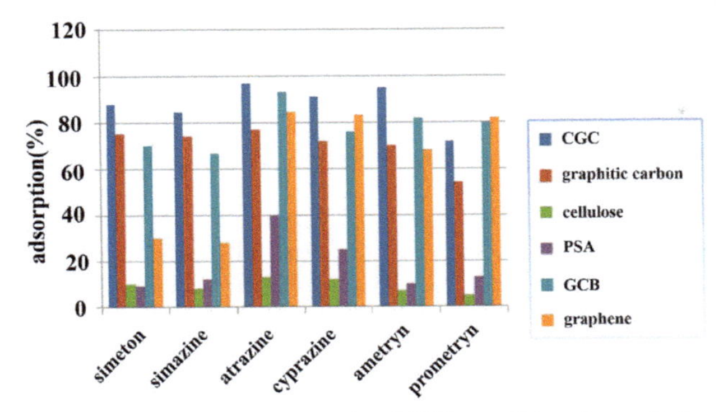

FIG. 9.7 Comparison of the adsorption capacity of CGC, graphitic carbon, cellulose, PSA, GCB, and graphene toward six triazine pesticides [59].

green with no effect on the environment due to less amount of graphene use in these composites. Apart from the synthesis procedures described in this chapter, methods like wet spinning, electrospinning, and freeze-drying can also be used to produce cellulose-graphene nanocomposites in the form of hydrogels, beads, nanofibers, etc. These nanocomposites have been used in different applications such as bioadsorbents for removal of pollutants and heavy metal ions, in biomedical field for drug delivery, scaffolding materials, fibers, immobilization of proteins/enzymes, etc. In addition, they have been used in biosensing, energy applications to make solar cells, and in electrochemistry field to make electrodes. In spite of the significant applications of these nanocomposites, it is important to consider the biodegradability of cellulose which in general causes premature degradation. Also, high-temperature poses a challenge to handle these nanocomposites during studies due to low thermal stability in most cases. However, improvement is possible with new hybrid nanocomposite materials as discussed in this chapter. Moreover, it is important to notice that some reports suggest the release of nanomaterials like Ag, Cu, montmorillonite nanoclay, Ag-zeolites from host matrix in some cases. As there is not enough research on the leakage of graphene from polymer host matrices, more studies need to be done on these hybrid structures for safe usage in different applications.

Reference

[1] Plesa I, Notingher PV, Schlogl S, Sumereder C, Muhr M. Properties of polymer composites used in high-voltage applications. Polymers (Basel) 2016;8(5):173.

[2] Camargo PHC, Satyanarayana KG, Wypych F. Nanocomposites: synthesis, structure, properties and new application opportunities. Mater Res 2009;12:1–39.

[3] Hussain F, Hojjati M, Okamoto M, Gorga RE. Review article: polymer-matrix nanocomposites, processing, manufacturing, and application: an overview. J Compos Mater 2006;40 (17):1511–75.

[4] Xiong R, Grant AM, Ma R, Zhang S, Tsukruk VV. Naturally-derived biopolymer nanocomposites: interfacial design, properties and emerging applications. Mater Sci Eng R Rep 2018;125:1–41.
[5] Sumrith N, Rangappa SM, Dangtungee R, Siengchin S, Jawaid M, Pruncu CI. In: Sanyang M, Jawaid M, editors. Biopolymers-based nanocomposites: properties and applications. Cham: Springer; 2019. p. 255–72.
[6] Mittal V. Polypropylene-layered silicate nanocomposites: filler matrix interactions and mechanical properties. J Thermoplast Compos Mater 2016;20(6):575–99.
[7] Mittal V. Mechanical and gas permeation properties of compatibilized polypropylene–layered silicate nanocomposites. J Appl Polym Sci 2008;107(2):1350–61.
[8] Favier V, Cavaille JY, Canova GR, Shrivastava SC. Mechanical percolation in cellulose whisker nanocomposites. Polym Eng Sci 1997;37(10):1732–9.
[9] Rouf TB, Kokini JL. Biodegradable biopolymer–graphene nanocomposites. J Mater Sci 2016;51(22):9915–45.
[10] Dideikin AT, Vul AY. Graphene oxide and derivatives: the place in graphene family. Front Phys 2019;6:149.
[11] Seabra AB, Bernardes JS, Favaro WJ, Paula AJ, Duran N. Cellulose nanocrystals as carriers in medicine and their toxicities: a review. Carbohydr Polym 2018;181:514–27.
[12] Duran N, Lemes AP, Seabra AB. Review of cellulose nanocrystals patents: preparation, composites and general applications. Recent Pat Nanotechnol 2012;6(1):16–28.
[13] Trache D, Hussin MH, Haafiz MK, Thakur VK. Recent progress in cellulose nanocrystals: sources and production. Nanoscale 2017;9(5):1763–86.
[14] Mariano M, El Kissi N, Dufresne A. Cellulose nanocrystals and related nanocomposites: review of some properties and challenges. J Polym Sci B Polym Phys 2014;52(12):791–806.
[15] Klemm D, Kramer F, Moritz S, Lindstrom T, Ankerfors M, Gray D, et al. Nanocelluloses: a new family of nature-based materials. Angew Chem 2011;50(24):5438–66.
[16] Moon RJ, Martini A, Nairn J, Simonsen J, Youngblood J. Cellulose nanomaterials review: structure, properties and nanocomposites. Chem Soc Rev 2011;40(7):3941–94.
[17] Brinchi L, Cotana F, Fortunati E, Kenny JM. Production of nanocrystalline cellulose from lignocellulosic biomass: technology and applications. Carbohydr Polym 2013;94(1):154–69.
[18] Domingues RM, Gomes ME, Reis RL. The potential of cellulose nanocrystals in tissue engineering strategies. Biomacromolecules 2014;15(7):2327–46.
[19] Peng BL, Dhar N, Liu HL, Tam KC. Chemistry and applications of nanocrystalline cellulose and its derivatives: a nanotechnology perspective. Can J Chem Eng 2011;89(5):1191–206.
[20] Lagerwall JPF, Schütz C, Salajkova M, Noh J, Hyun Park J, Scalia G, et al. Cellulose nanocrystal-based materials: from liquid crystal self-assembly and glass formation to multifunctional thin films. NPG Asia Mater 2014;6(1):e80.
[21] Lin N, Huang J, Dufresne A. Preparation, properties and applications of polysaccharide nanocrystals in advanced functional nanomaterials: a review. Nanoscale 2012;4(11):3274–94.
[22] Eichhorn SJ. Cellulose nanowhiskers: promising materials for advanced applications. Soft Matter 2011;7(2):303–15.
[23] Habibi Y, Lucia LA, Rojas OJ. Cellulose nanocrystals: chemistry, self-assembly, and applications. Chem Rev 2010;110(6):3479–500.
[24] Oliveira FB, Bras J, Pimenta MTB, Curvelo AAS, Belgacem MN. Production of cellulose nanocrystals from sugarcane bagasse fibers and pith. Ind Crop Prod 2016;93:48–57.
[25] Kontturi E, Meriluoto A, Penttila PA, Baccile N, Malho JM, Potthast A, et al. Degradation and crystallization of cellulose in hydrogen chloride vapor for high-yield isolation of cellulose nanocrystals. Angew Chem 2016;55(46):14455–8.

[26] Tang LR, Huang B, Ou W, Chen XR, Chen YD. Manufacture of cellulose nanocrystals by cation exchange resin-catalyzed hydrolysis of cellulose. Bioresour Technol 2011;102(23):10973–7.
[27] Satyamurthy P, Jain P, Balasubramanya RH, Vigneshwaran N. Preparation and characterization of cellulose nanowhiskers from cotton fibres by controlled microbial hydrolysis. Carbohydr Polym 2011;83(1):122–9.
[28] Chen X, Deng X, Shen W, Jiang L. Controlled enzymolysis preparation of nanocrystalline cellulose from pretreated cotton fibers. Bioresources 2012;7:4237–48.
[29] Novo LP, Bras J, García A, Belgacem N, Curvelo AAS. Subcritical water: a method for green production of cellulose nanocrystals. ACS Sustain Chem Eng 2015;3(11):2839–46.
[30] Yu H, Qin Z, Liang B, Liu N, Zhou Z, Chen L. Facile extraction of thermally stable cellulose nanocrystals with a high yield of 93% through hydrochloric acid hydrolysis under hydrothermal conditions. J Mater Chem A 2013;1(12):3938.
[31] Mohd Amin KN, Annamalai PK, Morrow IC, Martin D. Production of cellulose nanocrystals via a scalable mechanical method. RSC Adv 2015;5(70):57133–40.
[32] Cheng M, Qin Z, Liu Y, Qin Y, Li T, Chen L, et al. Efficient extraction of carboxylated spherical cellulose nanocrystals with narrow distribution through hydrolysis of lyocell fibers by using ammonium persulfate as an oxidant. J Mater Chem A 2014;2(1):251–8.
[33] Cao X, Ding B, Yu J, Al-Deyab SS. Cellulose nanowhiskers extracted from TEMPO-oxidized jute fibers. Carbohydr Polym 2012;90(2):1075–80.
[34] Lin N, Dufresne A. Nanocellulose in biomedicine: current status and future prospect. Eur Polym J 2014;59:302–25.
[35] Kim J-H, Shim BS, Kim HS, Lee Y-J, Min S-K, Jang D, et al. Review of nanocellulose for sustainable future materials. Int J Precision Eng Manuf Green Technol 2015;2(2):197–213.
[36] George J, Sabapathi SN. Cellulose nanocrystals: synthesis, functional properties, and applications. Nanotechnol Sci Appl 2015;8:45–54.
[37] Carpenter AW, de Lannoy CF, Wiesner MR. Cellulose nanomaterials in water treatment technologies. Environ Sci Technol 2015;49(9):5277–87.
[38] Miao C, Hamad WY. Cellulose reinforced polymer composites and nanocomposites: a critical review. Cellulose 2013;20(5):2221–62.
[39] Dufresne A. Nanocellulose: a new ageless bionanomaterial. Mater Today 2013;16(6):220–7.
[40] Ambrosi A, Chua CK, Bonanni A, Pumera M. Electrochemistry of graphene and related materials. Chem Rev 2014;114(14):7150–88.
[41] Yang G, Li L, Lee WB, Ng MC. Structure of graphene and its disorders: a review. Sci Technol Adv Mater 2018;19(1):613–48.
[42] Mansuriya BD, Altintas Z. Graphene quantum dot-based electrochemical immunosensors for biomedical applications. Materials 2019;13(1):96.
[43] Azman NHN, Mamat Mat Nazir MS, Ngee LH, Sulaiman Y. Graphene-based ternary composites for supercapacitors. Int J Energy Res 2018;42(6):2104–16.
[44] Aradhana R, Mohanty S, Nayak SK. Comparison of mechanical, electrical and thermal properties in graphene oxide and reduced graphene oxide filled epoxy nanocomposite adhesives. Polymer 2018;141:109–23.
[45] Potts JR, Dreyer DR, Bielawski CW, Ruoff RS. Graphene-based polymer nanocomposites. Polymer 2011;52(1):5–25.
[46] Kim C-J, Khan W, Kim D-H, Cho K-S, Park S-Y. Graphene oxide/cellulose composite using NMMO monohydrate. Carbohydr Polym 2011;86(2):903–9.
[47] Huang H-D, Liu C-Y, Li D, Chen Y-H, Zhong G-J, Li Z-M. Ultra-low gas permeability and efficient reinforcement of cellulose nanocomposite films by well-aligned graphene oxide nanosheets. J Mater Chem A 2014;2(38):15853–63.

[48] Jeon GW, An J-E, Jeong YG. High performance cellulose acetate propionate composites reinforced with exfoliated graphene. Compos Part B 2012;43(8):3412–8.
[49] Huang Y, Qin Y, Zhou Y, Niu H, Yu Z-Z, Dong J-Y. Polypropylene/graphene oxide nanocomposites prepared by in situ Ziegler-Natta polymerization. Chem Mater 2010;22(13):4096–102.
[50] Fim FC, Guterres JM, Basso NRS, Galland GB. Polyethylene/graphite nanocomposites obtained byin situpolymerization. J Polym Sci A Polym Chem 2010;48(3):692–8.
[51] Jang J, Kim M, Jeong H, Shin C. Graphite oxide/poly(methyl methacrylate) nanocomposites prepared by a novel method utilizing macroazoinitiator. Compos Sci Technol 2009;69 (2):186–91.
[52] Zhang H, Zhai D, He Y. Graphene oxide/polyacrylamide/carboxymethyl cellulose sodium nanocomposite hydrogel with enhanced mechanical strength: preparation, characterization and the swelling behavior. RSC Adv 2014;4(84):44600–9.
[53] Ji X, Xu Y, Zhang W, Cui L, Liu J. Review of functionalization, structure and properties of graphene/polymer composite fibers. Compos A: Appl Sci Manuf 2016;87:29–45.
[54] Zafar R, Zia KM, Tabasum S, Jabeen F, Noreen A, Zuber M. Polysaccharide based bionanocomposites, properties and applications: a review. Int J Biol Macromol 2016;92:1012–24.
[55] Shi H, Li W, Zhong L, Xu C. Methylene blue adsorption from aqueous solution by magnetic cellulose/graphene oxide composite: equilibrium, kinetics, and thermodynamics. Ind Eng Chem Res 2014;53(3):1108–18.
[56] Wan C, Li J. Graphene oxide/cellulose aerogels nanocomposite: preparation, pyrolysis, and application for electromagnetic interference shielding. Carbohydr Polym 2016;150:172–9.
[57] Xu M, Huang Q, Wang X, Sun R. Highly tough cellulose/graphene composite hydrogels prepared from ionic liquids. Ind Crop Prod 2015;70:56–63.
[58] Yadav M, Rhee KY, Jung IH, Park SJ. Eco-friendly synthesis, characterization and properties of a sodium carboxymethyl cellulose/graphene oxide nanocomposite film. Cellulose 2013;20 (2):687–98.
[59] Zhang C, Zhang RZ, Ma YQ, Guan WB, Wu XL, Liu X, et al. Preparation of cellulose/graphene composite and its applications for triazine pesticides adsorption from water. ACS Sustain Chem Eng 2015;3(3):396–405.
[60] Zhu W, Li W, He Y, Duan T. In-situ biopreparation of biocompatible bacterial cellulose/graphene oxide composites pellets. Appl Surf Sci 2015;338:22–6.
[61] Daniyal W, Fen YW, Abdullah J, Sadrolhosseini AR, Saleviter S, Omar NAS. Label-free optical spectroscopy for characterizing binding properties of highly sensitive nanocrystalline cellulose-graphene oxide based nanocomposite towards nickel ion. Spectrochim Acta A Mol Biomol Spectrosc 2019;212:25–31.
[62] Gao Y, Xu H, Cheng Q. Multiple synergistic toughening graphene nanocomposites through cadmium ions and cellulose nanocrystals. Adv Mater Interfaces 2018;5(10):1800145.
[63] Javanbakht S, Namazi H. Doxorubicin loaded carboxymethyl cellulose/graphene quantum dot nanocomposite hydrogel films as a potential anticancer drug delivery system. Mater Sci Eng C Mater Biol Appl 2018;87:50–9.
[64] Kafy A, Akther A, Zhai L, Kim HC, Kim J. Porous cellulose/graphene oxide nanocomposite as flexible and renewable electrode material for supercapacitor. Synth Met 2017;223:94–100.
[65] Montes S, Carrasco PM, Ruiz V, Cabañero G, Grande HJ, Labidi J, et al. Synergistic reinforcement of poly(vinyl alcohol) nanocomposites with cellulose nanocrystal-stabilized graphene. Compos Sci Technol 2015;117:26–31.
[66] Rasoulzadeh M, Namazi H. Carboxymethyl cellulose/graphene oxide bio-nanocomposite hydrogel beads as anticancer drug carrier agent. Carbohydr Polym 2017;168:320–6.

[67] Meng X, Wang S, Gao W, Han W, Lucia LA. Thermal pyrolysis characteristics and kinetic analysis of nanofibrillated cellulose/graphene oxide composites. Bioresources 2020;15:4851–65.

[68] Gan S, Zakaria S, Syed Jaafar SN. Enhanced mechanical properties of hydrothermal carbamated cellulose nanocomposite film reinforced with graphene oxide. Carbohydr Polym 2017;172:284–93.

[69] Wen Y, Wu M, Zhang M, Li C, Shi G. Topological design of ultrastrong and highly conductive graphene films. Adv Mater 2017;29(41):1702831.

[70] Feng Y, Zhang X, Shen Y, Yoshino K, Feng W. A mechanically strong, flexible and conductive film based on bacterial cellulose/graphene nanocomposite. Carbohydr Polym 2012;87(1):644–9.

[71] Zhang X, Liu X, Zheng W, Zhu J. Regenerated cellulose/graphene nanocomposite films prepared in DMAC/LiCl solution. Carbohydr Polym 2012;88(1):26–30.

[72] Han D, Yan L, Chen W, Li W, Bangal PR. Cellulose/graphite oxide composite films with improved mechanical properties over a wide range of temperature. Carbohydr Polym 2011;83(2):966–72.

[73] Wang B, Lou W, Wang X, Hao J. Relationship between dispersion state and reinforcement effect of graphene oxide in microcrystalline cellulose–graphene oxide composite films. J Mater Chem 2012;22(25):12859.

[74] Mahmoudian S, Wahit MU, Imran M, Ismail AF, Balakrishnan H. A facile approach to prepare regenerated cellulose/graphene nanoplatelets nanocomposite using room-temperature ionic liquid. J Nanosci Nanotechnol 2012;12(7):5233–9.

[75] Tian M, Qu L, Zhang X, Zhang K, Zhu S, Guo X, et al. Enhanced mechanical and thermal properties of regenerated cellulose/graphene composite fibers. Carbohydr Polym 2014;111:456–62.

[76] Palanisamy S, Ramaraj SK, Chen SM, Yang TC, Yi-Fan P, Chen TW, et al. A novel laccase biosensor based on laccase immobilized graphene-cellulose microfiber composite modified screen-printed carbon electrode for sensitive determination of catechol. Sci Rep 2017;7:41214.

[77] Wang F, Drzal LT, Qin Y, Huang Z. Multifunctional graphene nanoplatelets/cellulose nanocrystals composite paper. Compos Part B 2015;79:521–9.

[78] Liu Y, Zhou J, Tang J, Tang W. Three-dimensional, chemically bonded polypyrrole/bacterial cellulose/graphene composites for high-performance supercapacitors. Chem Mater 2015;27(20):7034–41.

[79] Sen I, Seki Y, Sarikanat M, Cetin L, Gurses BO, Ozdemir O, et al. Electroactive behavior of graphene nanoplatelets loaded cellulose composite actuators. Compos Part B 2015;69:369–77.

Chapter 10

Hybrid nanocomposites based on cellulose nanocrystals/nanofibrils: From preparation to applications

H Mohit[a], G Hemath Kumar[b], MR Sanjay[a], S Siengchin[a], and P Ramesh[c]
[a]*Natural Composites Research Group Lab, Department of Materials and Production Engineering, The Siridhorn International Thai-German Graduate School of Engineering, King Mongkut's University of Technology North Bangkok, Bangkok, Thailand,* [b]*Composite Research Center, Chennai, India,* [c]*Department of Production Engineering, National Institute of Technology, Tiruchirappalli, India*

10.1 Introduction to cellulose-based composites

Investigations into the advancement of renewable or sustainable polymeric composites have improved, explicitly concerning the progress of particular characteristics such as thermal stability and mechanical properties. Inorganic and organic nanofillers' utilization in composite laminate production has been beneficial, as these types of composites give the perception of ecological safety and greener surroundings [1, 2]. In this regard, when any specific material is under progress, the production and sustainability of environmentally friendly products from biodegradable and renewable sources are more desirable leading to opportunities, but still open for debate [3–6]. Several attempts have been made to develop and apply green-based polymer composites using plant cellulose fibers as alternatives for traditional inorganic particles such as carbon or glass [5]. Plant cellulose fibers have intrinsic characteristics exceeding conventional reinforcing fillers [7]. These materials are low-cost and biodegradable, lower weight, lower wear rate, superior mechanical characteristics, better thermal stability, enhanced carbon dioxide sequestration, and energy recovery [8, 9]. Nevertheless, the use of biodegradable polymer resins has been slowly introduced and needs some improvements because of the main problems associated to their characteristics, especially brittleness, lower gas permeability,

Cellulose Nanocrystal/Nanoparticles Hybrid Nanocomposites. https://doi.org/10.1016/B978-0-12-822906-4.00009-8

higher prices, and lower degradation temperature compared with other commercial plastic materials such as polyolefins, polyvinyl chloride, and polyethylene terephthalate [10].

To overcome the bio-based polymers' inherent physical limitations, the addition of nano-filler materials has been perceived as among the different routes to enhance their characteristics [8]. Cellulose obtained from plants is the most available material in nature, showing higher compatibility with the polymer chemistry. Under this condition, the nano-scale form is a promising component for the fabrication of polymer-based nanocomposite materials. Apart from this cost-efficient fabrication, nanocellulose has fascinating characteristics such as higher strength, lower density, extraordinary mechanical strength, and higher stiffness than most materials used today. This is why nanocellulose has been established as a potential renewable reinforcement nanoparticle in polymer-based laminates, with emerging hybrid laminates presenting higher overall characteristics than macro- and micro-based cellulose laminates. The design freedom of nanocellulose polymer laminates and the possibility to combine and characterize these laminates lead to their broad use in electronics, automotive, biotechnology, and packaging industries among different applications. Noticeably, the analysis of nanocellulose has propelled the acceptance of natural-based cellulose resins in various fields associated to sustainable nanocomposite materials. The larger surface area of nanocellulose allows stronger interactions with neighboring organic and polymeric components, water, living cells, and nanofillers [8].

Cellulose is the most abundant and broadly dispersed natural polymer, identified as a critical source of renewable material mainly in plant cellulose fiber. As per Onoja et al. [3], the biomass collected from oil palms such as fruit bunch, trunk, and frond involves a relatively higher cellulose amount. This biodegradable and low-cost resin is water-insoluble, fibrous, and resistant, with an annual total production of up to 10^{12} tons [11]. Since cellulose is biocompatible, non-polluting, renewable, and non-toxic, this led to several applications like fabrics and scaffold structures. Cellulose contains D-glucopyranose small molecules connected with glycosidic networks with polymerization degrees up to 15,000 in wood cellulose, but this value can be higher in indigenous plant cellulose fibers. Cellulose prevails in the plant cell walls and different sea animals, algae, bacteria, and tunicates [12]. The cell walls are based on carbohydrates and proteins. As it is one of the main constituent of plant cell walls, cellulose produces over 50% of botanic gardens' carbon concentration. These plants' cell walls are composed of three organic components: lignin, hemicellulose, and cellulose, which are in general with a ratio of 4:3:3. But, this proportion varies from resources like grass, softwood, and hardwood. The lignin and hemicellulose comprise nearly 40%–50% of plant wood, but cotton may have up to 90% cellulose content [13]. Cellulose can occur in two states as amorphous and crystalline. In the crystalline state, the cellulose chains are mainly composed of lignin, waxes, hemicellulose, trace elements, extractives, and arranged in a disordered aspect [13]. One of the main characteristics of cellulose is the presence of three hydroxyl groups at position 6, 3, and 2 in every glucose unit. The secondary alcohol's hydroxyl region is located at the 6th place, which is 10 times more reactive than those at the 3rd and 2nd position. Moreover, the hydroxyl group at the 2nd position is 2 times more reactive than the 3rd position [14]. The stiffness and hierarchical structure of the cellulose network is stabilized with intramolecular and intermolecular bonds. The intermolecular bond connects the hydroxyl group of the nearby glucose unit's ring oxygen and carbon [14].

Nanocellulose has been assigned as a pseudo-plastic material showing different fluid characteristics, even though it is usually more fluid-like under common cases. Nanocellulose is mainly classified as cellulose nanofibers and cellulose nanocrystals, both extracted from plant cellulose fiber and microbial cellulose obtained from bacterial micro-organisms. This commonly derived bulk cellulose includes both amorphous and crystalline portions, and the proportion of crystalline to amorphous parts may vary with the cellulose source [15]. The crystalline structure of the microfibrils can be derived from the consolidation of mechanical and chemical surface modifications [16, 17] to produce stiff rods or fillers made up of a cellulose network aligned in a perfect crystalline structure.

Cellulose nanocrystals are proposed as possible nanocomposite reinforcement materials and have been broadly investigated in recent years due to their relatively lower density, higher strength, and modulus [18]. Currently, cellulose nanocrystals are mainly extracted from lignocellulosic resources using top-down techniques, where wood is a primary source of cellulose fibers.

Furthermore, fierce competition between different fabrication industries, such as building, furniture producers, and pulp and paper sectors creates a high demand to provide all these industries with reasonably economic wood in more important amounts. On the other hand, non-woody plant cellulose fibers combine microfibrils of cellulose that are less tightly packed in the essential cell wall than in the trivial cell wall, which controls the production process [19].

Cellulose nanofibers can be obtained under different forms depending on the technique by which they are fabricated. The cellulose nanofibers' fundamental unit is the microfibril of cellulose, which is a single nanofiller [20]. Cellulose nanofibers have been broadly investigated for their notable mechanical strength, flexibility, chirality, thermostability, biodegradability, and potential blending and functionalization, with lower thermal expansion [21]. Cellulose nanofibers are produced by energy-intensive techniques such as mechanical defibrillation combined with homogenization, grinding, and refining. The demand for energy to make cellulose nanofibers can reach 30,000 kWh/ton with high-pressure homogenization as illustrated by Nakagaito and Yano [22]. Hence, many cellulose pre-surface modifications for cellulose nanofibers' preparation have been broadly studied to decrease the energy consumption, especially for natural fiber-based cellulose. Habibi et al. [23] examined the acid-hydrolyzed production of wood cellulose nanocrystals leading to a length and width of 50–500 nm and 3–5 nm respectively, and a crystallinity index of 54%–88%. On the other hand, mechanical refining to fabricate wood-based cellulose nanofibers gives length and width of 500–2000 nm and 4–20 nm respectively, with a crystallinity index of 46%–65% [24].

Bacterial cellulose is also termed microbial cellulose because it is produced from different types of bacteria in a pellicle's state under the interface of air and liquid. Its interphase features create a substitute for natural fiber-derived cellulose for different utilizations in medical science, food, and other industrial sectors [25]. The bacterial cellulose's fundamental structure in fibrils contains glucan networks with molecular formulations placed in intramolecular and intermolecular bonding. The fibril chain of bacterial cellulose is correlated with uniformly dispersed nanofibers creating a hydrogel sheet and developing a higher surface area. During the preparation, the glucose network proton fibers are recovered through the bacteria's cell wall and gathered together in nanofibers. The technique eventually tends to produce a spiral-like chain of cellulose [26]. Bacterial cellulose also differs from cellulose-based on the chemical and physical characteristics. Bacterial cellulose is of higher crystallinity and purity with good biological aspects [27]. These distinctive and valuable characteristics stem from bacterial cellulose's particular ultrastructure created from the pure cellulose ribbons [28]. Bacterial cellulose is designed to produce a diet food component and is used in different applications such as speaker diaphragms, makeup pas, medical pads, artificial skins, and paint thickeners [25].

Metal matrix composite based nanoparticles are natural minerals broadly applied as reinforcement nanofillers in different polymer laminates. The essential characteristics of these nanofillers are mechanical, barrier, and flame retardancy characteristics, which are important for their broad use in numerous industries such as automotive, pharmaceuticals, inks, and coating applications [29]. These components are commonly in the nano form of 1 nm in thickness, with a significant aspect ratio between 10 to 1000:1. From this range of aspect ratios, these nanoparticles are making better reinforcement. Even though metal matrix composite-based nanoparticles are non-renewable, this nanofiller has characteristics to be applicable for incorporation into other plastics or polymers by keeping the finished product as completely biodegradable [30].

The consolidation of nanocellulose and nanoclay was also studied in polymer-based laminates. It was found that poly(vinyl alcohol)/nanocellulose-based laminates exhibited enhanced thermomechanical characteristics (storage moduli from 3.5 to 7.9 GPa, glass transition temperature from 60 °C to 84 °C, and 2nd degradation temperature from 424 °C to 450 °C), decreased strain at break (from 6% to 1%), and oxygen transmission rate (from 0.04 to 0.005 cm^3 mm/(m^2 day atm)) above 30% of relative humidity with the addition of clay [31]. The addition of 1 wt% of cellulose nanocrystals with nanoclay in polylactic acid laminate consisting of 5 wt% of nanoclay tends to increase the strain at break (~10% to ~79%) for the laminates [32]. Similarly, cellulose nanocrystals' incorporation to the laminate tends to improve water diffusion (5.722 to 6.376 mm^2/s) and water absorption (up to 1.5%) based on the amount of cellulose nanocrystals addition [33, 34]. The consolidation of cellulose nanocrystals and nanoclay tend to enhanced the thermomechanical characteristics (glass transition temperature up to 90 °C and coefficient of thermal expansion as 25.1 μm/m °C) of maleic acid-modified polymers composites [35].

The arrival of nanotechnology led to the production of nanocellulose, which attracted an appreciable quantity of attention for utilization in advanced polymeric materials. The nanocellulose collected from plant cellulose fiber has changed bio-based materials for a range of applications. With their good optical characteristics, nano-scale, larger surface area, stiffness, higher crystallinity index, renewability, and biodegradability, cellulose nanoparticles were used for different interesting applications [36, 37]. The investigation reported here aimed at evaluating the potential synergistic effects as an outcome of combining nanocellulose and metal matrix nanocomposite fillers in polyester composites with a focus on characteristics to enhanced polyester materials and to develop them into an industrially suitable different polymer for structural applications.

10.2 Materials and methods

10.2.1 Materials

Polyester resin (Unsaturated polyester Isophthalic resin, HS code 39079190, IS9 grade) with cobalt octoate (accelerator) and methyl-ethyl-keto-peroxide (catalyst) as matrix materials were obtained from Sakthi Fiber Glass Inc, Chennai, India. The cellulose nanofibers were extracted from sugarcane bagasse which was collected from different juice shops. Aluminum (10 μm) and silicon carbide (150 μm) particles are commercially available from Carborundum Universal Limited, Kochi and Ganapathi Colors, Chennai, India, respectively. Sodium chloride (NaCl), sodium hydroxide (NaOH), and distilled water were supplied from Lab Chemicals, Chennai. All chemicals were used as received.

10.2.1.1 Preparation of nanocellulose fiber from sugarcane bagasse

The collected sugarcane bagasse was washed with fresh tap water to remove the dirt and dust particles. The sugarcane bagasse was soaked in a salt solution (prepared from NaCl and water in the ratio of 10:1) for 48 h and treated with a 0.1-N NaOH aqueous solution for 4 h at 100 °C. Finally, the treated sugarcane bagasse was transferred to an industrial grinder to produce fine particles and separated using a 50-nm nano mesh sieve.

10.2.1.2 Synthesis of Al-SiC nanoparticles

Aluminum and silicon carbide were poured into an horizontal high energy ball mill and mixed to get fine nanoparticles. The milling process was carried out at

150 rpm, 10:1, and 240 min of milling speed, balls to powder weight ratio, and milling time, respectively. After the milling process, the ball-milled fine particles were separated using a nano mesh sieve of 70 nm.

10.2.1.3 Polyester composites fabrication

The hybrid polyester composites were fabricated by combining a resin solution of polyester, uniform dispersion of both sugarcane nanocellulose (5 wt%), and Al-SiC nanoparticles (5 wt%). An ultrasonicator probe was used to mix the nanofillers within the polyester matrix at 400 W and 120 min to get uniform particles dispersion. For post-curing, the accelerator and catalyst were poured into the resin mixture, stirred vigorously using a mechanical stirrer for 60 s, and rapidly transferred to the mold (250 mm (length) × 250 mm (width) × 5 mm (thickness)). After curing, the fabricated polyester composites were transferred to a heating furnace and post-cured at 60 ± 2 °C for 24 h.

10.2.2 Characterization

The tensile modulus and strength of all the fabricated polyester hybrid composites were determined from a Tinius Olsen Universal Testing Machine with a load cell of 20 kN. The tests were carried out according to ASTM D 3039 with a test speed of 1 mm/min.

Uniaxial compression tests on the polyester hybrid nanocomposites were performed on a Tinius Olsen Universal Testing Machine following ASTM D 695. The specimens (15 × 10 × 5 mm) were cut into a rectangular piece followed by cyclic compression loading between the flat plates under a crosshead speed of 10 mm/min.

The flexural characteristics were performed according to ASTM D 790. The flexural tests were performed using a three-point bending geometry (span of 60 mm) with rectangular samples having dimensions 100 × 20 × 5 mm. The sample deflection was determined from the crosshead position, and the test outcomes contain displacement, flexural strength, and modulus.

The impact strength was defined as the energy in Joules needed to break a specific volume of material. The test was conducted as per ASTM D 256 standard using samples with dimensions of (62.5 × 12.5 × 5 mm).

All the mechanical properties experiments were conducted five times at room temperature (24 ± 2 °C) and a relative humidity of 60%.

The hardness was determined to characterize the sample surface prone to cracking. A superficial hardness tester consists of a resolution and hardness range of 0.25 lm/grid and 1–2967 HV respectively, with an accuracy of ±3% to ± 4%. The experiments were performed with a diamond-based Vickers indenter with a load of 0.1 kg and 20 s of dwell time.

Thermogravimetric analysis was conducted using a TGA 4000 (Perkin Elmer) from 30 °C to 600 °C with a flow rate of 50 mL/min under a nitrogen atmosphere and a heating rate of 20 °C/min. A specimen of 10 mg was heated in an alumina crucible. As temperature was increased, the weight as a function of temperature was obtained.

A field emission scanning electron microscope (FE-SEM, SU8000, Hitachi) was applied to examine the polyester composite materials' tensile fracture surface under 15 kV of accelerating voltage.

DMA Q800 (TA instruments) was applied to examine the polyester hybrid nanocomposites' dynamic mechanical and thermal characteristics. The samples' dimension was 50 × 10 × 5 mm under flexural mode with a static force of 0.1 N, strain amplitude of 30 μm, and 1 Hz of sinusoidal frequency. The temperature was raised from 30 °C to 150 °C with a heating rate of 10 °C/min.

The polyester hybrid nanocomposites were further studied to get their Fourier Transform Infrared (FTIR) spectrum with a Perkin Elmer 2 spectrometer. The scan rate was 32 scans per minute for wavenumbers between 400 and 4000 cm^{-1}. The sugarcane nanocellulose and Al-SiC nanoparticles reinforced hybrid nanocomposites structure were analyzed with a Rigaku Ultima IV X-ray diffractometer (XRD) with a Cu-Kα value of 0.154 nm under 40 mA, 40 kV, and 2θ (5 to 80 degree) for the current, scanning voltage, and diffraction angle range, respectively.

10.3 Results and discussion

10.3.1 Characteristic curves

From the FTIR spectra of nanocellulose/Al-SiC hybrid polyester composites (Fig. 10.1), the main peaks at 1627, 1396, and 1727 cm^{-1} are assigned to C—C stretch, O—H stretch, and C-O stretch bending vibration, respectively. However, C—O (alkoxy) and C—O (polyester) stretch vibration peaks are present at 1054

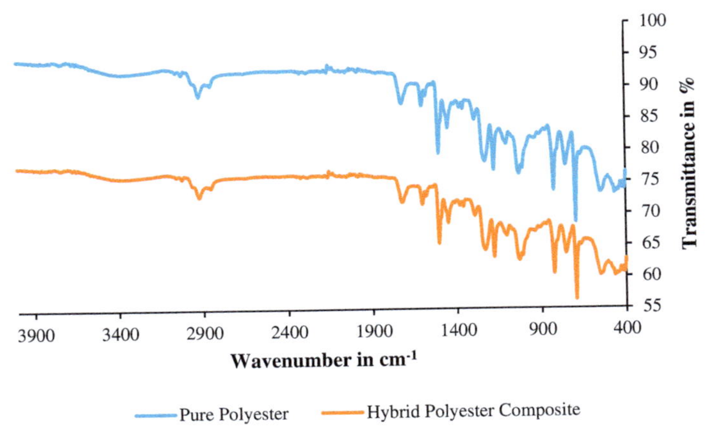

FIG. 10.1 FTIR curves of polyester composites.

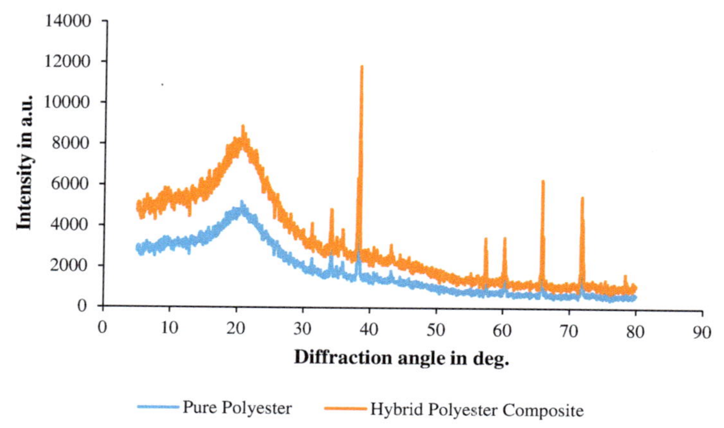

FIG. 10.2 XRD curves of polyester composites.

and 1216 cm^{-1}, respectively [38–40]. After adding nanocellulose and Al-SiC nanoparticles, the peaks' intensity corresponding to oxygenous groups are significantly reduced because of efficient reaction. The important peaks for —OH stretching in sugarcane nanocellulose/Al-SiC polyester are observed at 3332 cm^{-1} compared with the neat polyester. This is related to the destruction of the honeycomb structure in the Al-SiC nanoparticles with the addition of sugarcane nanocellulose and the production of stronger interfacial bonding between the matrix and nanofillers [41]. The uniform dispersion of the nanocellulose and Al-SiC nanoparticles was further studied with XRD as shown in Fig. 10.2. As observed from the XRD spectra, the neat polyester (powder) presents a free diffraction peak around 5 degree, which is relates to the fillers' d-spacing within the matrix. This strong peak for the hybrid polyester laminate is not present, showing that the Al-SiC nanoparticles are dispersed during processing and the Al-SiC nanoparticles in the laminates are highly exfoliated. Similar observations have been reported by other investigators [34]. XRD results show that the d-spacing of the nanocellulose/Al-SiC in the polyester matrix was considerably improved by combining with polyesters, even though a fully exfoliated state without nanofiller diffraction peak could not be produced in the hybrid composite. The crystal size of the polyester hybrid composites is estimated from the full width at half maximum of the Al-SiC nanoparticles (100) peaks which improved due to the Al-SiC addition. The anionic nature of the polyester hybrid composite can be related to the production of Al-SiC nanofillers structures in the laminate [42].

10.3.2 Mechanical properties

The tensile stress-strain plots, tensile strength, and modulus of the composites are shown in Fig. 10.3A and Table 10.1. The neat polyester laminate (without Al-SiC and sugarcane nanocellulose) has a tensile strength and modulus of

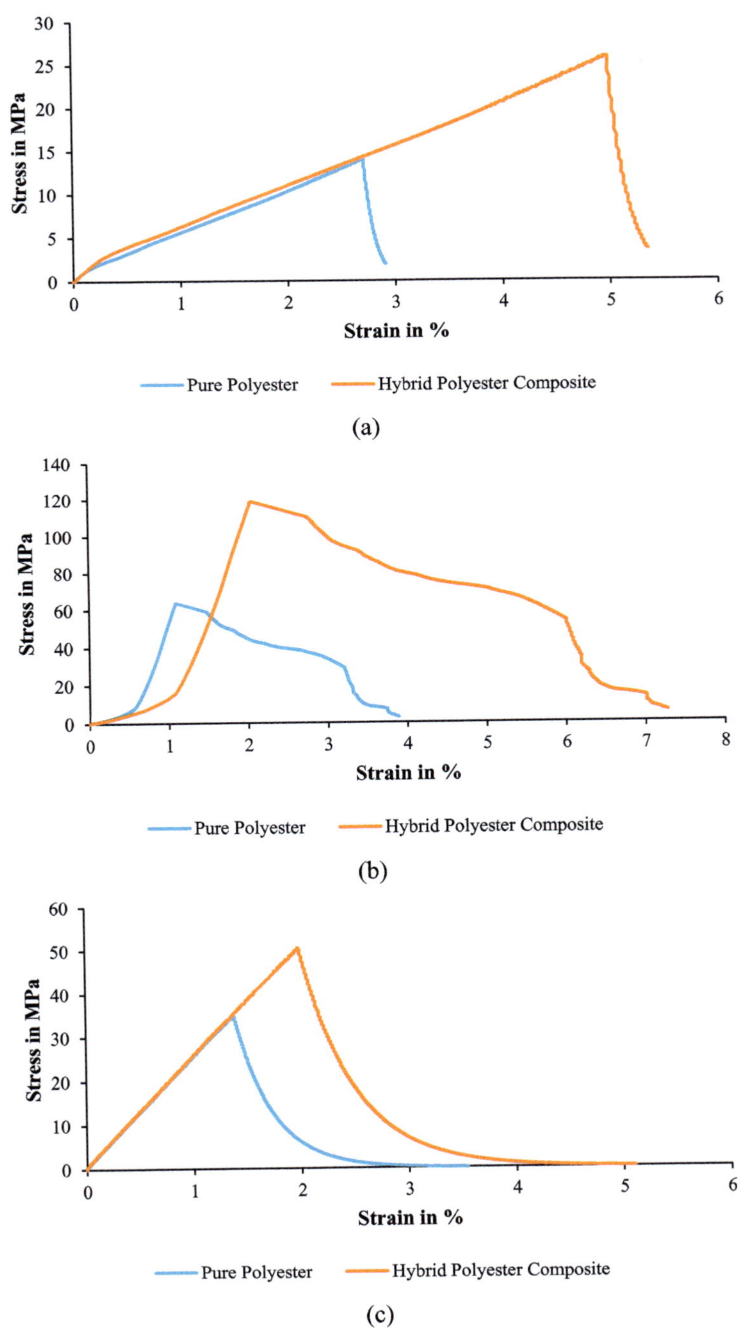

FIG. 10.3 Stress-strain graph of polyester composites (A) tensile, (B) compression, and (C) flexural.

TABLE 10.1 Properties of the newly developed polyester composites.

Property	Neat polyester	Hybrid polyester composite
Tensile strength (MPa)	12.8±0.4	23.7±0.8
Tensile modulus (GPa)	0.97±0.03	1.52±0.05
Compression strength (MPa)	62.3±2.2	116±4
Compression modulus (GPa)	0.77±0.03	1.47±0.06
Flexural strength (MPa)	32.5±1.3	46.9±1.9
Flexural modulus (GPa)	3.01±0.12	5.58±0.23
Impact strength (kJ/m^2)	0.05±0.01	1.25±0.05
Vickers hardness (–)	20.8±0.7	46.2±1.6
Glass transition temperature (°C)	60.0±1.9	72.5±2.4
Damping factor (–)	0.74±0.03	0.49±0.02

12.8 MPa and 0.97 GPa, respectively. These values are related to the high crystallinity index (around 60%) and lower aspect ratio of the polyester resin [43, 44]. The mechanical properties of the polyester hybrid composites were significantly enhanced compared to the Al-SiC and sugarcane nanocellulose fillers. The resulting composite had a higher tensile strength and modulus of 23.7 MPa and 1.52 GPa respectively, with a lower density of 2196 kg/m^3.

The higher tensile strength of the polyester hybrid composite is similar to recently published literature, but lower than other types of biobased inorganic/organic laminates [45, 46]. The effectiveness of Al-SiC nanofillers in the polyester hybrid composites seems to be lower than for other types of polymer composites [47–50]. Also, this high strength leads to a 5.4% elongation at break with 5 wt% of nanocellulose and Al-SiC nanofillers. The tensile modulus of 1.52 GPa for the polyester hybrid laminate indicates that each nanofiller was fully dispersed and discrete as reinforcement. This higher strength is apparently because of an effective nanolaminate influence from proper nanofillers dispersion at the nano level in the composite [51], producing more ionic interactions and hydrogen bonds at the interfaces between the matrix and nanocellulose/Al-SiC nanoparticles. The hydrogen bonds are the main bonding interactions between the matrix and nanofillers, leading to higher mechanical properties. These interactions can be observed from the FTIR spectra (Fig. 10.1).

Fig. 10.4 presents the SEM micrographs of the tensile fracture surface of the neat polyester and hybrid polyester composites. Dendrites are clearly observed

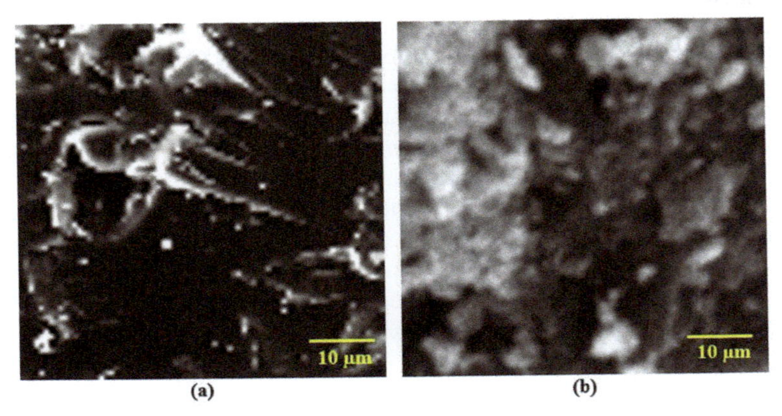

FIG. 10.4 SEM micrographs of polyester composites.

in Fig. 10.4A for the neat polyester due to the brittle nature of the laminate. The SEM micrographs of the polyester hybrid laminate shows the nanolayered and closely packed shapes of Al-SiC nanofillers in the vertical direction when combined with sugarcane nanocellulose, which played an important role in providing stronger bonding with the polyester resin. The hybrid polyester composite samples exhibit highly aligned and sheet-like structure of Al-SiC nanofillers and hence the production of nanolayered sugarcane nanocellulose structures (Fig. 10.4B) acting as a natural medium obtained by combining with higher crystallinity polyester resin as the matrix [42].

The compression strength and modulus of the neat polyester and polyester hybrid composites slightly improved, and the laminates were not flexible (Fig. 10.3B). But compared to the neat polyester, the polyester hybrid composites displayed substantial enhancement. However, the densities significantly influence the composites' strength with similar constituents leading to moduli of 0.77 and 1.77 GPa for the neat polyester and the polyester hybrid composites, respectively. After a 50% compression strain, the elastic relaxation of all the cross-linked composites improved compared to the neat specimen. This indicates that an effective network structures was created between the nanofillers. A flexural experiment (Fig. 10.3C) was performed to determine the flexibility of the composites. When 4.5 mm thick composites were bent to their highest curvature (2.25 cm) in a downward direction, damages were produced in both the neat and hybrid polyester samples; i.e. chemical bonds were destroyed during the bending process. Likewise, their greater surface area offers more reaction positions for the silanol and hydroxyl groups, hence producing dense cross-linked chains which can highly improve the resistance to bending, but at the cost of ductility loss [52].

Impact resistance of the composites may give information on the energy absorbed in relation to the strength while sustaining an Izod impact test and analyzing the fractured behavior of the exposed specimen. It can be seen from Table 10.1 that the impact strength improved by up to 1.25 kJ/m^2 by adding sugarcane nanocellulose and Al-SiC nanoparticles. The reason may be related to both nanoparticles in the polyester matrix being able to support the applied stresses by achieving good chemical bonding with the polymer. Here, the nanoparticles reinforced polyester matrix enables a broad space for the molecules to move and absorb energy under impact testing. Another reason is the excellent adhesion and higher surface area of Al-SiC nanoparticles. So these results confirm again the good resistance of the composites against flexural deformations, independent of the rate of applied force. From Table 10.1, it is easy to understand that the impact strength of the polyester laminates increases with the incorporation of Al-SiC nanoparticles and sugarcane nanocellulose as their incorporation tend to distribute the stresses more evenly [53]. It can be concluded that adequate transfer of loads among Al-SiC nanoparticles and sugarcane nanocellulose in the polyester matrix was able to produce advanced laminates to sustain higher mechanical stresses and rates as a function of the type and content of reinforcement. While comparing the impact strength characteristics to other samples like 1 and 1.5 wt% of TiO_2, the values also depend on particle pull-out thus explaining why 0.5 wt% is generally better leading to higher results than those of the 5 wt% Al-SiC nanoparticles. These results indicate that higher nanofillers loading may be the controlling parameter leading to lower interfacial bonding inside the polyester resin. Moreover, these experimental observations show that the impact strength is a function of the applied load direction with respect to delamination [54, 55].

The hardness measurements on the polyester laminates are shown in Table 10.1. Hardness for a polyester composite is difficult but important in materials testing. The present investigation reports on the Vickers hardness. It can be seen that the values depend on the sugarcane nanocellulose and Al-SiC nanoparticles content. The polyester hybrid laminate displayed a higher value of 46.2 HV compared to 20.8 HV for the neat matrix, which represents a 55% increase (Table 10.1). A similar increase was observed by Lim et al. [56] when hard SiC particles were used. This result displays a critical aspect here: the Vickers hardness is highly sensitive to the presence of nanofillers and their concentration. The hardness improves with the nanofillers concentration and decreases with particle size as reported by Zare [57].

10.3.3 Viscoelastic properties

The storage modulus offers some information with respect to the matrix/filler adhesion, rigidity, and stiffness of the polyester composites (Fig. 10.5A). It shows that the storage modulus of reinforced polyester hybrid laminates is higher than the neat polyester laminates below 30 °C because of the influence of reinforcement on the polyester resin with sugarcane nanocellulose and Al-SiC nanoparticles. Fig. 10.5A also shows that the storage modulus curve

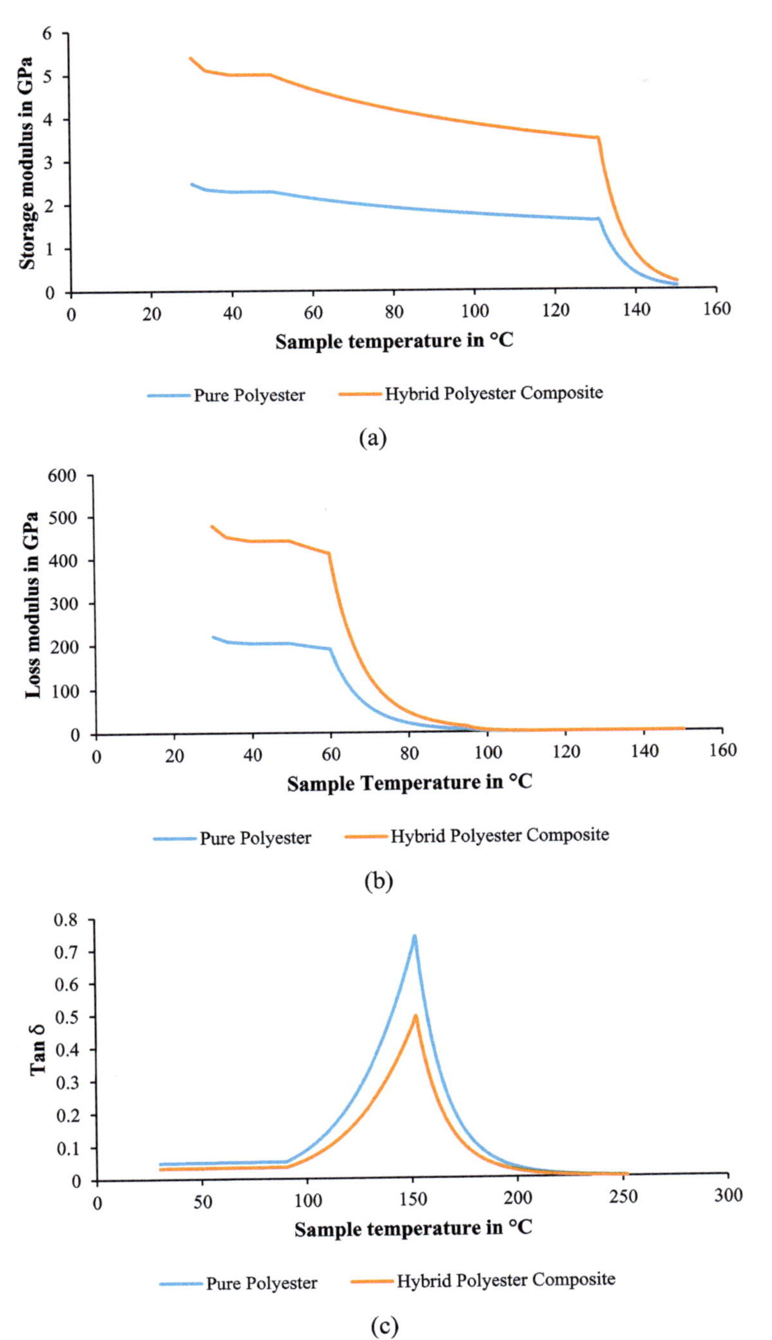

FIG. 10.5 Viscoelastic properties of polyester composites (A) storage modulus, (B) loss modulus, and (C) damping parameter.

can be decomposed into three different regions: glass transition, rubbery, and rigid. In the first region, the composite is stiff and tightly packed due to the fixed polymeric networks. The second region is associated to storage modulus decrease which is the glass transition temperature because of more mobility in the polymeric network [58]. For the third region, the rubbery region is achieved as the higher temperature improves molecular motion leading to limited changes in storage modulus. It is clear that increasing the temperature led to lower hybrid laminates' rigidity and stiffness. However, in the rubbery region, the storage modulus does not change much with temperature. This is especially the case when sugarcane nanocellulose and Al-SiC nanoparticles are present to limit molecular motion [59–61]. A similar result was reported by Keusch et al. [59] as the storage modulus is proportional to the interfacial adhesion bonding. An increase in the polyester laminate's storage modulus was observed which represented a balance between the packing arrangement of glass fibers and palm yarn, and excellent compatibility with the resin matrix. The reason may be that high compatibility leads to efficient stress transfer and higher interfacial bonding between the fiber and matrix during the cyclic loading process [62].

The loss modulus measures the dissipation of energy per cycle in the sinusoidal deformation when different schemes are compared with a specific strain. In principle, the viscous response of the polyester is susceptible to the molecular movement of the polymeric network. Fig. 10.5B depicts the loss modulus of the polyester hybrid composite reinforced by sugarcane nanocellulose and Al-SiC nanoparticles as a function of temperature. It can be observed that the loss modulus increased with the addition of both nanofillers. Comprehensively, the loss modulus increases and then decreases when going through the glass transition temperature. But a shift of the glass transition temperature towards higher temperature corresponds with the polymeric network's lower chain mobility due to the presence of the rigid nanoparticles and improved interfacial adhesion between the cellulose and Al-SiC nanoparticles with the polymer matrix. Here, the glass transition temperature from the loss modulus was found to increase from 60.0 °C to 72.5 °C when the nanoparticles were added. Based on recently published literature [63], the higher glass transition temperature can be assigned to the reduction in of polymeric chain mobility and better interaction between the polymer and fillers as observed when increasing the concentration of sugar palm fiber from 30 to 40 wt%. However, higher loss modulus is obtained as the fiber concentration increased due to higher energy dissipation and internal friction [64]. Similar results were achieved for oil palm nano filler/kenaf/epoxy hybrid nanocomposite [65]. Similar to the storage modulus curve, after the glass transition temperature region both curves overlapped together indicating that the presence of nanoparticles is less important at higher temperature.

The damping parameter represents the ratio between the loss modulus and storage modulus. This parameter can be related to the polyester's impact toughness [65]. The glass transition temperature can also be determined by the peak of the damping curve. This peak in the damping parameter curve is used to estimate the glass transition temperature of the polyester hybrid nanocomposite [66]. Compared with other plots, it presents a larger-range of molecular movements persistent in the rubbery region, based on the molecular structure or permanent deformation. Fig. 10.5C presents the damping factor for the neat and sugarcane nanocellulose/Al-SiC nanoparticles reinforced hybrid polyester composites. The presence of fillers can also be identified by the height of the damping factor curve. So fillers addition in the polyester resin leads to a lower damping parameter than for the neat polyester resin. Table 10.1 also presents the damping parameter and glass transition temperature collected from Fig. 10.5C. The damping parameter is lower for the polyester hybrid laminates which indicates that the stress transfer between the fillers and matrix is improved, and the filler-matrix interfacial bonding is increased [67]. The Al-SiC nanoparticles are an elastic material that can prevent energy dissipation and store energy, while the polyester is a viscoelastic material having both elasticity and viscosity. When the polyester is deformed, one portion of the input is dissipated as heat. It is thus expected that the addition of Al-SiC nanoparticles and sugarcane nanocellulose with an efficient loading of polyester will affected by the resin's energy dissipation. Idicula et al. [62] observed that materials can have lower damping values and glass transition temperature related to their load-bearing characteristics because of enhanced interfacial adhesion. Similar results were also found for hybrid laminates reinforced with glass fiber/palm yarn [68].

10.3.4 Thermal stability

The level of temperature can be associated with the degradation of the molecular structure of the polyester matrix which can be evaluated by thermogravimetric analysis. Fig. 10.6 shows the thermal properties of the neat and hybrid

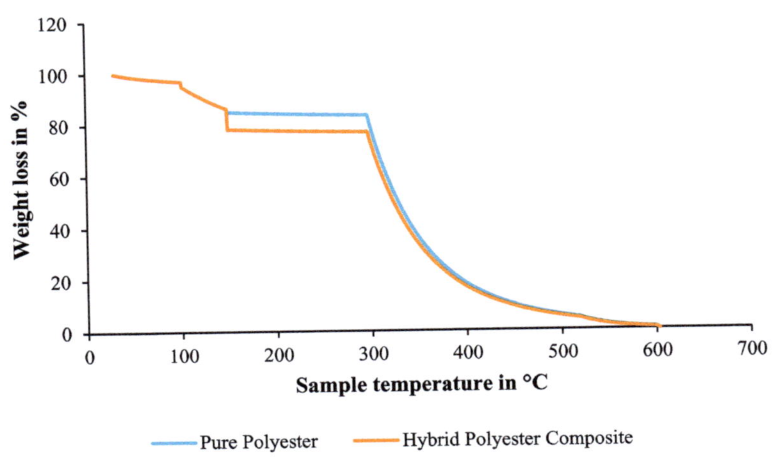

FIG. 10.6 Thermogravimetric analysis of polyester composites.

polyester laminates under a nitrogen atmosphere. It can be observed that the composite degradation occurs in three steps. The first process, between 30 °C and 190 °C, is related to volatile components of low molecular weight (mainly water). Then, the initial degradation temperature of polyester is about 250 °C with a remaining weight (residues) of about 60% (600 °C). The reason is the presence of sugarcane nanocellulose and Al-SiC nanoparticles in the polymer composite. The plots also show a second decomposition step around 300 °C with the highest degradation rate of 0.31 μg/min and the end of the second decomposition process occurs at 412 °C, where the composite is keeping 75% of its weight. The third decomposition process is featured by a peak of high decomposition rate (0.21 μg/min) around 520 °C. Similar results were observed by other investigators [69].

10.4 Applications of polyester hybrid composites

Firstly, cellulose nanofiber investigation is mainly focused in Europe, while cellulose nanocrystals are mainly studied in North America and Asia. In a same way, the fabrication of different types of nanocellulose is mainly performed in industrial centres before arriving on the markets. This section presents applications of sugarcane nanocellulose and Al-SiC nanoparticles reinforced polyester polymer composites in sensing, medical, and electronic applications. Progress in uses and needs are based on a complete knowledge of nanocellulose interaction via colloidal schemes. Nanocellulose fibers and Al-SiC nanoparticles are merely introduced into thin films and nanosheets which can be applied as protection materials [70]. Request for environmentally-friendly materials is leading to attempts tending to use a significant volume of these components. Also, for apathetic packaging, sugarcane nanocellulose and Al-SiC nanoparticles films can be applied for effective parts and maintained discharge. The combination of highly crystalline sugarcane nanocellulose and Al-SiC nanoparticles also offers a fixed surface with tuneable functional groups applicable for grafting and modification. These composites are presented as non-toxic. For biomedical use, such characteristics are very interesting and many research teams are recommending nanocellulose based hybrid composites for different medical applications [71].

Modified nanocellulose has been proposed in casted transportation of chemotherapeutical medicines [72]. It was observed that the structure of sugarcane nanocellulose is favorable for the transport of folic acid to aim at cancer or tumor in the brain. Zoppe et al. [73] established cellulose nanocrystals-based schemes as energetic inhibitors and recommended that cellulose nanocrystals be used for other viruses, especially for HIV. Also, for transportation, cellulose nanocrystals provide possibility for biosensing and detection. For illustration, biosensors based on cellulose nanocrystals by peptide amalgamation were proposed to identify the human neutrophil elastase [74, 75]. For biomedical applications, where sugarcane nanocellulose and Al-SiC nanoparticles reinforced polyester materials are essential, sugarcane nanocellulose fiber is the selected material. The reason may be the smoothness, higher surface area, decreased porosity, and benign behavior. These composite films have been explored as a possible component for biosensors. These types of biosensors are generally fabricated from proteins or binding peptides to the polymer. Cellulose nanofiber components stimulated by National Health Service/Ehlers-Danlos syndromes (NHS/EDS) systems have been presented to bond to BSA (bovine serum albumin), offering non-porous cellulosic sheets for diagnostics [76]. Another method to fabricate a biosensor for Immunoglobulin G (IgG) and BSA identification has been explored by stimulating the cellulose nanofiber surface from a grafting process. Many beneficial characteristics have already been reported, and possibility as aiding materials are forecasted to produce cellulose nanofibers dedicated to the advancement of bioactive interfaces.

Moreover, sugarcane nanocellulose and Al-SiC nanoparticles have properties enabling the production of electronics devices. Presently, cellulose nanofibers were shown to work as carbon-based nanoparticles for anode materials in sodium-ion batteries [77]. Furthermore, it was observed that carbon fibers generated from cellulose nanofibers have higher reversibility, which is the capacity of cycling and fast charging. Production of cellulose nanofibers based electroactive laminates by coating cellulose fibrils with polymers has been established by Nyström et al. [78]. The cellulose laminate was produced with a straightforward technique of coating the fibrils in an aqueous solution and then processed with filtration and drying. The output laminate was electroactive, conductive, and adequate for electrochemically controlled separation and energy storage. The fabricated hybrid composite can be applied as an arrangement to fabricate electronic materials [79]. Considering the benefits of close packing of sugarcane nanocellulose and Al-SiC nanoparticles leading to lower porosity, the resulting composite film showed enhanced conductivity compared to the conventional sheet modified by conductive materials such as gold and silver particles [80]. These cellulosic conductive materials are useful components for advanced flexible electronics with the advantage of suitable chemical and thermal stability and lightweight.

In principle, the development of opaque composite films and flexible electronics has been explored. By combining conductive materials, nanocellulose can improve the conductive laminate production for supercapacitor applications [81]. Similarly, cellulose nanofibers based supercapacitors have been fabricated with graphene nanoparticles [82]. These composites have also been used in light-emitting diodes by coating them with indium tin oxide [83]. When used to reinforce conductive thin films, cellulose nanofibers enable organic light emitting diode (OLED) utilization where biocompatibility is needed. These hybrid polyester composites can be applied as flexible field-effect transistors (organic) [84]. The above-explained developments in flexible electronics feature sugarcane nanocellulose as an active material and coating for flexible packaging applications [85, 86]. Bio-uses have been developed in the field of catalysis process. These materials can be established as a unique base of copper ions to produce catalytic composite using the click reaction [87]. Similarly, gold particles can be used in reinforced cellulose nanofibers to fabricate laminates with specific catalytic activity [88, 89].

10.5 Conclusion

Sugarcane nanocellulose and Al-SiC nanoparticles reinforced polyester hybrid composites were fabricated from ultrasonic-assisted wet layup methods. From the samples produced, morphological, mechanical, viscoelastic, and thermal characteristics were investigated with respect to their potential utilization. The results of the study are:

1. The tensile strength of the polyester matrix improved from 12.8 to 23.7 MPa, while the flexural strength and modulus increased from 32.5 to 46.9 MPa and 3.01 to 5.58 GPa, respectively.
2. The composite micrographs indicate a good interfacial adhesion bonding between the nanofillers and polyester matrix leading to improved mechanical and thermal properties.
3. The impact strength increased from 0.05 to 1.25 kJ/m^2. The polyester hybrid composite's fracture mechanism shows that crack initiation in hybrid polyester may potentially be one of the critical material aspects controlling the toughening effect in the newly developed laminates.
4. The storage and loss modulus were higher for polyester hybrid composites when compared with the neat polyester sample. The newly developed composites exhibited better filler interactions, which led to good compatibility between the polyester matrix and fillers decreasing the damping factor of the laminates.
5. The results of the thermal analysis showed that the degradation temperature, char residue, and thermal stability were improved with sugarcane nanocellulose and Al-SiC nanoparticles addition.

Acknowledgment

The work was financed by Thailand Science Research and Innovation Fund and King Mongkut's University of Technology North Bangkok (KMUTNB), Thailand with Contract no. KMUTNB-BasicR-64-16.

References

[1] Mohit H, Sanjay MR, Vinod K, Dhakal HN, Siengchin S. A comprehensive review on mechanical, electromagnetic radiation shielding, and thermal conductivity of fibers/inorganic fillers reinforced hybrid polymer composites. Polym Compos 2020;41(10):3940–65.

[2] Yashas GTG, Sanjay MR, Jyotishkumar P, Siengchin S. Natural fibers as sustainable and renewable resource for development of eco-friendly composites: a comprehensive review. Front Mater 2019;6:1–14.
[3] Onoja E, Chandran S, Razak FIA, Mahat NA, Wahab RA. Oil palm (Elaeis guineensis) biomass in Malaysia: the present and future prospects. Waste Bio Val 2018;10:2099–117.
[4] Sanjay MR, Siengchin S. Editorial corner—a personal view. Exploring the applicability of natural fibers for the development of biocomposites. Express Polym Lett 2021;15(3):193.
[5] Sanjay MR, Siengchin S, Dhakal HN. Green-composites: eco-friendly and sustainability. Appl Sci Eng Prog 2020;13(3):183–4.
[6] Sanjay MR, Siengchin S, Jyotishkumar P, Jawaid M, Pruncu CI, Khan A. A comprehensive review of techniques for natural fibers as reinforcement in composites: preparation, processing and characterization. Carbohydr Polym 2019;207(1):108–21.
[7] Ramesh M, Deepa C, Kumar LR, Sanjay MR, Siengchin S. Life-cycle and environmental impact assessments on processing of plant fibers and its bio-composites: a critical review. J Ind Text 2020;1–25.
[8] Elias N, Chandren S, Attan N, Mahat NA, Razak FIA, Jamalis J, et al. Structure and properties of oil palm-based nanocellulose reinforced chitosan nanocomposite for efficient synthesis of butyl butyrate. Carbohydr Polym 2017;176:281–92.
[9] Vinod A, Sanjay MR, Siengchin S, Jyotishkumar P. Renewable and sustainable biobased materials: an assessment on biofibers, biofilms, biopolymers and biocomposites. J Clean Prod 2020;258:120978.
[10] Sanchez-Garcia MD, Lagaron JM. On the use of plant cellulose nanowhiskers to enhance the barrier properties of polylactic acid. Cellulose 2010;17:987–1004.
[11] Suhas V, Gupta VK, Carrott PJM, Singh R, Chaudhary M, Kushwaha S. Cellulose: a review as natural, modified and activated carbon adsorbent. Bioresour Technol 2016;216:1066–76.
[12] Kargarzadeh H, Mariano M, Huang J, Lin N, Ahmad I, Dufresne A, et al. Recent developments on nanocellulose reinforced polymer nanocomposites: a review. Polymer 2017;132:368–93.
[13] Ezeilo UR, Zakaria II, Huyop F, Wahab RA. Enzymatic breakdown of lignocellulosic biomass: the role of glycosyl hydrolases and lytic polysaccharide monooxygenases. Biotechnol Biotechnol Equip 2017;31:647–62.
[14] Prabu LS. Nanocellulose bio-nanomaterial: a review. Bioequi Bioavail Int J 2017;1(1):1–7. 000106.
[15] Newman RH, Hemmingson JA. Carbon-13 NMR distinction between categories of molecular order and disorder in cellulose. Cellulose 1995;2:95–110.
[16] George J, Sabapathi SN. Cellulose nanocrystals: synthesis, functional properties, and applications. Nanotechnol Sci Appl 2015;8:45–54.
[17] Mohit H, Selvan VAM. A comprehensive review on surface modification, structure interface and bonding mechanism of plant cellulose fiber reinforced polymer based composites. Compos Interf 2018;25(5-7):629–67.
[18] Ureña-Benavides EE, Ao G, Davis VA, Kitchens CL. Rheology and phase behavior of lyotropic cellulose nanocrystal suspensions. Macromolecules 2011;44:8990–8.
[19] Trache D, Hussin MH, Haafiz MKM, Thakur VK. Recent progress in cellulose nanocrystals: sources and production. Nanoscale 2017;9:1763–86.
[20] Yano H. Production of cellulose nanofibers and their applications. Annal High Perform Paper Soc Japan 2010;49:15–20.
[21] Menon MP, Selvakumar R, Kumar PS, Ramakrishna S. Extraction and modification of cellulose nanofibers derived from biomass for environmental application. RSC Adv 2017;7: 42750–73.

[22] Nakagaito AN, Yano H. The effect of morphological changes from pulp fiber towards nanoscale fibrillated cellulose on the mechanical properties of high-strength plant fiber-based composites. Appl Phys A Mater Sci Process 2004;81:1109–12.

[23] Habibi Y, Lucia LA, Rojas OJ. Cellulose nanocrystals: chemistry, self-assembly, and applications. Chem Rev 2010;110:3479–500.

[24] Zhao Y, Moser C, Lindstrom ME, Henriksson G, Li J. Cellulose nanofibers from softwood, hardwood, and tunicate: preparation–structure–film performance interrelation. ACS Appl Mater Interfaces 2017;9:13508–19.

[25] Rangaswamy B, Vanitha KP, Hugund BS. Microbial cellulose production from bacteria isolated from rotten fruit. Int J Polym Sci 2015;2:1–8.

[26] Maria LCS, Santos ALC, Oliveira PC, Valle ASS, Barud HS, Messaddeq Y, et al. Preparation and antibacterial activity of silver nanoparticles impregnated in bacterial cellulose. Polímeros 2010;20:72–7.

[27] Hungund BS, Gupta S. Production of bacterial cellulose from Enterobacter amnigenus GH-1 isolated from rotten apple. World J Microbiol Biotechnol 2010;26:1823–8.

[28] Krystynowicz A, Czaja W, Wiktorowska-Jezierska A, Gonclaves-Miskiewicz M, Turkiewicz M, Bielecki S. Factors affecting the yield and properties of bacterial cellulose. J Ind Microbiol Biotechnol 2002;29:189–95.

[29] Lee WF, Fu YT. Effect of montmorillonite on the swelling behaviour and drug-release behavior of nanocomposite hydrogels. J Appl Polym Sci 2003;89:3652–60.

[30] Ashori A, Nourbakhsh A. Effects of nanoclay as a reinforcement filler on the physical and mechanical properties of wood-based composite. J Compos Mater 2009;43:1869–75.

[31] Spoljaric S, Salminen A, Luong ND, Lahtinen P, Vartiainen J, Tammelin T, et al. Nanofibrillated cellulose, poly(vinyl alcohol), montmorillonite clay hybrid nanocomposites with superior barrier and thermomechanical properties. Polym Compos 2014;35:1117–31.

[32] Arjmandi R, Hassan A, Haafiz MKM, Zakaria Z. Partial replacement effect of montmorillonite with cellulose nanowhiskers on polylactic acid nanocomposites. Int J Biol Macromol 2015;81:91–9.

[33] Arjmandi R, Hassan A, Haafiz MKM, Zakaria Z, Islam MS. Effect of hydrolysed cellulose nanowhiskers on properties of montmorillonite/polylactic acid nanocomposites. Int J Biol Macromol 2016;82:998–1010.

[34] Trifol J, Plackett D, Sillard C, Szaboo P, Bras J, Daugaard AE. Hybrid poly(lactic acid)/nanocellulose/nanoclay composites with synergistically enhanced barrier properties and improved thermomechanical resistance. Polym Int 2016;65:988–95.

[35] Hong J, Kim DS. Preparation and physical properties of polylactide/cellulose nanowhisker/nanoclay composites. Polym Compos 2013;34:293–8.

[36] Lee H, Hamid SBA, Zain SK. Conversion of lignocellulosic biomass to nanocellulose: structure and chemical process. Sci World J 2014;631013:1–20.

[37] Yorseng K, Sanjay MR, Tengsuthiwat J, Harikrishnan P, Jyotishkumar P, Siengchin S, Moure MM. Information in United States Patents on works related to 'Natural Fibers': 2000-2018. Curr Mater Sci 2019;12(1):4–76.

[38] Guo Y, Bao C, Song L, Yuan B, Hu Y. In situ polymerization of graphene, graphite oxide, and functionalized graphite oxide into epoxy resin and comparison study of on-the-flame behaviour. Ind Eng Chem Res 2011;50:7772–83.

[39] Li D, Muller MB, Gilje S, Kaner RB, Wallace GG. Processable aqueous dispersions of graphene nanosheets. Nat Nanotechnol 2008;3:101–5.

[40] Zhou Q, Ma Y, Gao J, Zhang P, Xia Y, Tian Y, et al. Facile synthesis of graphene/metal nanoparticle composites via self-catalysis reduction at room temperature. Inorg Chem 2013;52:3141–7.

[41] Yang S, Xue B, Li Y, Li X, Xie L, Qin S. Controllable Ag-rGO heterostructure for highly thermal conductivity in layer-by-layer nanocellulose hybrid films. Chem Eng J 2020;383: 123072.
[42] Wu CF, Saito T, Fujisawa S, Fukuzumi H, Isogai A. Ultrastrong and high gas-barrier nanocellulose/clay-layered composites. Biomacromolecules 2012;13:1927–32.
[43] Shinoda R, Saito T, Okita Y, Isogai A. Relationship between length and degree of polymerization of tempo-oxidized cellulose nanofibrils. Biomacromolecules 2012;13:842–9.
[44] Fukuzumi H, Saito T, Iwata T, Kumamoto Y, Isogai A. Transparent and high gas barrier films of cellulose nanofibers prepared by TEMPO-mediated oxidation. Biomacromolecules 2009;10:162–5.
[45] Wang J, Cheng Q, Tang Z. Layered nanocomposites inspired by the structure and mechanical properties of nacre. Chem Soc Rev 2012;41:1111–29.
[46] Tseng C, Liou H, Tsai W. The influence of nitrogen content on corrosion fatigue crack growth behavior of duplex stainless steel. Mater Sci Eng A 2003;344:190–200.
[47] Tang Z, Kotov NA, Magonov S, Ozturk B. Nanostructured artificial nacre. Nat Mater 2003;2:413–8.
[48] Podsiadlo P, Kaushik AK, Arruda EM, Waas AM, Shim BS, Xu J. Ultrastrong and stiff layered polymer nanocomposites. Science 2007;318:80–3.
[49] Munch E, Launey ME, Alsem DH, Saiz E, Tomsia AP, Ritchie RO. Tough, bio-inspired hybrid materials. Science 2008;322:1516–20.
[50] Bonderer LJ, Studart AR, Gauckler LJ. Bioinspired design and assembly of platelet reinforced polymer films. Science 2008;319:1069–73.
[51] Paul DR, Robesson LM. Polymer nanotechnology: nanocomposites. Polymer 2008;49:3187–204.
[52] Jiang S, Zhang M, Jiang W, Xu Q, Yu J, Liu L, et al. Multiscale nanocelluloses hybrid aerogels for thermal insulation: the study on mechanical and thermal properties. Carbohydr Polym 2020;247:116701.
[53] Kavimani V, Prakash KS, Thankachan T, Udayakumar R. Synergistic improvement of epoxy derived polymer composites reinforced with Graphene Oxide (GO) plus Titanium dioxide (TiO2). Compos Part B Eng 2020;191:107911.
[54] Pokharel P, Pant B, Pokhrel K, Pant HR, Lim J, Lee DS. Effects of functional groups on the graphene sheet for improving the thermomechanical properties of polyurethane nanocomposites. Compos Part B Eng 2015;78:192–201.
[55] Mustata F, Tudorachi N, Rosu D. Thermal behavior of some organic/inorganic composites based on epoxy resin and calcium carbonate obtained from conch shell of Rapana thomasiana. Compos Part B Eng 2012;43:702–10.
[56] Lim KS, Mariatti M, Kamarol M, Ghani ABA, Halim HS, Bakar AA. Properties of nanofillers/crosslinked polyethylene composites for cable insulation. J Vinyl Addit Technol 2018;25(S1): E147–54.
[57] Zare Y. Study of nanoparticles aggregation/agglomeration in polymer particulate nanocomposites by mechanical properties. Compos Part A Appl Sci Manuf 2016;84:158–64.
[58] Nishar H, Sreekumar PA, Francis B, Yang W, Thomas S. Morphology, dynamic mechanical and thermal studies on poly(styrene-co-acrylonitrile) modified epoxy resin/glass fiber composites. Compos Part A Appl Sci Manuf 2007;38:2422–32.
[59] Keusch S, Haessler R. Influence of surface treatment of glass fibers on the dynamic mechanical properties of epoxy resin composites. Compos Part A Appl Sci Manuf 1999;30:997–1002.
[60] Koronis G, Silva A, Fontul M. Green composites: a review of adequate materials for automotive applications. Compos Part B Eng 2013;44(1):120–7.

[61] Nurazzi NM, Khalina A, Sapuan SM, Ilyas RA, Rafiqah SA, Hanafee ZM. Thermal properties of treated sugar palm yarn/glass fiber reinforced unsaturated polyester hybrid composites. J Mater Res Technol 2020;9(2):1606–18.
[62] Idicula M, Malhotra SK, Joseph K, Thomas S. Dynamic mechanical analysis of randomly oriented intimately mixed short banana/sisal hybrid fiber reinforced polyester composites. Compos Sci Technol 2005;65:1077–87.
[63] Rashid B, Leman Z, Jawaid M, Ghazali MJ, Ishak MR. Dynamic mechanical analysis of treated and untreated sugar palm fiber-based phenolic composites. Bioresources 2017;12: 3448–62.
[64] Manoharan S, Suresha B, Ramadoss G, Bharath B. Effect of short fiber reinforcement on mechanical properties of hybridphenolic composites. J Mater 2014;2014:1–9.
[65] Saba N, Paridah MT, Abdan K, Ibrahim NA. Dynamic mechanical properties of oil palm nano filler/kenaf/epoxy hybrid nanocomposites. Constr Build Mater 2016;124:133–8.
[66] Lau K, Hung P, Zhu M, Hui D. Properties of natural fiber composites for structural engineering applications. Compos Part B Eng 2018;136:222–33.
[67] Jawaid M, Khalil HPSA, Hassan A, Dungani R, Hadiyane A. Effect of jute fiber loading on tensile and dynamic mechanical properties of oil palm epoxy composites. Compos Part B Eng 2013;45:619–24.
[68] Nurazzi MN, Ilyas RA, Sapuan SM, Abdan K. Mechanical properties of sugar palm yarn/ woven glass fiber reinforced unsaturated polyester composites: effect of fiber loadings and alkaline treatment. Polymers 2019;10:665–75.
[69] Ribeiro B, Corredor JAR, Costa ML, Botelho EC, Rezende MC. Multifunctional characteristics of glass fiber-reinforced epoxy polymer composites with multiwalled carbon nanotube buckypaper interlayer. Polym Eng Sci 2020;60(4):740–51.
[70] Syverud K, Stenius P. Strength and barrier properties of MFC films. Cellulose 2009;16:75–85.
[71] Lam E, Male KB, Chong JH, Leung ACW, Luong JHT. Applications of functionalized and nanoparticle-modified nanocrystalline cellulose. Trends Biotechnol 2012;30:283–90.
[72] Dong S, Cho H, Lee YW, Roman M. Synthesis and cellular uptake of folic acid conjugated cellulose nanocrystals for cancer targeting. Biomacromolecules 2014;15:1560–7.
[73] Zoppe JO, Ruottinen V, Ruotsalainen J, Ronkko S, Johansson L, Hinkkanen A, et al. Synthesis of cellulose nanocrystals carrying tyrosine sulfate mimetic ligands and inhibition of alphavirus infection. Biomacromolecules 2014;15:1534–42.
[74] Edwards JV, Prevost N, French A, Concha M, DeLucca A, Wu Q. Nanocellulose-based biosensors: design, preparation, and activity of peptide-linked cotton cellulose nanocrystals having fluorimetric and colorimetric elastase detection sensitivity. Engineering 2013;5:20–8.
[75] Edwards JV, Prevost N, Sethumadhavan K, Ullah A, Condon B. Peptide conjugated cellulose nanocrystals with sensitive human neutrophil elastase sensor activity. Cellulose 2013;20:1223–35.
[76] Orelma H, Filpponen I, Johansson L, Osterberg M, Rojas OJ, Laine J. Surface functionalized nanofibrillar cellulose (NFC) film as a platform for immunoassays and diagnostics. Biointerphases 2012;7:1–12.
[77] Luo W, Schardt J, Bommier C, Wang B, Raznik J, Simonsen J, et al. Carbon nanofibers derived from cellulose nanofibers as a long-life anode material for rechargeable sodium-ion batteries. J Mater Chem A 2013;1:10662–6.
[78] Nyström G, Mihranyan A, Razaq A, Lindstorm T, Nyholm L, Stomme M. A nanocellulose polypyrrole composite based on microfibrillated cellulose from wood. J Phys Chem B 2010;114:4178–82.

[79] Huang J, Zhu H, Chen Y, Preston C, Rohrbach K, Cumings J, et al. Highly transparent and flexible nanopaper transistors. ACS Nano 2013;7:2106–13.
[80] Hsieh M, Kim C, Nogi M, Suganuma K. Electrically conductive lines on cellulose nanopaper for flexible electrical devices. Nanoscale 2013;5:9289–95.
[81] Kang YJ, Chun S, Lee S, Kim B, Kim JH, Chung H, et al. All-solid-state flexible supercapacitors fabricated with bacterial nanocellulose papers, carbon nanotubes, and triblock-copolymer ion gels. ACS Nano 2012;6:6400–6.
[82] Gao K, Shao Z, Li J, Wang X, Peng X, Wang W, et al. Cellulose nanofiber–grapheme all solid-state flexible supercapacitors. J Mater Chem A 2013;1:63–7.
[83] Legnani C, Vilani C, Calil VL, Barud HS, Quirino WG, Achete CA, et al. Bacterial cellulose membrane as flexible substrate for organic light emitting devices. Thin Solid Films 2008;517:1016–20.
[84] Valentini L, Bon B, Cardinali M, Fortunati E, Kenny JM. Cellulose nanocrystals thin films as gate dielectric for flexible organic field-effect transistors. Mater Lett 2014;126:55–8.
[85] Li F, Biagioni P, Bollani M, Maccagnan A, Piergiovanni L. Multi-functional coating of cellulose nanocrystals for flexible packaging applications. Cellulose 2013;20:2491–504.
[86] Koga H, Azetsu A, Tokunaga E, Saito T, Isogai A, Kitaoka T. Topological loading of Cu(I) catalysts onto crystalline cellulose nanofibers for the Huisgen click reaction. J Mater Chem 2012;22:5538–42.
[87] Jagadeesh P, Yashas GTG, Madhu P, Sanjay MR, Siengchin S. Effect of natural filler materials on fiber reinforced hybrid polymer composites: an overview. J Nat Fiber 2020;1–16.
[88] Azetsu A, Koga H, Isogai A, Kitaoka T. Synthesis and catalytic features of hybrid metal nanoparticles supported on cellulose nanofibers. Catalysts 2011;1:83–96.
[89] Koga H, Tokunaga E, Hidaka M, Umemura Y, Saito T, Isogai A. Topochemical synthesis and catalysis of metal nanoparticles exposed on crystalline cellulose nanofibers. Chem Commun 2010;46:8567–9.

Chapter 11

Mechanical modeling of hybrid nanocomposites based on cellulose nanocrystals/nanofibrils and nanoparticles

Fatima-Zahra Semlali Aouragh Hassani[a], Zineb Kassab[b], Mounir El Achaby[b], Rachid Bouhfid[a], and Abou el Kacem Qaiss[a]
[a]*Moroccan Foundation for Advanced Science, Innovation and Research (MAScIR), Composites and Nanocomposites Center, Rabat Design Center, Rabat, Morocco,* [b]*Materials Science and Nano-Engineering (MSN) Department, Mohammed VI Polytechnic University (UM6P), Ben Guerir, Morocco*

11.1 Introduction

The addition of nanoscale fillers into a polymer matrix enables to obtain materials with improved or new properties such as thermal, mechanical, electrical, optical, and barrier properties [1–3]. The particularity of nanocomposites, compared to conventional composites, is to present a very large amount of interface per unit volume and short inter-particle distances. Thus, the physical phenomena brought into play on the surface will play a significant role on the overall materials mechanical behavior. The interaction forces between the charges, even small, can lead to charge aggregation phenomena, which can go up to the sample's scale (percolation). Percolation of the reinforcements can provide significant increase in the nanocomposites mechanical properties [1, 2, 4].

Recently, the society's environmental awareness led to the use of biomass-derived materials that are more easily degradable after their life cycle and therefore more environmentally friendly. Thus, the use of cellulosic fillers is attracting particular attention to valorize these materials to increase their added value in different applications such as pharmaceutical, food, paper, and composites manufacturing [1, 4, 5].

Nanocellulose is a form of nano-structured cellulose which can be typically used in the form of cellulose nanocrystals, cellulose nanofibrils or cellulose hairy nanocrystals. However, monofunctional cellulose nanoparticles or fibrils

Cellulose Nanocrystal/Nanoparticles Hybrid Nanocomposites. https://doi.org/10.1016/B978-0-12-822906-4.00003-7

can improve only some properties of a host polymer, but cannot provide additional functional properties such as antibacterial and UV-resistance properties [5, 6].

Recently, polymer nanohybrids based on nanocellulose attracted considerable attention in the field of materials science and tissue engineering because of their versatility [7–9]. Compared to neat polymers, nano-hybrid materials based on cellulose and other reinforcement (inorganic, metallic or carbon allotropes) have shown to significantly improve the nanocomposite performance in terms of high strength, stiffness, thermal stability, and reduced water absorption properties, which cannot be achieved in binary systems (a single type of reinforcement) [8]. The advantages of cellulose reinforcement complete those that are missing in the other type of reinforcement, providing a good balance between performance and cost [10]. However, from an experimental point of view, it is very difficult to characterize the structure and control the production of hybrid polymer nanocomposites. The development of such materials remains largely empirical and a more accurate prediction of their properties is not yet possible. Therefore, the modelling of mechanical properties will play an increasingly important role in the prediction and design of material properties and will guide the experimental work such as synthesis and characterization [10, 11]. For example, the prediction of the viscoelastic behavior of a heterogeneous material from the properties of its constituents, their volume fraction, and the developed morphology, is a subject widely studied in the literature. But the models available are generally based on classical mechanical approaches, such as phenomenological or homogenization methods [10, 11].

This chapter firstly provides an overview of the wide range of nanocomposite reinforcements, in particular cellulose, and their influence on the material mechanical properties of polymer nanocomposites. It then focuses on the different methods to produce cellulose-based nano-hybrid materials and presents some of their mechanical properties. Finally, the main mechanical models developed for nanocomposite hybrid materials are presented to predict properties of interest.

11.2 Nanocomposites reinforcement

Reinforcement (or reinforcing filler) is any immiscible body enabling to modify the mechanical, electrical or thermal properties of a polymer. Three main types of reinforcements are available: organic reinforcements (flax fibers, carbon fibers, graphene, etc.), inorganic reinforcements (mica, glass fibers, silica, etc.), and metallic reinforcements (boron fibers, aluminum, etc.). These reinforcements can be in the form of particles, short or long fibers, woven or not. Each type of reinforcement has its own advantages (UV aging resistance, antimicrobial, electromagnetic, etc.) and disadvantages (agglomeration, high viscosity, fiber impregnation problems, etc.). A classification of these nanoparticles is presented next.

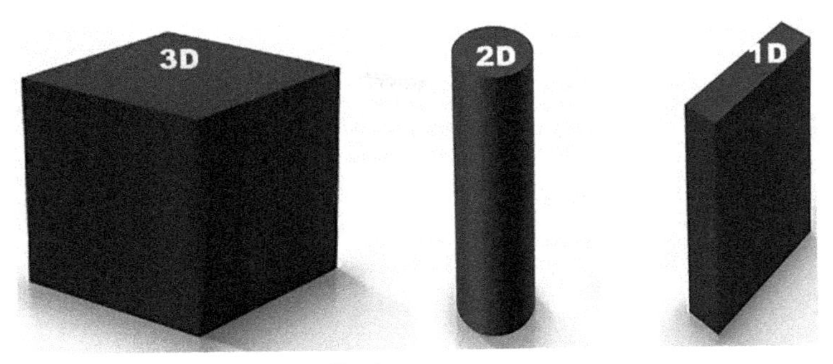

FIG. 11.1 Nano-reinforcements classification.

11.2.1 Nano-reinforcements classification

In the context of this study, only particulate reinforcements are presented. They can be classified according to their geometry (Fig. 11.1) [12].

11.2.1.1 3D geometry reinforcement

These reinforcements are isodimensional; i.e. the three dimensions are of a nanometer order (less than 100 nm). In most cases, these reinforcements have the advantage of being easily produced and inexpensive (silica beads). They are, inter alia, used to improve brightness, mechanical properties [13] or to modify the material rheology. They are found in various applications such as packaging [14], medical [15], and tire industry [16].

11.2.1.2 2D geometry reinforcement

Two dimensions are of the nanometer order, thus forming a fibrillar structure such as carbon nanotubes. Such reinforcements lead to materials having exceptional properties, in particular their rigidity. Despite their high cost, these reinforcements are very interesting for electronic applications thanks to their high conductivity [17], excellent mechanical properties [18], and gas barrier properties for packaging applications [19].

11.2.1.3 1D geometry reinforcement

Only one of the three dimensions is in the nanometer range. These particles are of nanometric thickness with lateral dimensions ranging from a few tens of nanometers to several microns. Among the most common 1D nanoparticles are clays (montmorillonites) and graphene. This type of reinforcement has excellent mechanical and thermal properties and is electrically conductive in the case of graphene. Even if the reinforcements provide mechanical strength and rigidity to the final composite, the intrinsic properties of the matrix also play a very important role in the final composite properties.

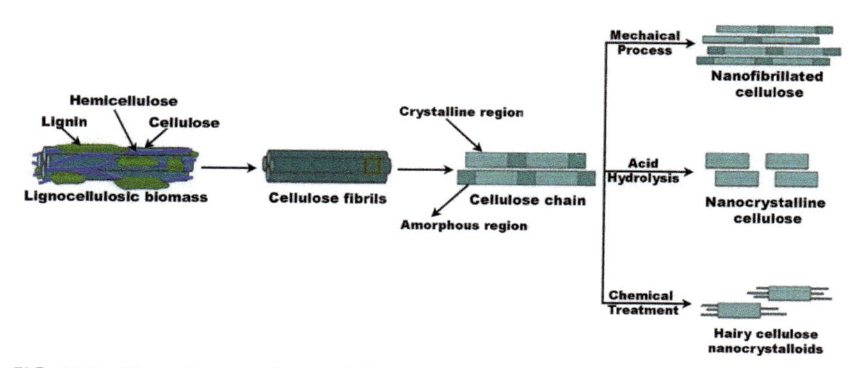

FIG. 11.2 The main steps of nanocellulose extraction from lignocellulosic biomass.

11.2.2 Nanocomposites based on cellulose reinforcement

Cellulose is a semi-crystalline biopolymer from the polysaccharide family. It is the most abundant biological molecule on earth forming the rigid skeleton of a wide variety of species including plants (with a content varying from 15% to 99%) and certain algae. It is also found in animals and in particular in the outer shell of marine animals (tunicates) [19, 20].

11.2.2.1 Cellulose classification

There are currently three types of cellulose nanoparticles (Fig. 11.2). The first one is a nanofiber whose structure corresponds to a succession of amorphous and crystalline parts (CNF). The second corresponds to a cellulose nanocrystal (CNC), while the third one has a crystalline part and an amorphous part at each end (CHNC).

Cellulose nanofibers (CNF)

The industrial cellulose fiber consists of several coaxial layers superposed on each other [21]. More precisely, these layers consist of cellulose nanofibers attached by hemicelluloses and hydrogen bonds. According to the scientific literature, these nanofibers have different name such as "cellulose nanofibers," "microfibrillated cellulose," "nanofibrillated cellulose," etc. To obtain cellulose nanofibers, oxidation is generally carried out using a reagent such as TEMPO (2,2,6,6-tetramethylpiperidine-1-oxyl) which is a radical able of selectively reacting on the C_6 carbon of the glucose group. This reaction allows the hydroxyl group to oxidize to an aldehyde. It is then again oxidized with NaClO to form a carboxyl group (COOH) [1]. In aqueous solutions, the carboxyl group is in its COO^- carboxylate form enables the separation of the nanofibers by electrostatic repulsion following a mechanical treatment.

Another option to reduce the energy required to separate the nanofibers from each other is an enzyme treatment coupled with mechanical refining [22].

In general, the size of the nanofibers manufactured, regardless of the process used to separate them, mainly depends on the fiber source. They are generally from 500 nm to a few microns in length with only a few nanometers in diameter. Due to their very high aspect ratio (length over diameter) and their great flexibility, these nanofibers tend to easily entangle with each other.

Cellulose nanocrystals (CNC)

Nanocrystalline cellulose corresponds to the crystalline part of the nanofiber. It is generally obtained following hydrolysis using sulfuric [21, 22] or hydrochloric acid as in the pulp and paper industry [21, 23]. This hydrolysis dissolves the amorphous part to release the crystalline part. Depending on the source of the wood pulp used, the crystalline part may have a different aspect ratio following the hydrolysis step. For example, cellulose nanocrystals from tunicate (marine species) are known to have an aspect ratio of around 100, while those from conventional wood have an aspect ratio of about 10 [24]. Hydrolysis with sulfuric acid leads to the grafting of sulphate esters groups (OSO_3^-) on the CNC surface which therefore become negatively charged in aqueous suspension [25]. Hydrolysis with hydrochloric acid does not change the chemical structure and therefore leaves the CNC uncharged. Habibi et al. [26] studied the oxidation with TEMPO on cellulose crystals preparation with hydrochloric acid. This oxidation enabled to increase the CNC stability in aqueous media due to the presence of carboxylate groups (COO^-) on their surface. The TEMPO agent was able to introduce a charge on the CNC surface, while hydrochloric acid did not.

Recently, phosphoric acid was proposed as a chemical reagent to extract nanocrystalline cellulose [27]. The authors compared the hydrolysis step by using hydrochloric, sulfuric, and phosphoric acid. It was shown that nanocrystals resulting from the hydrolysis with hydrochloric and phosphoric acid had better thermal resistance than those treated with sulfuric acid. Also, the CNC stability in solvents such as water, dimethylsulfoxide (DMSO) or dimethylformamide (DMF), did not seem to be affected. So, this study showed that sulfuric acid could be replaced by other reagents such as phosphoric acid, to extract nanocrystals.

Cellulose hairy nanocrystals (CHNC)

Cellulose nanocrystals are produced by a two-step oxidation of cellulose fibers. The first step consists in oxidizing the hydroxyl groups of the C_2 and C_3 carbons of the glucose units to aldehyde using periodate ions (IO_4^-). The second step is an oxidation using hypochlorite ions (ClO^-) in the presence of hydrogen peroxide (H_2O_2) to transform the aldehyde functions (R-COH) into carboxylic acid (COOH). The chemical process is such that a cellulose nanocrystal containing a central crystalline part as well as an amorphous part at each end is obtained. As the crystals have amorphous chains at their ends, an increase in steric hindrance results. Furthermore, as these amorphous chains are charged with carboxyl groups, this allows the presence of electrostatic repulsions. An electrostatically and sterically stabilized nanocrystals, known as “electrosterically stabilized nanocrystals of cellulose (ECNC)”, is obtained [28].

It is also possible to extract the cellulose nanocrystals by keeping the C_2 and C_3 carbons in their aldehyde form. To do this, the suspension after its first oxidation with periodate ions is heated as described by Yang et al. [28]. These particles are called "sterically stabilized nanocrystals of cellulose or SCNC." Chen et al. [29] showed that the size of the crystalline part is a function of the reaction time. For short times (26 h), the crystalline part is about 590 nm in length against 240 and 100 nm for reaction times of 42 h and 84 h, respectively. The authors proposed a chemical mechanism based on the fact that the reaction front transforms the crystalline part into an amorphous region.

It is also possible to confer a positive charge on the filament nanoparticles. For this, it is necessary to react a paper pulp previously oxidized by $NaIO_4$ with (2-hydrazinyl-2-oxoethyl)-trimethylazanium chloride which will cationize the C_2 and C_3 carbons. Subsequent heating allows the nanoparticles to be separated [30].

Filament nanoparticles were only discovered very recently. The scientific literature describing them is therefore very limited [25–36]. So, their structure is only hypothetical because no scientific work has yet confirmed this model.

11.2.2.2 Effects of nanocellulose on polymer mechanical properties

The properties of composites based on cellulose fibers are governed by different parameters: aspect ratio, volume fraction, orientation, and dispersion, as well as the fiber/matrix state (contact, interaction, adhesion, etc.). Each of these parameters will have a direct effect on the composites performance, so it is important to understand the mechanisms involved to maximize their effects.

Fiber aspect ratio

The composites mechanical performances are closely linked to the stress distribution within the material. When the composite is under stresses, the matrix is the first to be affected before transmitting these stresses to the reinforcement. The more efficiently the stresses are transferred from the matrix to the reinforcement, the better the mechanical properties of the composite. To optimize this stress management, it is necessary to maximize the fiber/matrix interactions while improving the adhesion quality. It has been shown that the higher the fiber aspect ratio (length/diameter, L/d), the better the stress transfer within the composite, especially above a critical L/d value [37]. Since cellulose fibers have very high aspect ratios and very large specific surface areas, this provides interesting reinforcing effects.

Fiber volume fraction

Another important parameter is the cellulose fibers volume fraction. In general, the higher the fiber content in the composite, the better the mechanical properties. Maintaining a high volume fraction with small fibers leads to an increase in the number of ends, but these ends act as crack initiators. Conversely, long fibers (above 9–10 mm) are difficult to process as they fold and entangle during mixing leading to high viscosity and agglomeration. In all cases, a very high volume fraction leads to low mechanical properties as above a critical volume fraction, the fibers tend to agglomerate resulting in poor dispersion (inhomogeneities). The formulation (content of all components) is also very important, especially at high fiber content so the shear forces induced during mixing must be sufficient to allow optimal mixing of the various constituents [37–39].

Fiber orientation

The fibers orientation has a significant effect on the composites mechanical properties. It is reported that the Young's modulus, Poisson's ratio, and tensile strength of materials increase as fiber orientation increases in the applied stress direction [37–39]. Cellulose fibers have a significant resistance in the length direction, while being weaker in the other directions. Hence, the fibers must be aligned in the length direction during composites processing to improve this effect.

Fiber dispersion

The cellulose fibers dispersion within the polymer matrix is a key element to take into account. Due to their polar nature, the fibers tend to agglomerate when mixed with a nonpolar matrix. The result is a heterogeneous composite with areas rich in cellulose fibers and others rich in matrix, which will adversely affect the composite performance. It is therefore imperative to have a good cellulose fibers dispersion. The fibers must be well separated from each other and each fiber must be coated by the matrix. Fiber/fiber interactions, such as hydrogen bonds, as well as fibers length, will directly influence their dispersion. So, the longer the fibers and the more intense the fiber/fiber interactions, the more they will entangle and agglomerate. The processing method is also important to limit these phenomena. The greater the shear forces and the mixing time, the better the dispersion. Conventionally, twin-screw extruders are used to achieve the best results. Nevertheless, physico-chemical treatments or chemical modification of the cellulose fibers are effective and commonly used in the literature [37, 40].

Fiber/matrix adhesion

The mechanical properties of cellulose fiber-reinforced composites are intimately dependent on the fiber/matrix interfacial adhesion [41, 42]. Stress transfer management within the composite is improved when the fiber/matrix interface is good. Since cellulose fibers are rich in hydroxyl groups, giving them a strong polarity and a marked hydrophilic behavior, they are of different nature compared with most thermoplastic matrices which are hydrophobic and nonpolar. This results in compatibility problems between both phases and a weak interfacial zone. In order to improve the fiber/matrix interfacial strength, it is common to use treatments aimed at reducing the fibers surface energy [8, 9]. The addition of a compatibilizing/coupling agent having intermediate properties between both constituents can also be used [8, 9].

Type of the fibers

All the parameters mentioned above (aspect ratio, volume fraction, orientation, dispersion, and interfacial adhesion) are highly influenced by the type of fiber used. The intrinsic properties of cellulose fibers, such as their mechanical properties, composition, water content or dimensions, will play a decisive role in the performance of composite materials [43, 44]. Thus, depending on the nature of these fibers, different physical and mechanical properties can be obtained as reported in Table 11.1 [43, 44].

In general, the cellulose fibers used as reinforcement in polymer composites significantly improve the mechanical properties of the neat matrix, which justifies their increasing use. The fibers stiffness enables to increase the Young's modulus, as well as the tensile strength of the composites under optimized conditions. In recent years, several studies focused on cellulose nanoparticles showing their superior mechanical properties compared to cellulose macro-fibers (MFC). Nevertheless, NFC and MFC have both important reinforcement potential leading to the production of very resistant (nano)composites.

TABLE 11.1 Physico-mechanical properties of some natural fibers [43, 44].

Fiber	Cellulose (wt%)	Tensile strength (MPa)	Young's modulus (GPa)	Density (g/cm^3)
Abaca	56–63	400	12	1.5
Bagasse	55.2	290	17	1.25
Bamboo	26–43	140–230	11–17	0.6–1.1
Flax	71	345–1035	27.6	1.5
Hemp	68	690	70	1.48
Jute	61–71	393–773	26.5	1.3
Kenaf	72	930	53	–
Coir	32–43	175	4-6	1.2

11.3 Cellulose based hybrid nanocomposites materials

Hybridization is a process in which two reinforcements (synthetic/nanofillers/natural/metallic) are incorporated into a single matrix or single reinforcements incorporated into a mixture of matrices to achieve better properties than each one used separately. These properties can be: high strength, rigidity, stiffness, thermal stability, reduced water absorption, etc., which cannot be achieved in conventional composite materials [8]. Several studies have shown that cellulose hybridization significantly improved the mechanical properties of polymer composites, giving designers the freedom to customize the composites and achieve more cost-effective properties that cannot be achieved in binary systems (one type of fiber dispersed in a single matrix). In addition, the benefits of cellulose reinforcement could complement those missing in other types of reinforcement. Therefore, a balance between performance and cost can be achieved through appropriate material design and formulation [10].

11.3.1 Manufacturing methods

Several methods have been used to produce polymer nanohybrids. Among them, the three main methods to manufacture hybrid nanocomposites based on cellulose are presented next.

11.3.1.1 Solution casting technique

Being particularly suitable for the synthesis of hybrid nanocomposites in small quantities, this technique was favored because of the limited availability and the high cost of manufacturing nanocomposites. This method consists of mixing the nanohybrids and the polymer dissolved in a solvent. The nanofiller dispersion is generally done using ultrasound, mechanical stirring, surfactant addition or a combination thereof. The solutions are mixed at or above room temperature. One of the main disadvantages of this method is the selection and removal of the solvent. Another drawback comes from the use of ultrasound, which, although very effective in dispersing nanocharges, can also destroy the macromolecular chains. One solution is to disperse the nanohybrids in solution using ultrasound before adding the polymer [8, 45, 46].

11.3.1.2 In situ technique

The manufacture of hybrid nanocomposites by an in-situ polymerization process overcomes the drawbacks of the previous methods while retaining the advantages. This process is divided into several stages. Initially, the charges are added to the liquid monomer. Given the low viscosity of the latter, the dispersion of the nanohybrids can be carried out by the use of ultrasound or even by mechanical agitation in turbulent regimes [47]. Once the charges are well dispersed in the matrix, the polymerization begins. When the polymerization is completed, all unreacted molecules are extracted (vacuum degassing). The first advantage of this method is that it does not require the use of solvents which are expensive and difficult to remove/recover. The second advantage is that it enables to obtain a good state of dispersion because the nanofillers are introduced into the low viscosity monomer, thus resulting in a strong improvement in the nanohybrids composites properties [45, 48].

11.3.1.3 Melt blending technique

This method is the simplest and industrially applied. In this case, the polymer and the nanohybrids are incorporated in a heated mixer (extruder). The shear stresses caused by the rotation of the screw(s) is the determining factor to control the dispersion level. The nanofillers can thus distribute within the molten polymer. The disadvantage of this method is that viscosity substantially increases with the nano-reinforcement content [8, 45]. The shaping of composite materials with a polymer matrix requires the implementation of very precise processes to obtain a quality material. The synthesis of a thermosetting polymer is more complex than that of a thermoplastic. The second parameter to take into account is the reinforcement dispersion within the matrix. The dispersion of nanohybrids is certainly the fundamental step in the composite manufacturing process [8, 45, 48].

11.3.2 Hybrid nanocomposites mechanical properties

The bond between the matrix and the reinforcements is a fundamental physicochemical phenomenon which is created during the composite preparation. It is the matrix/reinforcement interface which will be decisive in terms of the final properties of the composite, and in particular from a mechanical point of view [49]. It should be remembered that for most polymer matrices, compatibility with the nanofillers is low due to their very different chemical structures. In fact, the adhesion between a nanoparticle and a polymer is poor due to the too high surface tensions involved [50]. This drawback also prevents an efficient and homogeneous dispersion of the nanoparticles inside the matrix. Several studies were carried out to understand and improve these interfacial phenomena [49]. It is mainly a question of working on the level of cohesion between all the constituents inside the composite. This can be achieved by modifying the nanofillers surface before their insertion into the matrix.

Several nanocomposites are made from polymer matrices in which the nanoparticles are dispersed. The incorporation of a hybrid reinforcement based on cellulose within polymeric materials enables to modify the mechanical [51], thermal [49, 50], electrical [52] or magnetic [53] properties, and thus broadening their field of application. Different types of nanofiller can be mixed with cellulose and used as reinforcements in polymer composites. The main ones are:

- Inorganic materials (spheres, fibers and platelets),
- Metallic materials (spheres),
- Carbon allotropes (spheres, fibers and platelets).

11.3.2.1 Polymer hybrid nanocomposites based on cellulose/inorganic materials

Polymer composites made from inorganic nanoparticles and organic polymers represent a new class of materials which have more effective properties compared to those of their micro-particle counterparts [54]. Inorganic particles provide mechanical and thermal stability, as well as new functionalities depending on their chemical nature, structure, size, and crystallinity of inorganic nanoparticles (silica, transition metal oxides, nanocells, metal phosphates, metal chalcogenides, and nanometallic). Inorganic particles provide better mechanical, thermal, magnetic, electronic, and optical properties [55].

The most important processes used for the preparation of inorganic nanocomposites based on polymers are: intercalation of nanoparticles in the polymer from a solution, in-situ intercalation polymerization, intercalation in fusion, direct mixing of polymers and particles, chemical synthesis, in situ polymerization, and sol-gel processes. However, combining cellulose and inorganic particles, such as clays, can be interesting in producing high-technology devices [56].

Perotti et al. [56] developed a novel material based on the hybridization of bacterial cellulose (BC) and Laponite clay (LC) with different inorganic/organic ratios (m/m) in order to evaluate possible modifications in the mechanical properties of BC after clay incorporation. The hybrid nanocomposites were prepared using the contact of never-dried membranes of BC with a previous dispersion of clay particles in water, while the mechanical properties were evaluated using a dynamic mechanical analyzer (DMA) equipped with a film tension clamp at 27 °C under a preload force of 0.01 N until rupture. The mechanical results clearly showed that increasing the clay content (from 0% to 30%) in the cellulose nanocomposites led to higher tensile strength (from 164 to 227 MPa) and elastic modulus (from 11.6 to 21 GPa). It was suggested that strong hydrogen bonding interactions between the inorganic and organic parts hindered the flow of fibrils over a wide area, which increased the force required to break the structure or reduced the maximum stress. The proposed schematic representation of the nanocomposite formation is shown in Fig. 11.3. Initially, disk shaped Laponite particles exfoliate in the presence of water at room temperature due to an osmotic swelling mechanism. After mixing the polymer and the expanded clay suspension, strong interactions between the inorganic platelets and the hydroxyl groups of the bacterial cellulose was produced leading to the formation of a thin homogeneous film after drying.

11.3.2.2 Polymer hybrid nanocomposites based on cellulose/metallic materials

Nanocomposites based on metal/polymer materials combine the properties of each component. They are considered as promising systems for advanced functional applications [57]. As a result of metallic nanoparticles incorporation into the polymer, a new generation of materials with unique electrical, optical or

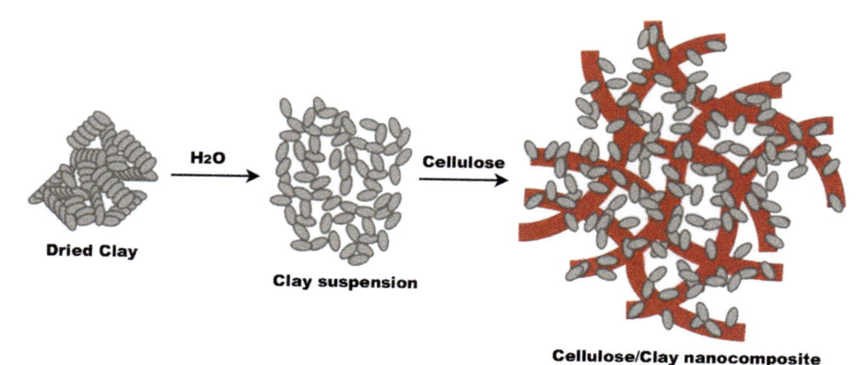

FIG. 11.3 Cellulose/clay hybrid nanocomposite formation.

mechanical properties was developed making them attractive for applications in optics [58], photoimaging and modeling [59], sensor design [2], catalysis [60], and antimicrobial coatings [61]. The search for new methods of preparing metallic nanocomposites has been strongly stimulated because of their attractive properties and their promising applications. Ionic and nano forms of metals can be applied to cellulose nanoparticles to provide antibacterial properties. The adsorption process is the conventional method to introduce metal ions into cellulose through ion exchange interactions between the metal ions and the H+ of the cellulose hydroxyl groups. The incorporation of nanometals into cellulose can be prepared either by the dispersion of metal nanoparticles in a polymerizable formulation (or in a polymer matrix), or by in situ polymerization to avoid agglomeration [7]. Abbate dos Santos et al. [62] manufactured hybrid materials made of poly(lactic acid) (PLA) and poly(lactic acid)/poly(lactic acid-glycol) (PLA/PLGA) mixture using cellulose nanocrystals (CNC) and/or organophilic silica (R972) in the form of nanoparticles. The CNC were obtained by acid hydrolysis of commercially available microcrystalline cellulose (MCC). The materials were produced in films by solution casting. The organophilic silica was incorporated at 3% by weight, while the CNC were added at 3% and 5% by weight. Two sets of films were obtained. The first was prepared using only PLA as the matrix and the second was prepared using PLA/PLGA mixtures. The mechanical properties of the films were evaluated according to ASTM D882 at 22 °C. The results showed that grafting sulphate groups onto the cellulose surface improved the nanoparticle dispersion by reducing the interaction between the side chains of the nanoparticles, but this change was not sufficient to promote an increase in tensile strength. The highest tensile strength (36.4 MPa) was achieved when both types of nanoparticles were added simultaneously (PLA/5CNC/R972), which indicates a better stress transfer from the matrix to the reinforcements. In contrast to the reduction in tensile strength observed in PLA films with CNC addition, the presence of PLGA improved the system homogeneity and the matrix-nanoparticle interaction, leading to an increase in tensile strength (from 24.5 to 29 MPa). For PLA/PLGA/5CNC/R972, the addition of 5 wt% of cellulose was also found to substantially improve the tensile strength (from 25 to 31.5 MPa). Similarly, for the tensile modulus, the PLA/5CNC/R972 and PLA/PLGA/5CNC/R972 films exhibited the highest values (2388 and 2346 MPa respectively) compared to the other composites. Higher tensile moduli were related to the degree of crystallinity which was significantly influenced by intermolecular bonding within the materials. Higher crystallinity led to tensile modulus increase. Also, when both PLA and PLGA are present, the PLGA copolymer promoted better interaction between the cellulose nanoparticles and PLA.

11.3.2.3 Polymer hybrid nanocomposites based on cellulose/carbon allotropes

Carbon nanofillers, such as nanotubes and graphene, have excellent properties due to their high mechanical strength (Young's modulus of 1 TPa) and high aspect ratio [63]. Graphene and its polymeric nanocomposite derivatives have shown high potential applications in the fields of electronics, aerospace, automotive, defense, and green energy industries due to their exceptional reinforcing effect in composites. To take full advantage of their properties, the integration of individual graphene in polymer matrices is essential. Compared to carbon nanotubes, graphene has a higher surface-to-volume ratio making it potentially more favorable for improving mechanical, electrical, thermal, gas permeability, and microwave absorption properties [64]. Carbon nanotubes (CNT) are considered as unique elements promoting the development of different polymer composites due to their outstanding properties such as high electrical conductivity ($\sim 10^6$ to 10^7 S/m), high tensile strength (50 GPa), and low density (1.3 g/cm^3). These characteristics make them useful in a wide range of industrial applications such as: electrical engineering, aerospace and electronics applications [63, 65]. Three main mechanisms of interaction of the polymer matrix with the carbon are involved: micro-mechanical interconnection, chemical bonding between the nanotubes and the matrix, and low van der Waals adhesion between the filler and matrix.

Carbon allotropes were shown to be effective by significantly improving the mechanical properties of polymers. They are suitable as fillers to improve the properties of cellulose based materials [66]. Gan et al. [66] manufactured homogeneous and highly ordered membranes reinforced with cellulose carbamate (CC) and graphene oxide (GO) using a simple solution mixing method. To investigate the effect of GO content on the tensile properties of CC-GO membranes, tensile tests were performed with a crosshead speed of 5 mm/min. The results clearly showed that increasing the GO content into the CC membrane significantly improved the tensile modulus (from 0.67 to 4.16 GPa) and tensile strength (from 26.4 to 50.9 MPa), but progressively decreased the elongation at break (from 4.8% to 2.1%). These improvements were attributed to the homogeneous dispersion of GO nanosheets at the molecular level in the CC matrix and the strong interaction between GO (which contains oxygen functions promoting hydrophilicity) and CC through hydrogen bonding. However, lower elongation at break is related to the brittle nature of the GO sheets.

11.4 Mechanical modeling of hybrid nanocomposites based on cellulose

The development of hybrid nanocomposite materials can be time consuming and costly. So, the prediction of the viscoelastic behavior of a heterogeneous material from the properties of its constituents, their volume fraction, and the developed morphology, is of high importance to reduce the experimental testing time and cost, as well as improving their optimization. To determine the effective properties of a nanohybrid composite material, several methods have been proposed such as phenomenological and homogenization models [67].

The advantages of such an approach is twofold [67]:

- At the structural measurement level, it is interesting to take into account the real behavior of the materials to predict the behavior of the composites,
- At the fundamental level, it allows the experimental results to be compared with micro-macro-mechanical models, with a possibility to validate specific hypotheses made such as the morphology (aggregation, percolation, etc.).

11.4.1 Phenomenological models

Several semi-empirical models have been used to determine the elastic properties of nanocomposite materials. Among these models, the main ones are [10, 39, 40]:

The rule of mixture (ROM) or Voigt model:

$$E_c = V_f E_f + V_m E_m \tag{11.1}$$

Inverse Rule of Mixture (IROM) or Reuss model:

$$E_c = \frac{E_f\, E_m}{V_f\, E_m + V_{fm}\, E_f} \tag{11.2}$$

Hirsch model:

$$E_c = X\left(V_f E_f + V_m E_m\right) + (1 - X)\frac{E_f\, E_m}{V_f\, E_m + V_{fm}\, E_f} \tag{11.3}$$

Tsai-Pagano model:

$$E_c = \frac{3}{8}\left(V_f E_f + V_m E_m\right) + \frac{5}{8}\left(\frac{E_f E_m}{V_f E_m + V_{fm} E_f}\right) \tag{11.4}$$

where, E_c, E_f, and E_m represent respectively the composite, fiber, and matrix elastic modulus. V_f and V_m are the fiber and matrix volume fraction, while X is a parameter (between 0 and 1) related to the stress transfer between the fibers and the matrix.

The average property of the hybrid nanocomposite materials is then given by the general rule of mixtures approximation as [67]:

$$\begin{aligned} \text{Discrete}: P_c &= V_f P_f + V_m P_m + \ldots = \sum V_i P_i \\ \text{Continuum}: P_c &= \int P(x) dV \end{aligned} \tag{11.5}$$

11.4.2 Homogenization models

The studies leading to the formulation of the laws of heterogeneous materials behavior have progressed considerably over the recent decades. The objective consisted initially in obtaining the elastic behavior of the material as a function of those of the constituent phases. Subsequently, these models were developed for the simulation of the nonlinear behavior (plasticity, damage, etc.). At the beginning of the XXth century, Voigt [68] and Reuss [69], and a little later Hill [70], laid the foundations for micromechanical modeling. The idea is to predict the overall behavior of a heterogeneous material from the morphology and the behavior of each of the constituents. Thus, the definition of a method called "homogenization" enables to establish a link between the macroscopic behaviors seen as homogeneous from the behavior of the different constituent phases (Fig. 11.4). These models are based on solving a scale transition problem by numerical or analytical methods. Thus, the approach requires a well-defined field of application, hence the use of the concept of a representative volume element (RVE), which is a crucial phase in the process of changing scales. This RVE, which is considered as the basic elementary cell of the material under study, must be large enough to contain all the heterogeneous phases of the microstructure while being small enough to be considered homogeneously stressed under a specific mechanical state.

Several homogenization techniques have been developed within this general framework which differs in the way of representing the microstructure. In fact, the choice of a suitable model is directly linked to the morphology of the material.

11.4.2.1 *Voigt and Reuss limiting cases*

The limits of Voigt [68] and Reuss [69] correspond to a first approximation enabling to determine the high-low limits for the elastic properties of any

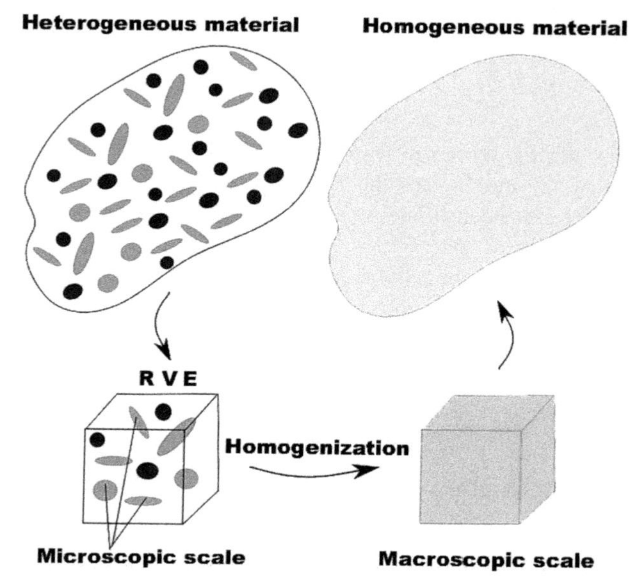

FIG. 11.4 Schematic representation of the homogenization principle.

heterogeneous material. The main interest is their simplicity and the fact that they make it possible to define the maximum and minimum values of the elastic properties. On the other hand, they do not allow microstructural parameters to be taken into account in a precise manner; only the stiffness tensors of the different phases and their volume fractions come into play. The geometry and the spatial distribution are not taken into account, so the localization tensors "*B*" (Voigt bound) and of concentration "*A*" (Reuss bound) are reduced to the identity tensor "*I*" [67]:

$$A = I = B \tag{11.6}$$

Voigt limit:

The Voigt bound, also called the deformation approach, assumes that the deformation is constant within all the phases of the material. The deformation at any point (ε) is equal to the macroscopic deformation applied (E), so it can be written as [67]:

$$\underline{\underline{\varepsilon_i}} = \underline{\underline{E}} \tag{11.7}$$

Reuss limit:

This is actually the Voigt's conjugate model, also called the stress approach. It assumes that the stress (σ) is equal to the macroscopic stress (Σ) at all points of the RVE. This allows to write the stress in each phase as [67]:

$$\underline{\underline{\sigma_i}} = \underline{\underline{\Sigma}} \tag{11.8}$$

Voigt and Reuss approximations constitute therefore a framework for the mechanical characteristics of the composite material [10, 63, 64]. They are used to determine the lower and upper limits of the equivalent elastic behavior. The other approaches are based on the important result established by Eshelby in 1957 [71], allowing a more precise description of the materials by adding microstructural parameters.

11.4.2.2 Eshelby approach

The Eshelby approach proposes a solution to solve the elementary problem of an ellipsoidal elastic inclusion (reinforcement) immersed in an isotropic infinite environment (matrix) [71]. This result is the starting point for several micromechanical models. This method links the free deformation that an inclusion outside the matrix could undergo with what it would undergo if it was embedded in the matrix. This relation is extended to treat heterogeneity whose properties, unlike inclusion, are different from those of the matrix.

Homogeneous inclusion of Eshelby

Let's consider an inclusion I immersed in an infinite isotropic environment of rigidity C_m. By assuming that the inclusion has the same mechanical properties as the environment in which it is immersed (matrix), one applies to the inclusion a deformation qualified free of stress (ε^L) equivalent to what it would undergo if it was isolated (out) of the matrix (elastic, plastic, thermal, etc.). However, the presence of the matrix phase generates an elastic deformation field in the matrix and in the inclusion. The objective is to determine an expression for this field at equilibrium, noted ε^i [67] (Fig. 11.5).

Eshelby proposed a methodology based on a series of virtual logical events enabling to determine an expression for the deformation of the inclusion [67].

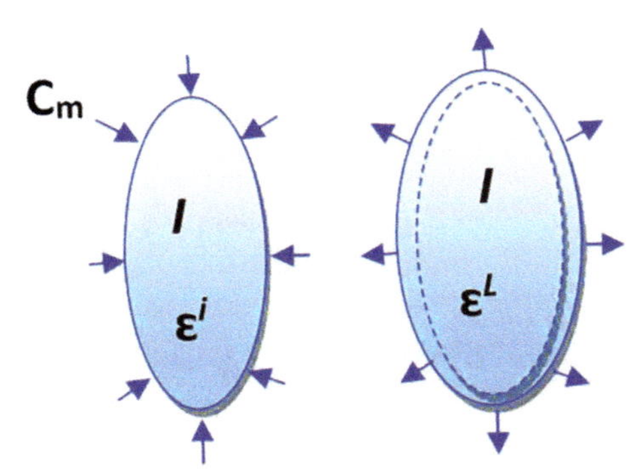

FIG. 11.5 Schematic representations of the homogeneous inclusion model of Eshelby.

He showed that the deformation undergone by the inclusion is linked to the free deformation via a tensor of order four, called the Eshelby tensor [71]. This tensor, whose general expression is based on Green's tensor, is explained in more details in the work of Mura [72], and can be written as:

$$\underline{\underline{\varepsilon^i}} = S^{esh} : \underline{\underline{\varepsilon^L}} = -\underline{\underline{P}} : \underline{\underline{\tau}} \tag{11.9}$$

where $\underline{\underline{\tau}} = -C_m : \underline{\underline{\varepsilon^L}}$ represents the polarization stress associated with the free strain and $\underline{\underline{P}}$ is the interaction tensor proposed by Hill whose expression is $\underline{\underline{P}} = S^{esh} : C_m{}^{-1}$.

The behavioral equations are therefore written as:

$$\begin{cases} \underline{\underline{\sigma_m}} = C_m : \underline{\underline{\varepsilon}} & \text{within the matrix} \\ \underline{\underline{\sigma^i}} = C_m : \left(\underline{\underline{\varepsilon^i}} - \underline{\underline{\varepsilon^L}}\right) = C_m : \left(S^{esh} - I\right) : \underline{\underline{\varepsilon^L}} = C_m : \underline{\underline{\varepsilon^i}} + \underline{\underline{\tau}} & \text{within the inclusion} \end{cases} \tag{11.10}$$

In the case of an imposed deformation (E^0) or of an imposed stress (Σ^0), the expressions become:

In the case of a stress imposed:

$$\underline{\underline{\sigma^i}} = \Sigma^0 + C_m : \left(S^{esh} - I\right) : \underline{\underline{\varepsilon^L}} \tag{11.11}$$

In the case of a deformation imposed:

$$\underline{\underline{\varepsilon^i}} = E^0 + S^{esh} : \underline{\underline{\varepsilon^L}} \tag{11.12}$$

An interaction law using the Hill influence tensor denoted C^* can be used. This tensor is symmetric, defined positive, and characterizes the response of the matrix phase to an imposed free deformation induced by an inclusion. It is exclusively dependent on the stiffness tensor C and the geometry of the inclusion leading to:

$$\underline{\underline{\sigma^i}} - \Sigma^0 = -C^* : \left(\underline{\underline{\varepsilon^i}} - E^0\right) \tag{11.13}$$

$$-C^* = C_m : \left(I - \left(S^{esh}\right)^{-1}\right) \tag{11.14}$$

Heterogeneous inclusion of Eshelby

Eshelby imposed that the stress tensor (τ), which appears in the case of the homogeneous problem, induced the same stress heterogeneity in the case of a heterogeneous problem. Thus, it shows that the heterogeneous problem can be solved in a similar way as the homogeneous problem [67] (Fig. 11.6).

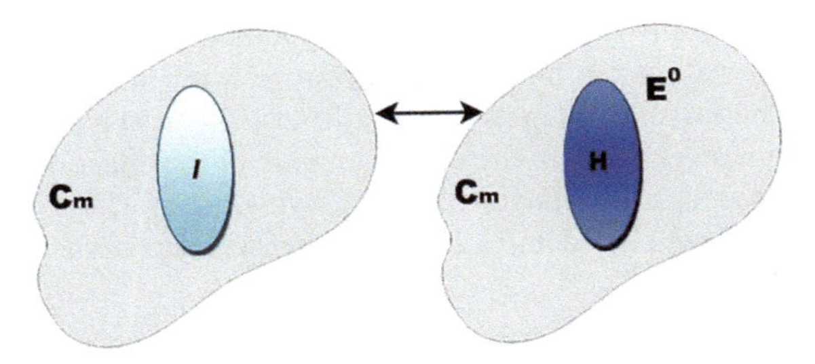

FIG. 11.6 Schematic representation of the heterogeneous inclusion model of Eshelby.

The behavioral equations in the case of the heterogeneous and homogeneous problem are written:

$$\begin{cases} \underline{\underline{\sigma}}^m = C_m : \underline{\underline{\varepsilon}}^m & \text{within the matrix} \\ \underline{\underline{\sigma}}^H = C_m : \underline{\underline{\varepsilon}}^H = C_m : \underline{\underline{\varepsilon}}^H + (C_H - C_m) : \underline{\underline{\varepsilon}}^H & \text{within the heterogeneity} \\ \underline{\underline{\sigma}}^i = C_m : \left(\underline{\underline{\varepsilon}}^i - \underline{\underline{\varepsilon}}^L \right) & \text{within the homogeneity} \end{cases} \tag{11.15}$$

where the index H refers to an ellipsoidal heterogeneity of deformation at equilibrium ε^H and of rigidity C_H immersed in an infinite matrix with a rigidity tensor C_m.

The equivalence between both problems gives an equality allowing to establish the expression of free deformation in the presence of heterogeneity.

$$\underline{\underline{\sigma}}^H = \underline{\underline{\sigma}}^i \Leftrightarrow (C_H - C_m) : \underline{\underline{\varepsilon}}^H = -C_m : \underline{\underline{\varepsilon}}^L \Leftrightarrow \underline{\underline{\varepsilon}}^L = -{C_m}^{-1} : (C_H - C_m) : \underline{\underline{\varepsilon}}^H \tag{11.16}$$

During a deformation imposed on the contour, ε^L is replaced by its expression and the expression of the deformation within heterogeneity is given by:

$$\underline{\underline{\varepsilon}}^H = E^0 + S^{esh} : \underline{\underline{\varepsilon}}^L \Leftrightarrow \underline{\underline{\varepsilon}}^H = \left(1 + S^{esh} : {C_m}^{-1} : (C_H - C_m)\right)^{-1} : E^0 \tag{11.17}$$

This expression can be written according to Hill's interaction tensor as:

$$\underline{\underline{\varepsilon}}^H = (1 + P : (C_H - C_m))^{-1} : E^0 \tag{11.18}$$

11.4.2.3 *Self-consistent model*

This model is applied when a continuous phase, such as the matrix, cannot be distinguished. Otherwise, none of the phases "*i*" has a predominant role. Each constituent is seen as being an inclusion which is embedded in a homogeneous equivalent medium (HEM). It generally applies to polycrystalline metallic materials made up of grains nested one inside the other [68, 69].

Therefore, the same formulation from the methods of dilute solutions with an Eshelby tensor estimated for a homogeneous isotropic matrix corresponding to the heterogeneous material in its overall stiffness C^{AC} is applied. The problem in this formulation comes from the implicit nature of the equations because the localization tensors A_i depend on the effective rigidity C^{AC}, which in turn depends on the tensors A_i. Identification is therefore done numerically by iterative methods as [68–70]:

$$A_i = \left(1 + S^{esh} : \left(C^{AC}\right)^{-1} : \left(C^i - C^{AC}\right)\right)^{-1} \quad (11.19)$$

$$C^{AC}_{eff} = \Sigma^n_{i=0} C^i f_i : \left(I + S^{esh} : \left(C^{AC}\right)^{-1} : \left(C^i - C^{AC}\right)\right)^{-1} \quad (11.20)$$

In the case of a mixture of fiber and matrix, the expression becomes:

$$C^{AC}_{eff} = C_m + \Sigma^n_{i=0}\left(C^i - C_m\right) f_i : \left(I + S^{esh} : \left(C^{AC}\right)^{-1} : \left(C^i - C^{AC}\right)\right)^{-1} \quad (11.21)$$

However, the self-coherent model generally overestimates the stiffness of materials with a continuous phase.

11.5 Conclusion

Hybrid composites based on regenerated cellulose (cellulose nanocrystals, cellulose nanofibers, and hairy cellulose nanocrystals) are increasingly used in several technical applications as they offer a number of improved properties and various advantages over traditional composite materials. In fact, it is known from the literature that mixing cellulose with a reinforcement of inorganic, metallic or carbon allotropes results in a significant increase in strength and elastic modulus. These improvements are mainly attributed to the matrix/reinforcement interface adhesion, which is crucial for the final material properties. However, the development of nanocomposite hybrid materials can be time consuming and expensive. So, predicting the viscoelastic behavior of a heterogeneous material from the properties of its constituents can reduce the time and cost of experimental trials. In order to determine the actual properties of a nano-hybrid composite material, several methods have been proposed, such as phenomenological and homogenization models. However, the choice of a suitable model must be directly related its morphology to avoid overestimating the material's rigidity.

References

[1] Isogai A, Saito T, Fukuzumi H. TEMPO-oxidized cellulose nanofibers. Nanoscale 2011;3:71–85.

[2] Shenhar R, Rotello VM. Nanoparticles: scaffolds and building blocks. Acc Chem Res 2003;36:549–61.

[3] Ouarhim W, Semlali Aouragh Hassani F, Qaiss A, et al. Rheology of polymer nanocomposites. In: Thomas S, Sarathchandran, Chandran CN, editors. Rheology of polymer blends and nanocomposites theory, modelling and applications. 1st ed. Elsevier; 2019. p. 73–96.
[4] Abdul Khalil H, Davoudpour Y, Sri Aprilia NA, et al. Nanocellulose-based polymer nanocomposite: isolation, characterization and applications. In: Thakur VK, editor. Nanocellulose polymer nanocomposites: fundamentals and applications. 1st ed. Scrivener Publishing; 2014. p. 273–309.
[5] Nechyporchuk O, Belgacem MN, Bras J. Production of cellulose nanofibrils: a review of recent advances. Ind Crop Prod 2016;93:2–25.
[6] Azizi Samir MAS, Alloin F, Dufresne A. Review of recent research into cellulosic whiskers, their properties and their application in nanocomposite field. Biomacromolecules 2005;6:612–26.
[7] Shankar S, Oun AA, Rhim JW. Preparation of antimicrobial hybrid nano-materials using regenerated cellulose and metallic nanoparticles. Int J Biol Macromol 2018;107:17–27.
[8] Saba N, Jawaid M, Asim M. Recent advances in nanoclay/natural fibers hybrid composites. In: Mohammad J, el Kacem QA, Rachid B, editors. Nanoclay reinforced polymer composites: natural fibre/nanoclay hybrid composites; 2016. p. 1–28.
[9] Semlali Aouragh Hassani F, Ouarhim W, Raji M, et al. N-Silylated benzothiazolium dye as a coupling agent for polylactic acid/date palm fiber biocomposites. J Polym Environ 2019;27:2974–87.
[10] Islam MS, Hasbullah NAB, Hasan M, et al. Physical, mechanical and biodegradable properties of kenaf/coir hybrid fiber reinforced polymer nanocomposites. Mater Today Commun 2015;4:69–76.
[11] Zeng QH, Yu AB, Lu GQ. Multiscale modeling and simulation of polymer nanocomposites. Prog Polym Sci 2008;33:191–269.
[12] Coll MW, et al. Les nanotechnologies. Paris: Edition Dunod; 2003.
[13] Blivi AS, Benhui F, Bai J, et al. Experimental evidence of size effect in nano-reinforced polymers: case of silica reinforced PMMA. Polym Test 2016;56:337–43.
[14] Yu Z, Li B, Chu J, et al. Silica in situ enhanced PVA/chitosan biodegradable films for food packages. Carbohydr Polym 2018;184:214–20.
[15] Chen F, Hableel G, Zhao ER, et al. Multifunctional nanomedicine with silica: role of silica in nanoparticles for theranostic, imaging, and drug monitoring. J Colloid Interface Sci 2018;521:261–79.
[16] Semlali Aouragh Hassani FZ, Ouarhim W, El Achaby M, et al. Recycled tires shreds based polyurethane binder: production and characterization. Mech Mater 2020;144:1–10.
[17] Mirmohammadi SA, Sadjadi S, Bahri-Laleh N. Electrical and electromagnetic properties of CNT/polymer composites. In: Carbon nanotube-reinforced polymers: from nanoscale to macroscale. Elsevier Inc; 2018. p. 233–58.
[18] Pantano A. Mechanical properties of CNT/polymer. In: Rafiee R, editor. Carbon nanotube-reinforced polymers: from nanoscale to macroscale. 1st ed. Elsivier; 2018. p. 201–32.
[19] Cui Y, Kundalwal SI, Kumar S. Gas barrier performance of graphene/polymer nanocomposites. Carbon 2016;98:313–33.
[20] Dufresne A. Cellulose nanomaterial reinforced polymer nanocomposites. Curr Opin Colloid Interface Sci 2017;29:1–8.
[21] Monti A, El Mahi A, Jendli Z, et al. Mechanical behaviour and damage mechanisms analysis of a flax-fibre reinforced composite by acoustic emission. Compos Part A Appl Sci Manuf 2016;90:100–10.

[22] Pääkko M, Ankerfors M, Kosonen H, et al. Enzymatic hydrolysis combined with mechanical shearing and high-pressure homogenization for nanoscale cellulose fibrils and strong gels. Biomacromolecules 2007;8:1934–41.
[23] De Souza Lima MM, Borsali R. Rodlike cellulose microcrystals: structure, properties, and applications. Macromol Rapid Commun 2004;25:771–87.
[24] Elazzouzi-Hafraoui S, Nishiyama Y, Putaux JL, et al. The shape and size distribution of crystalline nanoparticles prepared by acid hydrolysis of native cellulose. Biomacromolecules 2008;9:57–65.
[25] Beck S, Méthot M, Bouchard J. General procedure for determining cellulose nanocrystal sulfate half-ester content by conductometric titration. Cellulose 2015;22:101–16.
[26] Habibi Y, Chanzy H, Vignon MR. TEMPO-mediated surface oxidation of cellulose whiskers. Cellulose 2006;13:679–87.
[27] Frost BA, Johan FE. Isolation of thermally stable cellulose nanocrystals from spent coffee grounds via phosphoric acid hydrolysis. J Renew Mater 2020;8:187–203.
[28] Yang H, Alam MN, van de Ven TGM. Highly charged nanocrystalline cellulose and dicarboxylated cellulose from periodate and chlorite oxidized cellulose fibers. Cellulose 2013;20:1865–75.
[29] Chen D, van de Ven TGM. Morphological changes of sterically stabilized nanocrystalline cellulose after periodate oxidation. Cellulose 2016;23:1051–9.
[30] Sheikhi A, Yang H, Alam MN, et al. Highly stable, functional hairy nanoparticles and biopolymers from wood fibers: towards sustainable nanotechnology. J Vis Exp 2016;2016:1–10.
[31] Yang H, Sheikhi A, Van De Ven TGM. Reusable green aerogels from cross-linked hairy nanocrystalline cellulose and modified chitosan for dye removal. Langmuir 2016;32:1–29.
[32] Yang H, van de Ven TGM. Preparation of hairy cationic nanocrystalline cellulose. Cellulose 2016;23:1791–801.
[33] Hosseinidoust Z, Alam MN, Sim G, et al. Cellulose nanocrystals with tunable surface charge for nanomedicine. Nanoscale 2015;7:16647–57.
[34] Lenfant G, Heuzey MC, van de Ven TGM, et al. A comparative study of ECNC and CNC suspensions: effect of salt on rheological properties. Rheol Acta 2017;56:51–62.
[35] Lenfant G, Heuzey MC, van de Ven TGM, et al. Intrinsic viscosity of suspensions of electrosterically stabilized nanocrystals of cellulose. Cellulose 2015;22:1109–22.
[36] Safari S, Sheikhi A, van de Ven TGM. Electroacoustic characterization of conventional and electrosterically stabilized nanocrystalline celluloses. J Colloid Interface Sci 2014;432:151–7.
[37] Belgacem MN, Gandini A. Monomers, polymers and composites from renewable resources; 2008. p. 1–553.
[38] Sheikhi A, Safari S, Yang H, et al. Copper removal using electrosterically stabilized nanocrystalline cellulose. ACS Appl Mater Interfaces 2015;7:11301–8.
[39] Yang H, Chen D, van de Ven TGM. Preparation and characterization of sterically stabilized nanocrystalline cellulose obtained by periodate oxidation of cellulose fibers. Cellulose 2015;22:1743–52.
[40] Ben BS, Ben CR. Influence of fibre orientation and volume fraction on the tensile properties of unidirectional Alfa-polyester composite. Compos Sci Technol 2007;67:140–7.
[41] Semlali Aouragh Hassani F, Ouarhim W, Zari N, et al. Natural fiber-based biocomposites. In: Kumar K, Davim JP, editors. Biodegradable composites materials, manufacturing and engineering. 1st ed. De Gruyter; 2019. p. 49–79.
[42] Semlali Aouragh Hassani F, Ouarhim W, Zari N, et al. Injection molding of short coir fiber polypropylene biocomposites: prediction of the mold filling phase. Polym Compos 2019;40:4042–55.

[43] Semlali Aouragh Hassani F, Ouarhim W, Bensalah MO, et al. Mechanical properties prediction of polypropylene/short coir fibers composites using a self-consistent approach. Polym Compos 2019;40:1919–29.
[44] Felix JM, Gatenholm P. The nature of adhesion in composites of modified cellulose fibers and polypropylene. J Appl Polym Sci 1991;42:609–20.
[45] Alawar A, Hamed AM, Al-kaabi K. Characterization of treated date palm tree fiber as composite reinforcement. Compos Part B 2009;40:601–6.
[46] Faruk O, Bledzki AK, Fink HP, et al. Biocomposites reinforced with natural fibers: 2000-2010. Prog Polym Sci 2012;37:1552–96.
[47] Moniruzzaman M, Winey KI. Polymer nanocomposites containing carbon nanotubes. Macromolecules 2006;39:5194–205.
[48] Shalwan A, Yousif BF. In state of art: mechanical and tribological behaviour of polymeric composites based on natural fibres. Mater Des 2013;48:14–24.
[49] Ajayan PM, Tour JM. Materials science: nanotube composites. Nature 2007;447:1066–8.
[50] Gao J, Itkis ME, Yu A, et al. Continuous spinning of a single-walled carbon nanotube-nylon composite fiber. J Am Chem Soc 2005;127:3847–54.
[51] Okada A, Usuki A. The chemistry of polymer-clay hybrids. Mater Sci Eng C 1995;3:109–15.
[52] Knite M, Teteris V, Polyakov B, et al. Electric and elastic properties of conductive polymeric nanocomposites on macro- and nanoscales. Mater Sci Eng C 2002;19:15–9.
[53] Barnakov YA, Scott BL, Golub V, et al. Spectral dependence of Faraday rotation in magnetite-polymer nanocomposites. J Phys Chem Solids 2004;65:1005–10.
[54] Rong MZ, Zhang MQ, Shi G, et al. Graft polymerization onto inorganic nanoparticles and its effect on tribological performance improvement of polymer composites. Tribol Int 2003;36:697–707.
[55] Jeon IY, Baek JB. Nanocomposites derived from polymers and inorganic nanoparticles. Materials (Basel) 2010;3:3654–74.
[56] Perotti GF, Barud HS, Messaddeq Y, et al. Bacterial cellulose-laponite clay nanocomposites. Polymer (Guildf) 2011;52:157–63.
[57] Armelao L, Barreca D, Bottaro G, et al. Recent trends on nanocomposites based on Cu, Ag and Au clusters: a closer look. Coord Chem Rev 2006;250:1294–314.
[58] Jin R, Cao Y, Mirkin CA, et al. Photoinduced conversion of silver nanospheres to nanoprisms. Science 2001;294:1901–3.
[59] Tizazu G, Adawi AM, Leggett GJ, et al. Photopatterning, etching, and derivatization of self-assembled monolayers of phosphonic acids on the native oxide of titanium. Langmuir 2009;25:10746–53.
[60] Vriezema DM, Aragonès MC, Elemans JAAW, et al. Self-assembled nanoreactors. Chem Rev 2005;105:1445–89.
[61] Palza H. Antimicrobial polymers with metal nanoparticles. Int J Mol Sci 2015;16:2099–116.
[62] Abbate Dos Santos F, Bruno Tavares MI. Development of biopolymer/cellulose/silica nanostructured hybrid materials and their characterization by NMR relaxometry. Polym Test 2015;47:92–100.
[63] Filleter T, Beese AM, Roenbeck MR, et al. Tailoring the Mechanical Properties of Carbon Nanotube Fibers. Elsevier; 2013.
[64] Galpaya D, Wang M, Liu M, et al. Recent advances in fabrication and characterization of graphene-polymer nanocomposites. Graphene 2012;01:30–49.
[65] Meguid SA, Weng GJ. Micromechanics and nanomechanics of composite solids. Springer; 2017.

[66] Gan S, Zakaria S, Chia CH, et al. Physico-mechanical properties of a microwave-irradiated kenaf carbamate/graphene oxide membrane. Cellulose 2015;22:3851–63.
[67] Bouhfid N, Raji M, Boujmal R, et al. Numerical modeling of hybrid composite materials. In: Jawaid M, Thariq M, Saba N, editors. Modelling of damage processes in biocomposites, fibre-reinforced composites and hybrid composites. 1st ed. Elsivier; 2018. p. 57–101.
[68] Voigt W. Lehrbuch der Kristallphysik (mit Ausschluß der Kristalloptik). Springer; 1966.
[69] Reuss A. Berechnung des fliessgrenze von mischkristallen auf grund der prastizitätsbedingunug für einkristalle. Zeitschrift für angewandte Mathematik und Mechanik. ZAMM J Appl Math Mech 1929;9:49–58.
[70] Hill R. Elastic properties of reinforced solids: some theoretical principles. J Mech Phys Solids 1963;11:357–72.
[71] Shelby J. The determination of the elastic field of an ellipsoidal inclusion, and related problems. Proc R Soc A Math Phys Eng Sci 1957;241:376–96.
[72] Mura T, Cheng PC. The Elastic Field Outside an Ellipsoidal Inclusion. J Appl Mech 1977;44:591–4.

Index

Note: Page numbers followed by *f* indicate figures and *t* indicate tables.

I

L

M

R

S